NONLINEAR ANALYSIS RESEARCH TRENDS

NONLINEAR ANALYSIS RESEARCH TRENDS

INÈS N. ROUX
EDITOR

Nova Science Publishers, Inc.
New York

For permission to use material from this book please contact us:
Telephone 631-231-7269; Fax 631-231-8175
Web Site: http://www.novapublishers.com

LIBRARY OF CONGRESS CATALOGING-IN-PUBLICATION DATA

Nonlinear analysis research trends / Inès N. Roux (editor).
p. cm.
ISBN 978-1-60456-358-0 (hardcover)
1. Nonlinear theories--Research. 2. Mathematical analysis--Research. I. Roux, Inès N.
QA427.N637 2008
515--dc22 2007050804

Published by Nova Science Publishers, Inc. ✢ New York

CONTENTS

PREFACE

Nonlinear analysis is a broad, interdisciplinary field characterized by a mixture of analysis, topology, and applications. Its concepts and techniques provide the tools for developing more realistic and accurate models for a variety of phenomena encountered in fields ranging from engineering and chemistry to economics and biology. This new book presents the latest research from around the globe.

Chapter 1 proposes a new variational approach to find the best constants and optimal functions of the Sobolev inequalities. The author shows that the extremals for these inequalities are obtained via solutions of a nonlinear ordinary differential equation that is solved explicitly using a strategic change of variable. The author's method does not require the result by Bliss (Journal of London Math. Soc.; 5 (1930)).

In Chapter 2, the mathematical modelling is reported of current density distribution effects, leading to the localisation of cathodic current efficiency during the electroforming of Ni-diamond composites for applications in the fabrication of advanced tools for the cutting and shaping of metals and ceramics. Lack of control of such local effects during the electrochemical fabrication process is related to inconsistencies in the quality of the electroformed tools, caused by hydrogen incorporation and eventually resulting in hydrogen embrittlement failures. The mathematical model — based on the numerical integration of the Laplace equation with randomly selected Neumann and Dirichlet boundary conditions — has been validated with experimental results on the embedding of diamond particles of different sizes in a Ni matrix electroplated from a Watts bath. The dependence of thc cathodic current efficiency has been measured as a function of the particle size and compared with the numerical simulations of electrodeposition processes carried out with different fractions of Neumann-to-Dirichlet boundary conditions.

In Chapter 3, the authors study a large class of variational inequalities that can be used to deal with some highly nonlinear phenomena in electronics like switching and clipping. Using recession tools like recession cones and recession functions, they first introduce a new class of problem that the authors call a "semi-complementarity problem". Then they show that the study of a large class of variational inequalities of the second kind can be reduced to the one of a generalized eigenvalue problem governed by some semicomplementarity problem. A general spectral condition is then used to proceed to a qualitative mathematical analysis of the aforementioned class of variational inequalities. Finally, using their theoretical results, the authors present a suitable methodology for the formulation and the mathematical analysis of circuits in electronics specially those using semi-conductors like diodes and transistors.

Various examples are then presented showing that their theory is applicable to determine qualitative properties of such circuits. Moreover numerical simulations are given with the examples to support the theoretical results.

In Chapter 4 nonlinear mathematical problems of the interaction of electromagnetic fields with deformable media are studied. The models are based on combination of the Lamé and Maxwell systems coupled through so-called seismomagnetic effect. Several direct and associated with them inverse problems are studied. Then speaking about the inverse problems, electromagnetic and elastic characteristics of a medium are the subject of reconstruction. The values of physical fields are connected through electromagnetoelastic interactions. The authors consider the processes which are observed when elastic waves propagate in an elastic electroconducting medium. Variations of the seismic and electromagnetic fields in this case are called electromagnetoelastic waves. There are described different statements of mathematical model of the electromagnetoelastic interactions. Then the theoretical results of the analytical solution are discussed for various nonlinear direct and inverse problems for the equations of electromagnetoelasticity.

Predictive control has emerged as a very efficient method to control dynamical systems. The classical approach is to define a linear model of the system to control and then to solve successive on-line optimization problems to compute the best control inputs to apply. The approach is tractable as long as the linear model is sufficient to describe the system behaviour. However, as the system becomes more and more complex, nonlinear models have to be considered. Consequently, the corresponding optimization problems can hardly be exactly solved and there are crucial computational challenges to overcome to use such efficient control methods. That's why a new trend is the use of stochastic optimization so as to get well suited solutions with low computation times. In Chapter 5, the generic optimal control problem, together with the predictive control principles and their application to linear systems are firstly called up. A brief description of deterministic non linear optimization methods is also provided. Stochastic optimization algorithms are then described, with an analysis of their advantages and drawbacks for the solution of non linear optimization problems. Finally, two examples of the application of non linear predictive control in the area of Power Systems are presented.

In most cases, the conflict between stability and filtering prevents the search of the appropriate solution, and the stability/fitering problem may still unsolvable when the desired performances are high. Chapter 6 deals with robust observer-based stabilization of multiple-models systems subject to Lipschitzian nonlinearities. First, new sufficient and necessary linear matrix inequality conditions are provided for designing switching nonlinear observers. After the decomposition of the observer-based problem into two convex optimization problems, the observer-based control of systems under consideration is totally solved in an efficient numerical way. In case of presence of system and output uncertainties, a robust nonlinear observer-based scheme is developed to enfeeble the effects of uncertainties. In this case, a dynamic output feedback stabilization issue with a prescribed degree of uncertainty attenuation is considered. The design of the observer and the controller gains is conditioned by the solutions of three linear matrix inequalities. A numerical example is provided to demonstrate the potential of the author's results.

The existence of multiple nontrivial solutions of semilinear elliptic eigenvalue problems under Dirichlet- or Neumann boundary conditions has received much attention over the last decades. It is well known that the multiplicity of solutions basically depends on the value of the eigenvalue parameter (with respect to the eigenvalues of an associated linear eigenvalue problem), and the structure and growth conditions imposed on the nonlinearity involved. In recent years the study of multiple solutions for quasilinear elliptic problems involving the *p*-Laplacian has become a subject of growing interest. The main goal of Chapter 7 is to present new multiple solutions result for quasilinear elliptic variational problems involving the *p*-Laplacian which, in general, depend on two parameters, and whose nonlinearity may be even multi-valued and given in the form of Clarke's generalized gradient. The dependence of the problem under consideration on two parameters, which are studied in relation to the Fučik spectrum of the p-Laplacian, allows for a more detailed analysis of the multiplicity of solutions. The multi-valued problem considered here requires advanced tools from Nonsmooth Analysis. The authors' approach will be based on comparison principles for multivalued elliptic problems and variational methods for nonsmooth functionals such as, e.g., nonsmooth critical point theory and second deformation lemma for nonsmooth and locally Lipschitz functionals.

In Chapter 8 the authors study the initial boundary value problem of wave equations and reaction-diffusion equations with two nonlinear source terms of different sign. By introducing a family of potential wells W_δ and corresponding outside sets V_δ of W_δ the authors first obtain the invariant sets and vacuum isolating of solutions. Then they get the threshold result of global existence and nonexistence of solutions. Finally the authors prove the global existence of solutions for the problem with critical initial conditions.

Chapter 9 focuses on a coupled 2×2 system of mixed type for viscosity-capillarity with periodic initial-boundary condition in viscoelastic material. The main concern in the present study is the stationary solutions. It is shown that the stationary system has one or multiple stationary solutions depending on the value of the viscosities ε_1 and ε_2. In particular, the non-trivial solutions always present phase transitions. The criteria for the type of system responses (one or multiple solutions) are specified. Furthermore, the calculation formula for the number of multiple solutions is provided. Finally, numerical simulations on the prototypical system are reported to verify the theoretical results.

Chapter 10 deals with the two-dimensional Navier-Stokes equations in which the source term involves a Dirac delta function and describes the elastic reaction of the immersed boundary. The authors analyze the existence of the approximate solution with Dirac delta function approximated by differentiable function. They obtain the result via the Banach Fixed Point Theorem and the properties of the solutions to the Navier-Stokes equations of viscous incompressible fluids with periodic boundary conditions.

In: Nonlinear Analysis Research Trends
Editor: Inès N. Roux, pp. 1-13
ISBN 978-1-60456-358-0

Chapter 1

A NEW ODE APPROACH TO SHARP SOBOLEV INEQUALITIES

*M. Agueh**
Department of Mathematics and Statistics,
University of Victoria, PO. Box. 3045 STN CSC,
Victoria B.C., Canada, V8W 3P4

Abstract

We propose a new variational approach to find the best constants and optimal functions of the Sobolev inequalities. We show that the extremals for these inequalities are obtained via solutions of a nonlinear ordinary differential equation that we solve explicitly using a strategic change of variable. Our method does not require the result by Bliss (Journal of London Math. Soc.; 5 (1930)).

Keywords: Sobolev inequalities, best constants, optimal functions.

AMS Subject Classifications: 34A34, 39B62, 35J20.

1. Introduction

The present work deals with the Sobolev inequality

$$\|u\|_{L^{p^*}(\mathbb{R}^n)} \leq c(n,p)\,\|\nabla u\|_{L^p(\mathbb{R}^n)} \quad \forall u \in W^{1,p}(\mathbb{R}^n), \tag{1}$$

where $1 < p < n$, $p^* := \frac{np}{n-p}$ denotes the Sobolev exponent, $c(n,p)$ is a constant, and

$$W^{1,p}(\mathbb{R}^n) := \{u \in L^p(\mathbb{R}^n),\ \nabla u \in L^p(\mathbb{R}^n)\}.$$

The best constant $c(p,n)$ and optimal functions in (1) are first obtained independently by Aubin [4] and Talenti [10]. Their proofs are based on a symmetrization argument which

*E-mail address: agueh@math.uvic.ca. The author is supported by a grant from the Natural Science and Engineering Research Council of Canada.

helped them reduce the problem to a 1-dimensional space, and then on a 1-dimensional calculus of variations which was previously studied by Bliss [5].

Recently, Cordero-Erausquin et al presented another proof of (1) using a mass transportation approach [7]. In [7], the authors established a duality between the functional $E(u) = \int_{\mathbb{R}^n} |\nabla u(x)|^p \, dx$ and a certain energy functional on probability densities on $\mathbb{R}^n$, from which they derived the sharp constant and extremals in (1).

In this paper, we investigate the link between these two approaches. We show that in fact, the mass transportation approach suggests a change of variable which helps to solve explicitly the nonlinear ordinary differential equation – leading to the sharp constant and extremals in (1) – obtained from the approach by Aubin [4] and Talenti [10] without using the result by Bliss [5]. Then, this leads to a new ODE approach to the sharp Sobolev inequalities that we briefly describe below:

1.1. Aubin and Talenti's Approach [4, 10]

Following [4, 10], the best constant, $c(p,n)$, and extremals in (1) can be obtained from the solutions of the variational problem

$$\sup\left\{I\left(u(x)\right) := \frac{\|u\|_{L^{p^*}(\mathbb{R}^n)}}{\|\nabla u\|_{L^p(\mathbb{R}^n)}} : \quad u \in C_c^1(\mathbb{R}^n)\right\}, \tag{2}$$

or equivalently

$$\sup\{\|u\|_{L^{p^*}(\mathbb{R}^n)} : \quad u \in C_c^1(\mathbb{R}^n), \ \|\nabla u\|_{L^p(\mathbb{R}^n)} = 1\}. \tag{3}$$

If u^* denotes the Schwarz symmetrization of $|u|$ (i.e., u^* is nonnegative, radially symmetric, nonincreasing, such that the level sets of u^* has the same measure as the level sets of $|u|$ and they are balls centered at the origin), then by Cavalieri and Pòlya-Szegö principles, we have that $I(u^*) \geq I(u)$, so that – up to replacing u by u^* –, (2) reduces to the 1-dimensional variational problem,

$$\sup\left\{J\left(u(r)\right) := \frac{\left(\int_0^\infty r^{n-1}|u(r)|^{p^*} \, dr\right)^{1/p^*}}{\left(\int_0^\infty r^{n-1}|u'(r)|^p \, dr\right)^{1/p}} : \quad u(r) \in K\right\}, \tag{4}$$

or equivalently

$$\sup\left\{\int_0^\infty r^{n-1}|u(r)|^{p^*} \, dr; \quad u(r) \in K, \ \int_0^\infty r^{n-1}|u'(r)|^p \, dr = C\right\}, \tag{5}$$

where $C > 0$ is some constant, and K is the set of all nonnegative and nonincreasing functions, $u : [0,\infty) \to [0,\infty)$, $(u(r) := u(x), \ r := |x|)$, that satisfy some conditions at ∞ (which essentially ensure that $\|u(x)\|_{L^{p^*}(\mathbb{R}^n)}$ and $\|\nabla u(x)\|_{L^p(\mathbb{R}^n)}$ are finite). At this stage, the authors recognize (4) or (5) as the 1-dimensional variational problem studied by Bliss in [5]. In [5], the value of (4) or (5) is computed along with the explicit formula of the maximizers. Therefore, one deduces the best constant and extremals in the Sobolev inequality (1) just by changing conveniently r to $|x|$ in the formulas of Bliss. As mentioned in [4, 10], the maximizers of (4) or (5) are the solutions of the ODE [1]

$$\left(r^{n-1}|u'(r)|^{p-1}\right)' - \lambda r^{n-1}u(r)^{p^*-1} = 0, \quad \lambda > 0, \tag{6}$$

which verify certain limit conditions at ∞ (as in $u(r) \in K$). Note that (6) is nothing but the p-Laplacian (or Laplace's if $p = 2$) equation,

$$\text{div}\left(|\nabla u(x)|^{p-2}\nabla u(x)\right) + \lambda u(x)^{p^*-1} = 0 \tag{7}$$

expressed in the radial form, $u(x) = u(r)$, $r = |x|$.

1.2. Mass Transportation Approach [7]

First, we observe that the best constant and extremals in (1) can be derived from the solutions of the variational problem

$$\inf\left\{E\left(u(x)\right) := \int_{\mathbb{R}^n} |\nabla u(x)|^p \, \mathrm{d}x : \quad u \in W^{1,p}(\mathbb{R}^n), \ \|u\|_{L^{p^*}(\mathbb{R}^n)} = 1\right\}, \tag{8}$$

where

$$W^{1,p}(\mathbb{R}^n) := \{u \in L^p(\mathbb{R}^n) : \quad \nabla u \in L^p(\mathbb{R}^n)\}.$$

In [7], the authors established a duality between $E(u)$ and an energy functional on probability density functions on $\mathbb{R}^n$, from which they obtained an explicit formula of the solutions to (8). This duality is based on the notion of displacement convexity [9], that is, the convexity of functionals defined on the space $\mathcal{P}(\mathbb{R}^n)$ of probability densities on $\mathbb{R}^n$ (with finite second moment) equipped with the Wasserstein metric d_2 (see [11] for the definition). Indeed, let ρ_0 and ρ_1 be in $\mathcal{P}(\mathbb{R}^n)$, and denote by T the optimal map that pushes ρ_0 forward to ρ_1 (i.e. $T_{\#}\rho_0 = \rho_1$) in the Monge-Kantorovich problem [11] for the quadratic cost $p(x-y) = \frac{|x-y|^2}{2}$. For $t \in [0,1]$, define $T_t = (1-t)\text{id} + tT$. Then the path $[0,1] \ni t \mapsto \rho_t = (T_t)_{\#}\rho_0$ is the minimal path joining ρ_0 and ρ_1 in $(\mathcal{P}(\mathbb{R}^n), d_2)$, in the sense that $d_2(\rho_0, \rho_1) = d_2(\rho_0, \rho_t) + d_2(\rho_t, \rho_1)$, and for all other paths $\tilde{\rho}_t \in \mathcal{P}(\mathbb{R}^n)$ joining ρ_0 and ρ_1, we have that $d_2(\rho_0, \rho_1) < d_2(\rho_0, \tilde{\rho}_t) + d_2(\tilde{\rho}_t, \rho_1)$. Let $F : [0, \infty) \to \mathbb{R}$ be such that $F(0) = 0$ and $x \mapsto x^n F(x^{-n})$ is convex and non-increasing. Then $\mathcal{F}(\rho) = H^F(\rho) := \int_{\mathbb{R}^n} F\left(\rho(x)\right) \, \mathrm{d}x$ is displacement convex [9], i.e., $[0,1] \ni t \mapsto \mathcal{F}(\rho_t) \in \mathbb{R}$ is convex for all pairs (ρ_0, ρ_1) in $\mathcal{P}(\mathbb{R}^n)$. A direct consequence is the following energy inequality (here we assume for simplicity that ρ_0 and ρ_1 have compact supports in $\mathbb{R}^n$):

$$\begin{aligned} H^F(\rho_1) - H^F(\rho_0) &\geq \left[\frac{d}{dt} H^F(\rho_t)\right]_{t=0} \\ &= \int_{\mathbb{R}^n} F'(\rho_0) \left.\frac{\partial \rho_t}{\partial t}\right|_{t=0} \mathrm{d}x \\ &= -\int_{\mathbb{R}^n} F'(\rho_0) \text{div}\left(\rho_0(T - \text{id})\right) \mathrm{d}x \\ &= \int_{\mathbb{R}^n} \rho_0 \nabla\left(F'(\rho_0)\right) \cdot (T - \text{id}) \, \mathrm{d}x, \end{aligned} \tag{9}$$

[1] In [4], the author identified – when $p = 2$ – (6) as a simple case of Emden-Fowler equations whose solutions are known explicitly. From that, he was able to guess the solutions to (4) for the general case, by modifying conveniently the formula for $p = 2$.

where the last equality is obtained by an integration by parts, and the second one can be understood as follows. Since $\rho_t = (T_t)_{\#}\rho_0$, we have for any smooth test function φ,

$$\int_{\mathbb{R}^n} \varphi(y)\rho_t(y)\,\mathrm{d}y = \int_{\mathbb{R}^n} \varphi\left(T_t(x)\right)\rho_0(x)\,\mathrm{d}x.$$

Differentiating the above expressions with respect to t, we have

$$\frac{d}{dt}\int_{\mathbb{R}^n} \varphi(y)\rho_t(y)\,\mathrm{d}y = \int_{\mathbb{R}^n} \nabla\varphi\left(T_t(x)\right)\cdot(T(x)-x)\rho_0(x)\,\mathrm{d}x,$$

that is,

$$\int_{\mathbb{R}^n} \varphi(y)\frac{\partial\rho_t(y)}{\partial t}\bigg|_{t=0}\mathrm{d}y = \int_{\mathbb{R}^n} \nabla\varphi(x)\cdot(T(x)-x)\rho_0(x)\,\mathrm{d}x = -\int_{\mathbb{R}^n} \varphi(x)\mathrm{div}\left(\rho_0(T-\mathrm{id})\right)\mathrm{d}x,$$

from which we obtain

$$\frac{\partial\rho_t}{\partial t}\bigg|_{t=0} = -\mathrm{div}\left(\rho_0(T-\mathrm{id})\right).$$

For a rigorous proof of (9), we refer to [2]. For convenience, we will use the following formulation of [1]:

$$-H^F(\rho_1) \le -H^{F+nP_F}(\rho_0) - \int_{\mathbb{R}^n} \rho_0\nabla\left(F'(\rho_0)\right)\cdot T(x)\,\mathrm{d}x, \tag{10}$$

where $P_F(x) = xF'(x) - F(x)$. This follows directly from (9) by simply rewriting the right hand side term as

$$\begin{aligned}\int_{\mathbb{R}^n} \rho_0\nabla\left(F'(\rho_0)\right)\cdot(T-\mathrm{id})\,\mathrm{d}x &= \int_{\mathbb{R}^n} \rho_0\nabla\left(F'(\rho_0)\right)\cdot T(x)\,\mathrm{d}x - \int_{\mathbb{R}^n} \nabla\left(P_F(\rho_0)\right)\cdot x\,\mathrm{d}x\\ &= \int_{\mathbb{R}^n} \rho_0\nabla\left(F'(\rho_0)\right)\cdot T(x)\,\mathrm{d}x + nH^{P_F}(\rho_0),\end{aligned}$$

where an integration by parts is used in the last term above.

By the Young inequality

$$-\nabla\left(F'(\rho_0)\right)\cdot T(x) \le \frac{|T(x)|^{p'}}{p'} + \frac{|\nabla F'(\rho_0)|^p}{p}, \quad \left(\frac{1}{p}+\frac{1}{p'}=1\right), \tag{11}$$

(10) gives

$$-H^F(\rho_1) \le -H^{F+nP_F}(\rho_0) + \frac{1}{p}\int_{\mathbb{R}^n} \rho_0|\nabla F'(\rho_0)|^p\,\mathrm{d}x + \frac{1}{p'}\int_{\mathbb{R}^n} \rho_0(x)|T(x)|^{p'}\,\mathrm{d}x,$$

i.e.,

$$-H^F_{|x|^{p'}/p'}(\rho_1) \le -H^{F+nP_F}(\rho_0) + \frac{1}{p}\int_{\mathbb{R}^n} \rho_0|\nabla F'(\rho_0)|^p\,\mathrm{d}x, \tag{12}$$

where we use that $T_{\#}\rho_0 = \rho_1$. Furthermore, if $\rho_0 = \rho_1$, then $T = \mathrm{id}$ and equality holds in (10). Then equality holds in (12) if it holds in the Young inequality (11). This occurs when $\rho_0 = \rho_1$ satisfies on it support

$$\nabla\left(F'\left(\rho_0(x)\right) + \frac{|x|^{p'}}{p'}\right) = 0.$$

We have then established the following duality:

$$\sup\Big\{-H^F_{|x|^{p'}/p'}(\rho_1):\quad \rho_1\in\mathcal{P}(\mathbb{R}^n)\Big\}$$
$$=\inf\Big\{J(\rho_0):=-H^{F+nP_F}(\rho_0)+\frac{1}{p}\int_{\mathbb{R}^n}\rho_0|\nabla F'(\rho_0)|^p\,\mathrm{d}x:\quad \rho_0\in\mathcal{P}(\mathbb{R}^n)\Big\},\tag{13}$$

and an optimal function $\rho_0=\rho_1:=\rho_\infty$ in both problems is a solution to the equation

$$\nabla\left(F'\left(\rho_\infty(x)\right)+\frac{|x|^{p'}}{p'}\right)=0.\tag{14}$$

In particular, choosing $F(x)=-nx^{1-1/n}$ and $\rho_0=u^{p^*}$, (13)-(14) shows that

$$\begin{aligned}\inf\Big\{E(u)&=\int_{\mathbb{R}^n}|\nabla u|^p\,\mathrm{d}x:\quad u\in W^{1,p}(\mathbb{R}^n),\ \|u\|_{L^{p^*}(\mathbb{R}^n)}=1\Big\}\\&=-p\left(\frac{n}{p(n-1)}\right)^p H^F_{|x|^{p'}/p'}(\rho_\infty),\end{aligned}\tag{15}$$

and u_∞ defined by

$$\nabla\left(F'\left(u_\infty^{p^*}(x)\right)+\frac{|x|^{p'}}{p'}\right)=0,\quad \rho_\infty:=u_\infty^{p^*}\in\mathcal{P}(\mathbb{R}^n)\tag{16}$$

is a minimizer of (8).

1.3. A New ODE Approach

We observe that the Euler-Lagrange equation to (8) is the PDE (7), whose radial form is the ODE (6) obtained in [4, 10], where $p\lambda$ represents here the Lagrange multiplier for the constraint $\|u\|_{L^{p^*}(\mathbb{R}^n)}=1$. Hence, (15)-(16) state that the function $u(r)=u(x)$, $r=|x|$, defined by (16) solves (6) and (7), or in other words, solutions to (6) satisfy

$$\frac{r^{p'}}{p'}=\frac{|x|^{p'}}{p'}=\gamma-F\left(u^{p'}(x)\right):=H\left(u(x)\right)=H\left(u(r)\right),\tag{17}$$

for some constant γ. Therefore, (17) suggests the change of variable $H\left(u(r)\right)=\frac{r^{p'}}{p'}$ to solve (6). Using this change of variable, we can rewrite (6) as the first order nonlinear ODE in $G'(t)=\frac{|H'(t)|^{p-2}H'(t)}{p-1}$,

$$\frac{\lambda p}{p^*}t^{p^*}G''(t)+\lambda(p-1)t^{p^*-1}G'(t)+\frac{p(n-1)}{p-1}G''(t)\int_0^t\frac{\mathrm{d}\tau}{G'(\tau)}=-n,\quad t\in(0,u(0)).\tag{18}$$

Setting $G''(t)\int_0^t\frac{\mathrm{d}\tau}{G'(\tau)}=k$ (for some constant k), (18) reduces to the linear first oder ODE in G',

$$\frac{\lambda p}{p^*}t^{p^*}G''(t)+\lambda(p-1)t^{p^*-1}G'(t)=-n-\frac{kp(n-1)}{p-1}\tag{19}$$

whose solutions can be easily obtained in closed forms. Then from the solutions G' to (18), we can derive a solution $u(r)$ to (6) via the relations $G'(t) = \frac{|H'(t)|^{p-2}H'(t)}{p-1}$ and (17). Finally, we can compute the best constant and extremals in (1) using (8).

We emphasize that this new approach neither requires a priori any information on the solutions of ODE (6), nor does it require any knowledge of mass transportation theory. It solely relies on the change of variable (17) and on symmetrization arguments which allow us to reduce PDE (7) to ODE (6). A similar technique can be used to derive the best constants and optimal functions of certain Gagliardo-Nirenberg inequalities, see [3]. Finally let us mention that other proofs of the sharp Sobolev inequalities are also available in the literature; for e.g., see [8, 6].

Throughout the paper, $\|u\|_p$ will denote the L^p-norm of a function u defined on $\mathbb{R}^n$.

In the next section, we study a related variational problem to the Sobolev inequality (1), and in section 3., we compute the best constant and optimal functions in (1) from the solutions of this problem.

2. An Optimization Problem

This section is devoted to the study of the optimization problem (8) and to the derivation of the sharp Sobolev inequality (1) from the solutions of (8).

Theorem 1. *Let $1 < p < n$ and set $p^* := np/(n-p)$. Then the variational problem*

$$\inf\left\{E(u) := \frac{1}{p}\int_{\mathbb{R}^n} |\nabla u(x)|^p\,dx : \quad u \in W^{1,p}(\mathbb{R}^n),\ \|u\|_{p^*} = 1\right\} \tag{20}$$

has a solution u_∞ which is nonnegative, radially symmetric and nonincreasing. Moreover, u_∞ is a solution of the PDE

$$\mathit{div}\left(|\nabla u(x)|^{p-2}\nabla u(x)\right) + \lambda u(x)^{p^*-1} = 0, \tag{21}$$

for a well-chosen $\lambda > 0$ (such that $\|u\|_{p^*} = 1$). Conversely, any nonnegative radially symmetric and nonincreasing solution $u \in W^{1,p}(\mathbb{R}^n)$ of (21) which satisfies the condition $\|u\|_{p^*} = 1$ solves (20).

Furthermore, the Sobolev inequality

$$\|u\|_{L^{p^*}(\mathbb{R}^n)} \leq c(n,p)\,\|\nabla u\|_{L^p(\mathbb{R}^n)} \quad \forall u \in W^{1,p}(\mathbb{R}^n), \tag{22}$$

holds, with the best constant

$$c(n,p) = \frac{1}{(pE(u_\infty))^{1/p}} \tag{23}$$

and optimal functions $u_{\sigma,\bar{x}}(x) = Cu_\infty(\sigma(x-\bar{x}))$, where $C \neq 0$, $\sigma > 0$ and $\bar{x} \in \mathbb{R}^n$ are arbitrary.

Proof. Let $u \in W^{1,p}(\mathbb{R}^n)$, $\|u\|_{p^*} = 1$, and let u^* denote the Schwarz symmetrization of $|u|$, that is, u^* is nonnegative, radially symmetric, nonincreasing, such that the level sets of u^* has the same measure as the level sets of $|u|$ and they are balls centered at the origin. By Cavalieri and Pòlya-Szegö principles, we have

$$\|u^*\|_{p^*} = \|u\|_{p^*} = 1, \quad \|u^*\|_p = \|u\|_p, \quad \int_{\mathbb{R}^n} |\nabla u^*|^p \le \int_{\mathbb{R}^n} |\nabla u|^p,$$

so that u^* is also admissible in (20) and $E(u^*) \le E(u)$. Therefore we can assume without loss of generality that the test functions in (20) are nonnegative, radially symmetric and nonincreasing. Let E_∞ denote the value of (20). Clearly E_∞ is finite because $E(u)$ is nonnegative. Now, let $(u_k)_k$ be a minimizing sequence in (20). We have that $u_k(x) = v_k(r), r = |x|$ and $\|u_k\|_{L^{p^*}(\mathbb{R}^n)} = 1$ for some nonincreasing function $v_k : [0,\infty) \to [0,\infty)$. Then $v_k(r) \to 0$ as $r \to \infty$. And since v_k is nonincreasing, we can assume without loss of generality that

$$v_k(r) \le 1, \quad \forall r \ge 0, \tag{24}$$

otherwise we replace $v_k(r)$ by $\tilde{v}_k(r) = v_k(r+r_0)$ where r_0 is such that $v_k(r) \le 1, \ \forall r \ge r_0$. By definition, $(u_k)_k$ is bounded in $W^{1,p}(\mathbb{R}^n)$, and then up to a subsequence, it converges weakly to a function $u_\infty \in W^{1,p}(\mathbb{R}^n)$ which obviously is nonnegative, radially symmetric and nonincreasing (because the u_k are). Then we can write that $u_\infty(x) = v_\infty(r)$, $r = |x|$, for some nonincreasing function $v_\infty : [0,\infty) \to [0,\infty)$. By weak lower-semicontinuity, we have that

$$E(u_\infty) \le \liminf_{k\to\infty} E(u_k) = E_\infty.$$

It remains to show that $\|u_\infty\|_{p^*} = 1$. For that, we will prove that $(u_k)_k$ converges strongly to u_∞ in $L^{p^*}(\mathbb{R}^n)$. Indeed, since the u_k are radial and bounded in $L^p(\mathbb{R}^n)$, there exists a constant $C > 0$ such that

$$C \ge \|u_k\|^p_{L^p(\mathbb{R}^n)} = n\omega_n \int_0^\infty r^{n-1} v_k(r)^p \, dr, \quad \forall k,$$

where ω_n denotes the Lebesgue measure of the unit ball in $\mathbb{R}^n$. Hence for all $R \ge 1$, we have

$$C \ge n\omega_n \int_0^R r^{n-1} v_k(r)^p \, dr \ge n\omega_n \int_0^R r^{n-1} v_k(R)^p \, dr = \omega_n R^n v_k(R)^p,$$

i.e.,

$$v_k(r) \le (C/\omega_n)^{1/p} r^{-n/p}, \quad \forall r \ge 1. \tag{25}$$

Set

$$h(r) := \alpha \min(1, r^{-n/p}), \quad \alpha := \max\left(1, (C/\omega_n)^{1,p}\right). \tag{26}$$

We have that

$$\begin{aligned}\int_0^\infty r^{n-1} h(r)^{p^*} \, dr &= \int_0^1 r^{n-1} h(r)^{p^*} \, dr + \int_1^\infty r^{n-1} h(r)^{p^*} \, dr \\ &\le \alpha^p \int_0^1 r^{n-1} \, dr + \alpha^p \int_1^\infty r^{n(1-p^*/p)-1} \, dr < \infty,\end{aligned}$$

because $n(1-p^*/p) = -np/(n-p) < 0$. Then $r^{n-1}h(r)^{p^*} \in L^1((0,\infty))$ and from (24)-(26), $r^{n-1}v_k(r)^{p^*} \leq r^{n-1}h(r)^{p^*}$ $\forall k$. Hence, the dominated convergence theorem gives that

$$\begin{aligned} 1 = \lim_{k\to\infty} \|u_k\|^{p^*}_{L^{p^*}(\mathbb{R}^n)} &= n\omega_n \lim_{k\to\infty} \int_0^\infty r^{n-1}v_k(r)^{p^*}\,\mathrm{d}r = n\omega_n \int_0^\infty \lim_{k\to\infty} r^{n-1}v_k(r)^{p^*}\,\mathrm{d}r \\ &= n\omega_n \int_0^\infty r^{n-1}v_\infty(r)^{p^*}\,\mathrm{d}r = \int_{\mathbb{R}^n} u_\infty(x)^{p^*}\,\mathrm{d}x = \|u_\infty\|^{p^*}_{L^{p^*}(\mathbb{R}^n)}. \end{aligned}$$

This shows that u_∞ is a minimizer of (20).

Clearly u_∞ is a solution of the Euler-Lagrange equation (21) to the variational problem (20), where λ is the Lagrange multiplier for the constraint $\|u\|_{p^*} = 1$.

To prove the reverse implication, let $\bar{u}_\infty \in W^{1,p}(\mathbb{R}^n)$ be a nonnegative radially symmetric solution of (21) such that $\|\bar{u}_\infty\|_{p^*} = 1$, and still denote by u_∞ the minimizer of (20) found above. There exists a sequence $(\bar{u}_k)_k$ in $C^1_c(\mathbb{R}^n)$ such that $\bar{u}_k$ converges weakly to $\bar{u}_\infty$ in $L^{p^*}(\mathbb{R}^n)$ and $\nabla\bar{u}_k$ converges weakly to $\nabla\bar{u}_\infty$ in $L^p(\mathbb{R}^n)$. Then, multiplying (21) by $\bar{u}_k$, we have after integration,

$$\int_{\mathbb{R}^n} |\nabla\bar{u}_\infty|^{p-2}\nabla u_\infty \cdot \nabla\bar{u}_k - \lambda \int_{\mathbb{R}^n} \bar{u}_\infty^{p^*-1}\bar{u}_k = 0.$$

Now using that $\bar{u}_\infty^{p^*-1} \in L^{(p^*)'}$ and $|\nabla\bar{u}_\infty|^{p-2}\nabla\bar{u}_\infty \in L^{p'}$ (where the $'$ denotes the conjugate exponent), we obtain after passing to the limit in the above equality that

$$\int_{\mathbb{R}^n} |\nabla\bar{u}_\infty|^p - \lambda \int_{\mathbb{R}^n} \bar{u}_\infty^{p^*} = 0,$$

that is,

$$\lambda = \int_{\mathbb{R}^n} |\nabla\bar{u}_\infty|^p,$$

because $\|\bar{u}_\infty\|_{p^*} = 1$. Similarly

$$\lambda = \int_{\mathbb{R}^n} |\nabla u_\infty|^p,$$

so that $E(u_\infty) = E(\bar{u}_\infty)$. And since by assumption u_∞ is a minimizer of (20), we deduce that $\bar{u}_\infty$ is also a minimizer of (20).

We easily show that (22) and (23) follow from (20) by writing

$$E(u_\infty) \leq E\left(\frac{u}{\|u\|_{p^*}}\right) = \frac{\|\nabla u\|_p^p}{p\|u\|_{p^*}^p}, \quad \forall\, 0 \neq u \in W^{1,p}(\mathbb{R}^n).$$

Finally, by definition, u_∞ is an optimal function in (22). And since (22) is invariant under translation $x \mapsto x - \bar{x}$, scaling $x \mapsto \sigma x$, and multiplication by a constant $u \mapsto Cu$, we deduce that $u_{\sigma,\bar{x}}(x) = Cu_\infty(\sigma(x-\bar{x}))$ is an optimal function in (22), for arbitrary $C \neq 0$, $\sigma > 0$ and $\bar{x} \in \mathbb{R}^n$. □

3. Sharp Sobolev Inequalities

Here, we compute a minimizer u_∞ of the variational problem (20), and we obtain the best constant and optimal functions of the Sobolev inequality (22) using Theorem 1.

Theorem 2. *Assume that* $1 < p < n$, *and set* $p' := p/(p-1)$. *Then*

$$u_\infty(x) = \frac{a}{(b + |x|^{p'})^{(n-p)/p}} \tag{27}$$

is a minimizer of the variational problem (20), where $a > 0$ *is arbitrary,*

$$b := \frac{\lambda a^{\frac{n(p-1)+p}{n-p}}}{n\left(\frac{a(n-p)}{p-1}\right)^{p-1}}, \tag{28}$$

and $\lambda > 0$ *is uniquely determined by* $\|u_\infty\|_{p^*} = 1$. *Therefore the Sobolev inequality (22) holds with the best constant given by (23), and the optimal functions*

$$u(x) = \frac{C}{(1 + \sigma|x - \bar{x}|^{p'})^{(n-p)/p}} \tag{29}$$

where $C \neq 0$, $\sigma > 0$ *and* $\bar{x} \in \mathbb{R}^n$ *are arbitrary.*

Proof. It follows the lines of the proofs of Theorem 3.1 and Corollary 3.4 in [3]. Indeed, let $u \in W^{1,p}(\mathbb{R}^n)$ be a nonnegative, radially symmetric and nonincreasing solution of PDE (21) such that $\|u\|_{p^*} = 1$. From Theorem 1, u is also a minimizer of the variational problem (20). Hence, there is a nonincreasing function $v : [0,\infty) \to [0,\infty)$, such that $\lim_{r\to\infty} v(r) = 0$, $\lim_{r\to\infty} v'(r) = 0$ and

$$v(r) = u(x), \quad r = |x|. \tag{30}$$

Note that $\lim_{r\to\infty} v(r) = 0$ and $\lim_{r\to\infty} v'(r) = 0$ because

$$\|u\|_{p^*}^{p^*} = n\omega_n \int_0^\infty r^{n-1} v(r)^{p^*}\, dr < \infty \quad \text{and} \quad \|\nabla u\|_p^p = n\omega_n \int_0^\infty r^{n-1}|v'(r)|^p\, dr < \infty.$$

Then (21) reads as

$$(p-1)|v'(r)|^{p-2}v''(r) + \frac{n-1}{r}|v'(r)|^{p-2}v'(r) + \lambda v(r)^{p^*-1} = 0. \tag{31}$$

To solve (31), we consider (as explained in the introduction) the change of variable

$$H\left(v(r)\right) = \frac{r^{p'}}{p'}, \quad \frac{1}{p} + \frac{1}{p'} = 1, \tag{32}$$

where H is some C^2-function on $(0, v(0))$, and we rewrite (31) in terms of the new variable H. Indeed, from the conditions on v, we have that $H'(t) \leq 0$ for all $t \in (0, v(0))$, $\lim_{t\to 0} H(t) = \infty$ and $\lim_{t\to\infty} H'(t) = -\infty$. Differentiating (32) with respect to r gives that

$$r = \left[H'\left(v(r)\right)v'(r)\right]^{\frac{1}{p'-1}} = \left[H'\left(v(r)\right)v'(r)\right]^{p-1}. \tag{33}$$

Differentiating again (33) with respect to r, we have

$$H'(v)^{p-2}H''(v)(v')^p + H'(v)^{p-1}(v')^{p-2}v'' = \frac{1}{p-1}, \tag{34}$$

i.e.,

$$|H'(v)|^{p-2}H''(v)|v'|^p + |H'(v)|^{p-2}H'(v)|v'|^{p-2}v'' = \frac{1}{p-1}. \tag{35}$$

Now, we express $|v'|^{p-2}v''$ and $|v'|^p$ in terms of v and $H(v)$. From (31),

$$|v'|^{p-2}v'' = \frac{-1}{p-1}\left(\frac{n-1}{r}|v'|^{p-2}v' + \lambda v^{p^*-1}\right), \tag{36}$$

and from (33),

$$\frac{|v'|^{p-2}v'}{r} = \frac{1}{|H'(v)|^{p-2}H'(v)}. \tag{37}$$

Combining (36) and (37), we have

$$|v'|^{p-2}v'' = \frac{-1}{p-1}\left(\frac{n-1}{|H'(v)|^{p-2}H'(v)} + \lambda v^{p^*-1}\right). \tag{38}$$

We multiply (38) by v' and integrate over (r,∞). Noticing that $\frac{d}{dr}\left(\frac{|v'(r)|^p}{p}\right) = |v'|^{p-2}v'v''$, we have

$$|v'|^p = -\frac{p}{p-1}\left(\frac{\lambda v^{p^*}}{p^*} - (n-1)\int_r^\infty \frac{v'(r)}{|H'(v'(r))|^{p-2}H'(v(r))}\mathrm{d}r\right),$$

i.e.

$$|v'|^p = -\frac{p}{p-1}\left(\frac{\lambda v^{p^*}}{p^*} + (n-1)\int_0^{v(r)} \frac{\mathrm{d}\tau}{|H'(\tau)|^{p-2}H'(\tau)}\right), \tag{39}$$

where we use the substitution $\tau = v(r)$ in the integral term. We insert (38) and (39) into (35) to obtain the ODE in H:

$$\begin{aligned}&\frac{\lambda p}{p^*}v^{p^*}|H'(v)|^{p-2}H''(v) + \lambda|H'(v)|^{p-2}v^{p^*-1}\\ &\quad + p(n-1)|H'(v)|^{p-2}H''(v)\int_0^{v(r)}\frac{\mathrm{d}\tau}{|H'(\tau)|^{p-2}H'(\tau)} = -n.\end{aligned} \tag{40}$$

For convenience, we set

$$G'(t) := \frac{|H'(t)|^{p-2}H'(t)}{p-1}, \tag{41}$$

and rewrite (40) as the first order nonlinear ODE in G':

$$\frac{\lambda p}{p^*}t^{p^*}G''(t) + \lambda(p-1)t^{p^*-1}G'(t) + p'(n-1)G''(t)\int_0^t \frac{\mathrm{d}\tau}{G'(\tau)} = -n, \quad \forall t \in (0, v(0)). \tag{42}$$

To solve (42), we rewrite it as the linear ODE

$$\frac{p}{p^*}t^{p^*}G''(t) + (p-1)t^{p^*-1}G'(t) = -\frac{n+kp'(n-1)}{\lambda}, \tag{43}$$

where

$$G''(t)\int_0^t \frac{\mathrm{d}\tau}{G'(\tau)} = k \tag{44}$$

is assumed to be constant. By a direct computation, (44) gives that

$$G'(t) = -Bt^{\frac{k}{k+1}} \tag{45}$$

for some arbitrary constant $B > 0$, and then (43) reads as

$$\left(p - 1 + \frac{pk}{p^*(k+1)}\right) t^{p^*-\frac{1}{k+1}} = \frac{n+kp'(n-1)}{\lambda B}, \quad \forall t. \tag{46}$$

It is easy to see that (46) holds for all t if and only if

$$\frac{pk}{p^*(k+1)} + p - 1 = 0 \quad \text{and} \quad \frac{n+kp'(n-1)}{\lambda B} = 0, \tag{47}$$

or

$$p^* - \frac{1}{k+1} = 0 \quad \text{and} \quad \frac{pk}{p^*(k+1)} + p - 1 = \frac{n+kp'(n-1)}{\lambda B}. \tag{48}$$

Case 1: If (47) holds, then $k = -p^*/p'$ and B is arbitrary. Therefore,

$$G'(t) = -Bt^{-p^*/p'}, \quad B > 0, \tag{49}$$

is a solution of (42). Combining (41) and (49), we have that

$$H'(t) = -\left((p-1)B\right)^{1/(p-1)} t^{-n/(n-p)}$$

is a solution of (40), i.e.,

$$H(t) = \frac{n-p}{n}\left((p-1)B\right)^{1/(p-1)} t^{-p/(n-p)} - \gamma, \tag{50}$$

for some constant γ. Next, we solve for $v(r)$ in (32) to have that

$$v(r) = \frac{a}{(b + r^{p'})^{(n-p)/p}}, \quad a := \left[\frac{p'(n-p)\left((p-1)B\right)^{1/(p-1)}}{n}\right]^{(n-p)/p}, \quad b := \gamma p'. \tag{51}$$

We substitute (51) into (31) to obtain that (51) solves (31), where b is given by (28) and $a > 0$ is arbitrary.

Case 2: If (48) holds, then $k = \frac{1}{p^*} - 1$ and $B = \frac{n-p}{\lambda p(p-1)}$. Therefore

$$G'(t) = -\frac{n-p}{\lambda p(p-1)} t^{1-p^*} \tag{52}$$

solves (42), i.e.,

$$H'(t) = -\left(\frac{n-p}{\lambda p}\right)^{1/(p-1)} t^{\frac{1-p^*}{p-1}}$$

or

$$H(t) = \frac{(n-p)(p-1)}{p^2}\left(\frac{n-p}{\lambda p}\right)^{1/(p-1)} t^{-\frac{p^2}{(n-p)(p-1)}} - \gamma \tag{53}$$

is a solution of (40), for some constant γ. As before, solving $v(r)$ in (32), we read that

$$v(r) = \frac{\lambda^{(p-n)/p^2}(p/n)^{(p-n)/p}}{(b + r^{p'})^{(n-p)(p-1)/p^2}}, \quad b := \gamma p'. \tag{54}$$

And after substituting (54) into (31), we easily see that (54) does not solve (31). Therefore, the only admissible solutions are the ones given by (51).

Finally, we combine (30) and (51) to show that (27) is a minimizer of problem (20). We conclude the proof by applying Theorem 1. □

Remark 1. From the proof of Theorem 2, we see that NOT every solution $H(t)$ to (40) gives a solution $v(r)$ to (31) via the change of variable (32). Hence the two ODEs (31) and (40) are not equivalent; in fact, the set of solutions of (31) is strictly included in the set of solutions of (40).

Acknowledgment

The author thanks D. Cordero-Erausquin for bringing his attention to this problem during the recent BIRS workshop on geometric inequalities.

References

[1] M. Agueh, N. Ghoussoub, and X. Kang. (2004). *Geometric inequalities via a general comparison principle for interacting gases*, *Geom. Funct. Anal.,* **14**, 215-244.

[2] M. Agueh. (2005). *Existence of solutions to degenerate parabolic equations via the Monge-Kantorovich theory*, *Adv. Differential Equations,* **10**, No. 3, 309-360.

[3] M. Agueh. (2007). *Sharp Gagliardo-Nirenberg inequalities via p-Laplacian type equations*, *Nonlinear Differential Equations Appl.,* (to appear).

[4] T. Aubin. (1976). *Problème isopérimétrique et espaces de Sobolev*, *J. Differential Geometry.,* **11**, 573-598.

[5] G. A. Bliss. (1930). An integral inequality, *Journal of London Math. Soc.* **5.**, 11, 40-46.

[6] S. Bobkov, and M. Ledoux. (2006). *From Brunn-Minkowski to sharp Sobolev inequalities*, Preprint.

[7] D. Cordero-Erausquin, B. Nazaret, and C. Villani. (2004). *A mass transportation approach to sharp Sobolev and Gagliardo-Nirenberg inequalities*, *Adv. Math.*, **182**, 307-332.

[8] E. Lieb. (1983). *Sharp constants in the Hardy-Littlewood-Sobolev and related inequalities*, *Annals of Mathematics.*, **118**, 349-374.

[9] R. J. McCann. (1994). *A convexity theory for interacting gases and equlibrium crystals*, Ph.D Thesis, Princeton University, 14, *AMS*, 1997.

[10] G. Talenti. (1976). *Best constants in Sobolev inequality*, *Ann. Mat. Pura Appl.* **110**, 353-372.

[11] S. T. Rachev, and L. Rüschendorf. (1998). *Mass transportation problems*, Vol I: Theory, Springer-Verlag, New-York.

In: Nonlinear Analysis Research Trends
Editor: Inès N. Roux, pp. 15-25
ISBN 978-1-60456-358-0

Chapter 2

Local Current Density Distribution Effects in the Electrodeposition of Composites

Benedetto Bozzini*
Dipartimento di Ingegneria dell'Innovazione,
Università del Salento, via Monteroni, I-73100 Lecce, Italy,
Ivonne Sgura†
Dipartimento di Matematica,
Università del Salento, via per Arnesano, I-73100 Lecce, Italy

Abstract

In this paper, the mathematical modelling is reported of current density distribution effects, leading to the localisation of cathodic current efficiency during the electroforming of Ni-diamond composites for applications in the fabrication of advanced tools for the cutting and shaping of metals and ceramics. Lack of control of such local effects during the electrochemical fabrication process is related to inconsistencies in the quality of the electroformed tools, caused by hydrogen incorporation and eventually resulting in hydrogen embrittlement failures. The mathematical model - based on the numerical integration of the Laplace equation with randomly selected Neumann and Dirichlet boundary conditions - has been validated with experimental results on the embedding of diamond particles of different sizes in a Ni matrix electroplated from a Watts bath. The dependence of the cathodic current efficiency has been measured as a function of the particle size and compared with the numerical simulations of electrodeposition processes carried out with different fractions of Neumann-to-Dirichlet boundary conditions.

1. Introduction

During the electrodeposition of metal matrix composites with dispersion of electrically insulating ceramic particles, it was shown that the local c.d. distribution can affect the properties of the coatings. In the past compositional effects in alloy plating were stressed [1], but

*E-mail address: benedetto.bozzini@unile.it
†E-mail address: ivonne.sgura@unile.it, Corresponding author

the effects are more general and can have a bearing on several functional characteristics of the electrodeposits. In the peculiar case of Ni-matrix deposits for tribological applications or tool fabrication, severe inconsistencies in the mechanical properties of composite coatings as well as in-service exfoliation of grinding wheels have been reported in the industrial practice. These mechanical problems can be explained on the basis of hydrogen incorporation in the coating, giving rise to embrittlement and adhesive failure of the coatings. The statistical nature of the realisation of a specific c.d. distribution during a plating batch as well as use of different granulometries of ceramics or otherwise identical plating processes gives rise to differences in the average cathodic efficiencies to which poorly predictable hydrogen incorporation levels can be connected. Lack of understanding and control on this factor might have a remarkable impact on the industrial acceptability of composite coatings, which have proved outstanding average performance of composite Ni-base electrodeposits in specific applications, but still exhibit cases of unpredictable anticipated failure which can be unacceptable for some high-technology or critical applications such as precision and automated machining. In the specific case of nickel-matrix nickel plating, the local and average current efficiencies are expected to be critical factor as far as the mechanical properties are concerned.

More specifically, for a given plating process, once the operating parameters (i.e. nominal c.d., ceramic size distribution and packing fraction of the ceramic in the cell) are fixed, the actual local c.d. distribution is still in principle a random function of the local, instantaneous packing geometry at the cathode. The actual deposit properties are therefore, at least in principle, a result of the latest, uncontrolled, factor. As a matter of fact, particle incorporation gives rise to an instantaneous variation of the degree of coverage of the cathode with insulating material. This fact, in turn, brings about the growth of metal with different catalytic activity and consequently sites with different activity for hydrogen reduction and incorporation. Of course, in the case of baths containing additives, the problem is still more complex, because the local, instantaneous variations of electrode activity gives rise to different local additive decomposition and incorporation rates.

It is consequently important to try and gain some degree of quantitative control over the expected dispersion of the properties as a function of the controllable plating parameters.

In this paper we propose a combination of mathematical modelling and experimental investigation endeavoring to put the problem of local c.d. distribution brought about by a random distribution of insulating particles on an electrode surface on a mathematical basis. Even though in this specific paper the focus is chiefly on the electrodeposition of composites, the same approach and modelling tools apply also for metal and alloy electrodeposition processes in the presence of insulating templates (of course either ceramic, such as in electrocatalysis applications or organics as in the case of plating through resin masks or self-assembled layers), of porous organic films, such as for electroplating in the presence of inhibitor films or thorough insulating polymer films or in corrosion processes going on through damaged or randomly porous insulating films (such as paints or conversion coatings).

2. Mathematical Modeling and Numerical Simulations

The mathematical model for the electrodeposition on an electrode surface can be formulated by means of the Laplace equation for the potential distribution $u(x,y)$ with $(x,y) \in \Omega$. For our purposes, we consider the rectangular domain $\Omega = [0,a] \times [0,b]$ where mixed boundary conditions are assigned and such that one of its sides represents the electrode surface. Let us consider

$$\begin{cases} \Delta u = 0 \qquad (x,y) \in \Omega \\ n^T c \nabla u + d\, u(x,y) = g(x,y) \quad (x,y) \in \delta\Omega, \end{cases} \tag{1}$$

where n is the normal unit vector and c and d are assigned constant values. The current distribution is obtained by considering the gradient $\nabla u(x,y)$ in Ω and in particular on the electrode boundary.

In particular, we fix zero Neumann conditions along the lines $(x,0)$ and (x,b) (up and bottom sides) and zero Dirichelet boundary condition along the right vertical side (a,y). Hence, we have

$$\frac{\partial u}{\partial n}(x,0) = \frac{\partial u}{\partial n}(x,b) = 0,\ 0 \le x \le a \qquad u(a,y) = 0,\ 0 \le y \le b.$$

Along the left side, that is on the segment $[0,b]$ we consider Dirichelet and Neumann conditions alternated in order to simulate the electrode surface with a given distribution of insulating particles.

For a fixed $b > 0$, we consider a given coverage θ of the electrode by N particles. As a first application we consider $\theta = 50\%$. This is formulated mathematically by assuming that $L_0 \equiv L_0^N = \bigcup_{i=1}^N D_i \cup \bigcup_{i=1}^N N_i$, where D_i is i-th empty zone and corresponds to imposed Dirichelet b.c., whereas N_i is the the zone occupied by the i-th insulating particle and corresponds to Neumann b.c..

The size of each zone is then determined by $h = \frac{b}{2N}$, but the distribution along L_0^N of the particles is not uniform. In fact, for a fixed N, all the possible combinations of N particles on $2N$ positions can be obtained in combinatoric way and then there exist say c_N possible configurations of L_0^N, that are given by the combinations $c_N = \begin{pmatrix} 2N \\ N \end{pmatrix} = \frac{(2N)!}{N!N!}$. Of course a large number of N would imply a prohibitively large number of possibilities.

The analytic solution of the Laplace equation (1) with Neumann or Dirichelet b.c is known only on very simple geometries, but, at our knowledge, the case of mixed b.c. is very complicated. For this reason we look for a numerical approximation of (1) and for this aim we use the *Partial Differential Equation (PDE) Toolbox* of *Matlab 6.5* [2] that is based on the finite element method. This choice allows us to find the numerical approximation of each PDE model (1) for a given set of boundary conditions L_0 by using all the facilities provided by this software environment (like mesh generation and adaption, geometry settings, etc.). Moreover, the Matlab function programming allows to implement easily all the choices of L_0 required and also to process a posteriori the numerical solutions of all the c_N PDE models solved to extract information about the current distribution on the electrode surface.

The first remark is that even for $N = 1$ at the interface between different b.c.s the solution of (1) is continuous but presents a discontinuity in the first derivative. Hence, when the current distribution along the boundary L_0 and at the corners of Ω is considered, a finite number of singularities is present.

Since we are interested in extracting information related to the d.c. along L_0, a very accurate numerical approximation is needed near this zone. For this reason each PDE model is approximated on a non uniform mesh M_N that is finer, that is with a big number of triangles N_T, in proximity of the side L_0. Because we expect that also the number of singularities increases with N, we consider a value of N_T sufficiently high for each N and that increases with the value of N. If we indicate as $p_N = p(N)$ the number of meshpoints along L_0, the criterion of choice for each p depends on the accuracy (and stability) required for the information, as will be explained in the following into more details.

Let $v_N = [v_1^N, \ldots, v_p^N]^T$ be the numerical approximation of the current distribution along L_0. Therefore, the information extracted for each N and for a fixed configuration of L_0 are the following:

i) $m_N = mean(v_N)$,

ii) $s_N = std(v_N)$,

iii) the approximation I_1 of the L_1-norm of the solution, that is the integral $I_1 \simeq \|v\|_1 := \frac{1}{b}\int_0^b |v(y)|dy$. $\|v\|_1$ represents the continuous mean value $< \bar{v} >$ of the continuous (unknown) current distribution $v(y) := \frac{\partial u}{\partial y}_{|x=0}$ along the electrode surface.

iv) the approximation I_2 of the following L_2-norm $I_2 \simeq \|v- < \bar{v} > \|_2 := \left(\frac{1}{b}\int_0^b |v(y)- < \bar{v} > |^2 dy\right)^{1/2}$.

Note that, by previous assumptions, the function $v(y)$ could be considered as a random variable, then the above norms correspond to the first and the second moment of the current distribution along the boundary. The following statistical analysis is based on this consideration.

We calculate I_1 and I_2 by means of the Composite Trapezoidal Rule using the space discretization and the numerical values of the current given by each PDE approximation. Note that, it is easy to prove that the $I_1 \equiv m_N$ and $I_2 \equiv s_N$ if the Rectangular instead of the Trapezoidal Rule is used. It is well known that the Trapezoidal Rule is a more precise (second order) quadrature approximation formula of the above integrals. We prefer this choice to have more accuracy in the calculation of the areas near the singularity zones.

We note that, if the number p of meshpoints along L_0 is not *sufficiently high*, the error in the approximations I_1 and I_2 of the integrals could be very high. For each fixed N, to determine the value p_N used in the following simulations, we have considered a set of meshes of increasing size $p^1 \leq p^2 \leq \cdots \leq p^m$ and we set $p_N := p^m$ such that the approximations $|I_1(p^m) - I_1(p^{m-1})| \leq$ 1e-2 and $|I_2(p^m) - I_2(p^{m-1})| \leq$ 1e-2 hold.

Moreover, as observed before, for each N, c_N different particle distributions are possible, that is c_N PDE models (1) have to be approximated. Let $v_N^{(i)}, i = 1, \ldots, c_N$ be all the numerical solutions obtained for all the possible particle distributions along L_0 and for the fixed mesh M_N.

Therefore, for a fixed N indeed we generate the following information:

i) $\mathbf{m_N} = [m_N^{(1)}, \ldots, m_N^{(c_N)}]^T, m_N^{(i)} = mean(v_N^{(i)}), i = 1, \ldots, c_N;$

ii) $\mathbf{s_N} = [s_N^{(1)}, \ldots, s_N^{(c_N)}]^T, s_N^{(i)} = std(v_N^{(i)}), i = 1, \ldots, c_N;$

iii) $\mathbf{I_1} = [I_1^{(1)}, \ldots, I_1^{(c_N)}]^T;$

iv) $\mathbf{I_2} = [I_2^{(1)}, \ldots, I_2^{(c_N)}]^T.$

The numerical results obtained for $N = 1, \ldots, 7$, on the rectangular domain $\Omega = [0, 10] \times [0, 12]$ ($b = 12$) are reported in Figure 1.

Moreover, to synthesize all these information, we calculate also the mean and the standard deviation of the above vectors. Hence, let us define

$\bar{\mathbf{m}}_\mathbf{N} = \frac{1}{c_N}\sum_{i=1}^{c_N} m_N^{(i)}$ and $\widetilde{\mathbf{m}}_\mathbf{N} = \sqrt{\frac{\sum_{i=1}^{c_N}(m_N^{(i)})^2}{c_N - 1}}$. Similarly, let us define $\bar{\mathbf{s}}_\mathbf{N}$, $\bar{\mathbf{I}}_\mathbf{1}$, $\bar{\mathbf{I}}_\mathbf{2}$ and , $\widetilde{\mathbf{s}}_\mathbf{N}$, $\widetilde{\mathbf{I}}_\mathbf{1}$, $\widetilde{\mathbf{I}}_\mathbf{2}$, respectively.

We observe that another piece of information can be useful, that is the number of effective conducting zones on L_0. In fact, among the c_N configurations only $N + 1$ of them are of length $h = b/2N$, in the other cases two or more conducting zones can be near each other and then they become equivalent to a conductors of twice or more length. We report also the percentage values

$$\% = \frac{\text{number of conducting zones}}{N}$$

to emphasize this situation and to show how the behavior of the above $means$ and the $std's$ values depend on these different distributions (always the 50% of the electrode surface is covered but with patches of different sizes and in many cases different from h).

In Figure 2, we show the same information as before but only for the configurations where N effective conductive zones of size h are present. The same trend as in Figure 1 can be observed, that is *for increasing values of N the mean value of current distribution increases*.

2.1. Electrode Coverage of $\theta = 25\%$ and $\theta = 75\%$ of Insulating Particles

For a fixed $b > 0$, we consider the coverage of the electrode by N particles such that $\theta = 25\%$. This is formulated mathematically by assuming that $L_0 \equiv L_0^N = \bigcup_{i=1}^{N} D_i \cup \bigcup_{i=1}^{3N} N_i$,

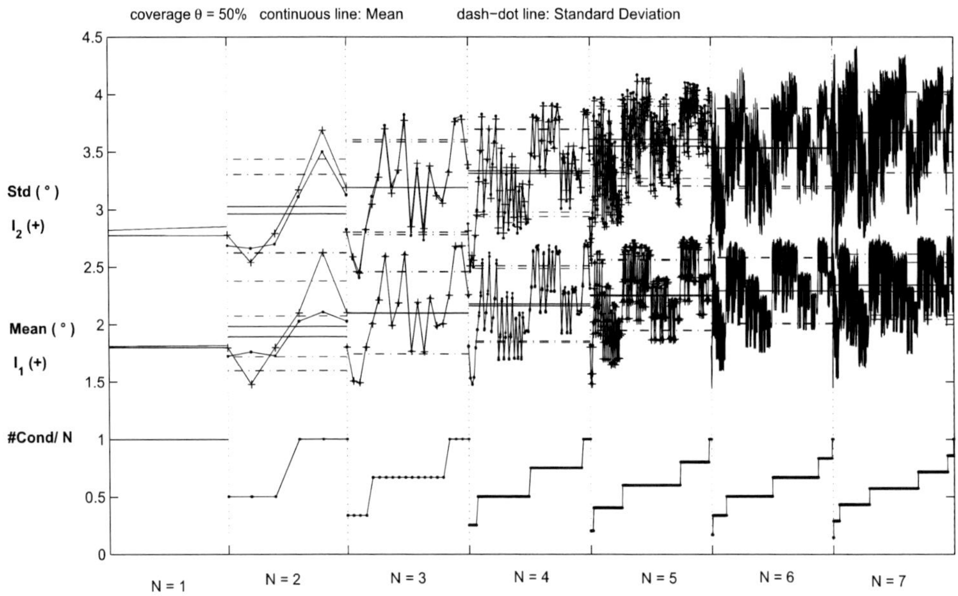

Figure 1. Coverage $\theta = 50\%$: at most N conductive zones of size $h = b/2N$ are present in L_0.

where D_i is the i-th empty zone and corresponds to imposed Dirichelet b.c., whereas N_i is the zone occupied by the i-th insulating particle and corresponds to Neumann b.c.. The size of each zone is then determined by $h = \frac{b}{4N}$, but again the distribution along L_0^N of the particles can be non uniform. In this case, for a fixed N, all the possible combinations of N particles on $4N$ possible locations on the electrode have to be considered. In fact, this situation corresponds to calculate all the ways in which N objects can be grouped (when the position does not matter) among $4N$ given objects. Hence they are given by the combinations $c_N = \begin{pmatrix} 4N \\ N \end{pmatrix} = \frac{(4N)!}{N!(3N)!}$. In the following simulation we consider the values of $N \leq 4$ since a larger value would imply $c_N > 10^4$ choices and a huge computational cost.

We remark that a similar analysis holds when the electrode coverage is of $\theta = 75\%$. In this case we can assume that $\bar{N} = 3N$ particles are present and that $L0 \equiv L_0^N = \bigcup_{i=1}^{\bar{N}} D_i \cup \bigcup_{i=1}^{N} N_i$. If we refer to the empty zones, the same number c_N of possible configurations as in the case of $\theta = 25\%$ is present. In Figures 3 and 4 we report the numerical results concerning the cases of coverages $\theta = 25\%$ and $\theta = 75\%$, respectively.

A remark holds for the percentage information in the case of $\theta = 75\%$. In fact, in this case we report c where c is the number of conducting zones only for $N = 1, 2$ (that is $\bar{N} = 3, 6$)

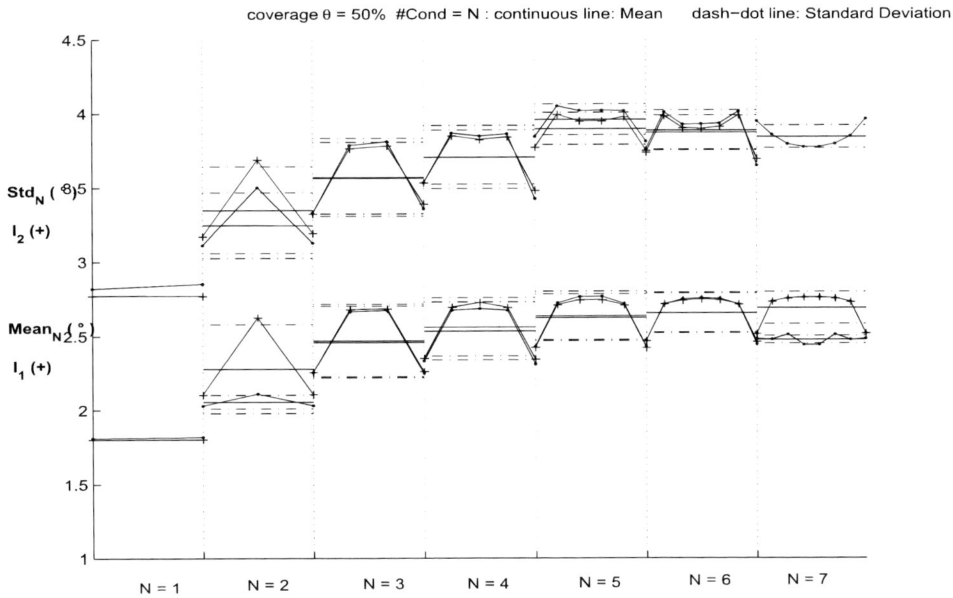

Figure 2. Coverage $\theta = 50\%$: exactly N conductive zones of size $h = b/2N$ are present in L_0, b is the length of the electrode.

3. Validation Experiments

A Watts-type nickel plating bath was used, with the following composition: $NiSO_4$ 300 g L^{-1}, $NiCl_2$ 50 g L^{-1}, H_3BO_4 30 g L^{-1}. An industrial process was simulated for the fabrication of the cutting layer of drills: graphite bars of diameter 0.5 cm were employed. Ni electroplating was used to fix three grades of diamond powders of diameters 50, 250 and 500 μm. Cylinders of diameter 2 cm and height 5 cm, closed at one end, were fabricated with porous PVC foil. The graphite bars were fixed along the axis of these cylinders and the remaining free space was filled with diamond powder. The assembly was plunged in the axial position of a 1 L beaker (Figure 5, position A), thermostated to 60$\pm$ 1°C. Fours soluble anodes, closed in fabric bags, were hung in diametrically opposite positions (Figure 5, positions B). A potential of 3V was applied between the cathodic bar and the anodes. Electrodeposition was carried out for 4 hours. 10 experiments were run with each type of diamond granulometry.

According to the models developed by the authors in [1, 3, 4] and assuming that the particles form a compact monolayer in contact with the electrode, the ceramic particles give rise to local screening of the cathode, whence a local current density distribution results, that tends to peak in the spaces left open by particle contact. In the ideal case of a hexagonal lattice of spherical insulating particles (Figure 6), the current density at the underlying cathode would be vanishing in position A, intermediate in B and maximum in C. It is of interest to understand the relationship between the local current density and the one would obtain with a free cathode surface in the absence of ceramic particles. Of course the hydrogen evolution reaction and the current efficiency η_c are affected by the current

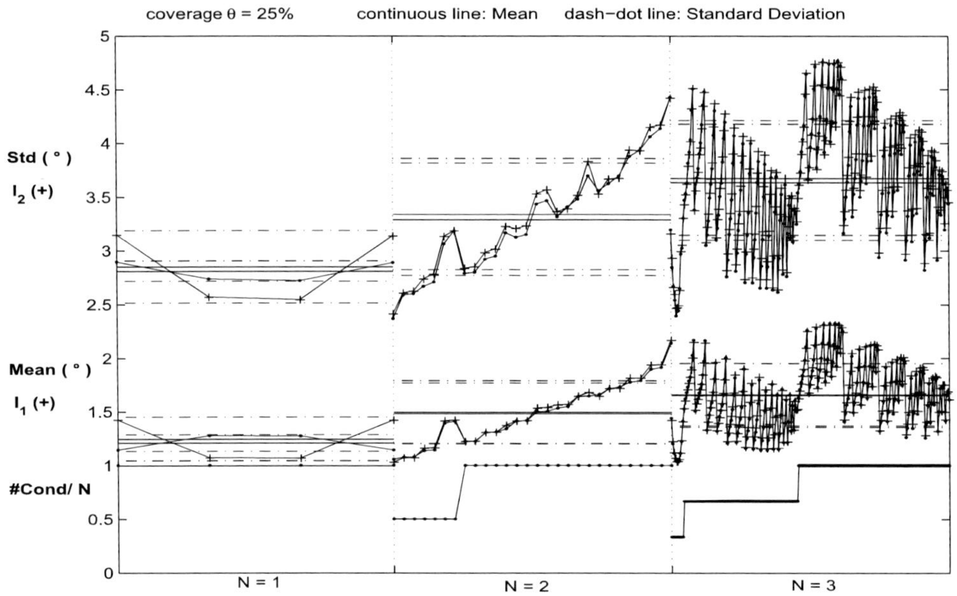

Figure 3. Coverage $\theta = 25\%$: at most N conductive zones of size $h = b/4N$ are present in L_0.

density distribution resulting from the presence of the particle layer at the cathode. Under the hypothesis of primary current density distribution - an acceptable approximation in the present case, where the electrolyte is essentially immobilised and the cathodic polarisation is relatively low - the nominal current density distribution was computed under the hypothesis of homogeneity. The average current density can be estimated rather straightforwardly by measuring η_c. This was done by dissolving the Ni binder in concentrated HNO_3; the composition of the solution thus obtained was analysed by VIS-UV absorption spectrometry. Compositional data, obtained by the wet-chemical method were compared with the integrated charge consumed in the process, to yield estimates of the cathodic current efficiency η_c; our results are reported in Table 1. Estimates of surface coverage with insulating material θ_{ins}, after [1] are also reported. Notwithstanding the experimental scatter, with the help of replicated experiments, a statistically meaningful trend could be measured. This result proves that the highest fraction of the variance among the estimated η_c values is explained by the granulometry of diamond. Other factors, such as preparation of the powder compact, geometrical irreproducibilities of the setup and lack of detailed control over bath chemistry during batch operation, contribute to a lesser extent.

Acceptable qualitative matching was obtained between the computations reported in Section 2 and the experimental results. Of course, the achievement of quantitative correspondence would be an ambitious task, implying strict control of the contact geometry between cathode and powder compact and computations including tertiary current density distribution, non linear boundary conditions with at least two electrochemical reactions and time-evolving geometry.

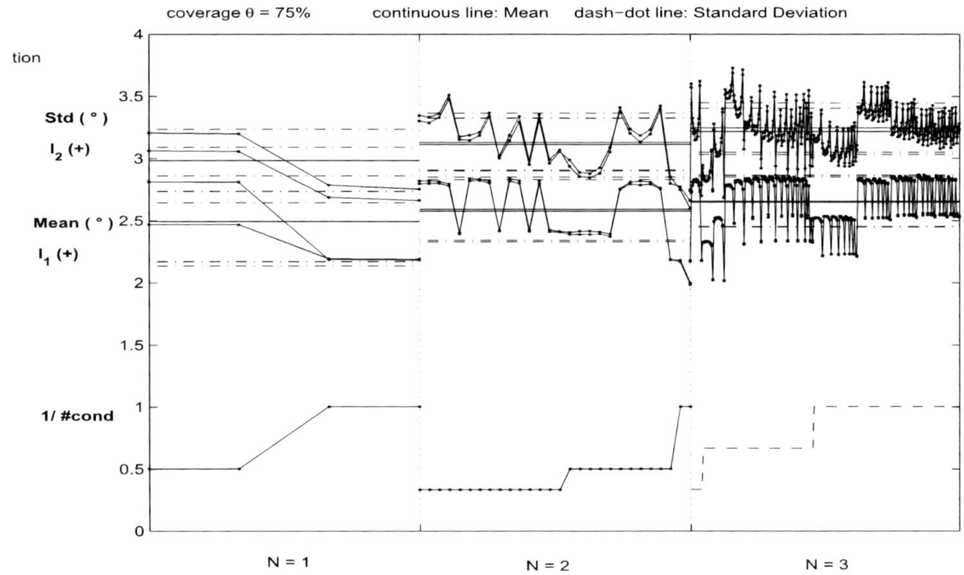

Figure 4. Coverage $\theta = 75\%$: at most N insulating zones ($\bar{N} = 3N$ conductive zones) of size at least $h = b/4N$ are present in L_0.

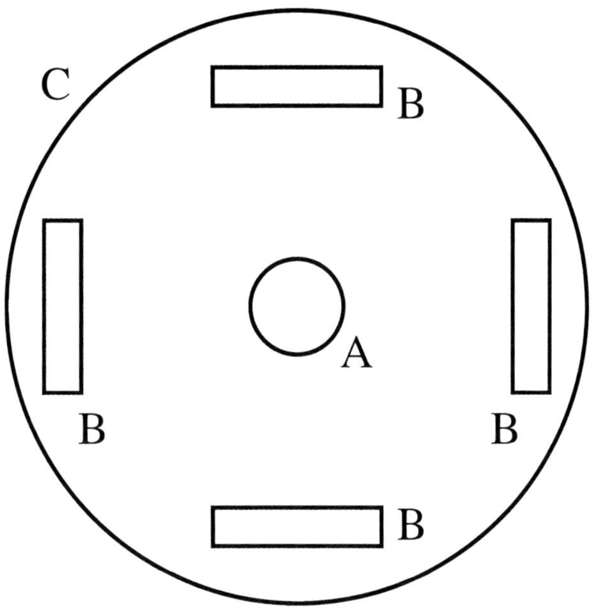

Figure 5. Schematic top view of the electrodeposition cell: A cathode, B anodes, C tank.

4. Conclusions

In this paper we show, with a simple numerical procedure and with a set of experiments close to industrial applications, that the granulometry of ceramic powders employed in the

Table 1. Estimated surface coverage with insulating particles θ_{ins} and measured cathodic current efficiency η_c and its standard deviation $\sigma(\eta_c)$, as a function of the particle diameter d_p.

$d_p/\mu m$	θ_{ins}	η_c	$\sigma(\eta_c)$
50	0.60	0.63	0.12
250	0.12	0.71	0.17
500	0.06	0.87	0.11

electrodeposition of metal matrix composites with electrically insulating ceramic dispersoids, has a bearing on the current density distribution. The local properties of the current density distribution, in turn, affect the ratio of rates of potentially concurrent electrochemical reactions. In the case of alloy plating, this phenomenon has been shown to give rise to compositional heterogeneity [1], in the case of electrodeposition systems with concurrent hydrogen evolution, like the one considered experimentally in this research, uncontrolled hydrogen uptake can occur since in some locations of the cathode, higher than average current densities can flow; typical practical precautions of limiting the average current density to a given value, compatible with the absence of hydrogen-embrittlement, can thus prove ineffective.

Notwithstanding the special case chosen for experimental demonstration, the results of this research can be generalised to all cases in which partial coverage of the cathode with insulating layers are relevant and more than one electrochemical reaction is expected.

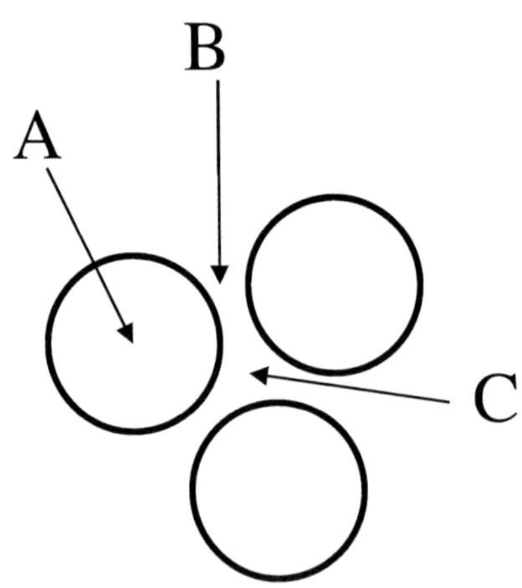

Figure 6. Schematic description of the current density distribution at an electrically conducting cathode covered with a cluster of spherical insulating particles.

References

[1] B. Bozzini, G. Giovannelli, P. L. Cavallotti, *J. Appl. Electrochem.* **29**, 685 (1999).

[2] Matlab 6.5, Release 13, PDE Toolbox, Reference users's guide.

[3] B. Bozzini, G. Giovannelli, P. L. Cavallotti, *J. Microsc. (Oxford)* **185**, 283 (1997).

[4] B. Bozzini, B. Brevaglieri, P.L. Cavallotti, G. Giovannelli, S. Natali. *Electrochim. Acta* **45**, 3431 (2000).

In: Nonlinear Analysis Research Trends
Editor: Inès N. Roux, pp. 27-97
ISBN 978-1-60456-358-0

Chapter 3

A Spectral Condition for Variational Inequalities: Theory, Methods and Applications in Electronics

***K. Addi*[*] *and D. Goeleven*[†]**
IREMIA, University of La Reunion,
97400 Saint-Denis, France

Abstract

In this paper, we study a large class of variational inequalities that can be used to deal with some highly nonlinear phenomena in electronics like switching and clipping. Using recession tools like recession cones and recession functions, we first introduce a new class of problem that we call a "semi-complementarity problem". Then we show that the study of a large class of variational inequalities of the second kind can be reduced to the one of a generalized eigenvalue problem governed by some semi-complementarity problem. A general spectral condition is then used to proceed to a qualitative mathematical analysis of the aforementioned class of variational inequalities. Finally, using our theoretical results, we present a suitable methodology for the formulation and the mathematical analysis of circuits in electronics specially those using semi-conductors like diodes and transistors. Various examples are then presented showing that our theory is applicable to determine qualitative properties of such circuits. Moreover numerical simulations are given with the examples to support the theoretical results.

Keywords: Variational inequalities, Complementarity problems, spectral condition, Recession functions, Recession cones, Set-valued ampere-volt characteristics in electronics.

1. Introduction

In this paper, we discuss a mathematical approach that can be used to derive conditions for the existence and uniqueness of a general class of variational inequalities of the second

[*]E-mail address: addi@univ-reunion.fr
[†]E-mail address: goeleven@univ-reunion.fr

kind, i.e. variational inequalities of the form: Find $u \in \mathbb{R}^n$ such that

$$\langle Mu + q, v - u \rangle + \Phi(v) - \Phi(u) \geq 0, \ \forall v \in \mathbb{R}^n, \tag{1}$$

where $M \in \mathbb{R}^{n \times n}$ is a real matrix, $q \in \mathbb{R}^n$ a vector and $\Phi : \mathbb{R}^n \to \mathbb{R} \cup \{+\infty\}$ a proper convex and lower semicontinuous function.

We develop an approach using recession tools and Brouwer degree theory to reduce the study of problem (1) to the one of the eigenvalue problem of the semi-complementarity problem:

$$\left\{ \begin{array}{l} z \in D(\Phi)_\infty \\ \\ Mz - \mu z \in D(\Phi_\infty)^* \\ \\ \langle (M - \mu I)z, z \rangle \leq 0, \end{array} \right. \tag{2}$$

where $D(\Phi)_\infty$ is the recession cone of the domain of Φ and $D(\Phi_\infty)^*$ is the dual cone of the domain of the recession function Φ_∞ of Φ.

Denoting by $\sigma_\infty(M, \Phi)$ the set of generalized eigenvalues of (M, Φ), i.e.

$$\sigma_\infty(M, \Phi) := \{\mu \in \mathbb{R} : \exists \ z \in \mathbb{R}^n \backslash \{0\} \text{ solution of } (2)\},$$

we show that this approach may be used to state a general spectral condition

$$\sigma_\infty(M, \Phi) \subset]0, +\infty[$$

that can be employed to discuss the qualitative properties of the solution(s) of Problem (1) for various important classes of matrices M and functions Φ and expose a general methodology that can be used to study problem (1) for various classes of possibly singular and possibly non symmetric matrices.

Then we show that problem (1) and our theoretical results are mandatory studying for the formulation and mathematical analysis of electrical devices and circuits in electronics like diodes, transistors, operational amplifiers, clipping circuits, slicers, sampling gates, four-diode bridge and full-wave rectifiers.

Such circuits are the basis of a large class of electrical engineering and industrial applications. To design such circuits it is common to use simulation softwares like SPICE [7]. Our approach could be a complement or an alternative for these circuits. For example, to simulate a circuit with diodes using SPICE it is necessary to replace diode by an equivalent circuit of 25 parameters. Our approach could be faster with more qualitative results if the application does not require a high diode modelling level.

Notation: For $x, y \in \mathbb{R}^n$, the notation $\langle x, y \rangle = \sum_{i=1}^{n} x_i y_i$ is used to denote the euclidean scalar product on $\mathbb{R}^n$ and $\|x\| = \sqrt{\langle x, x \rangle}$ to denote the corresponding norm. The identity mapping on $\mathbb{R}^n$ will be denoted by $id_{\mathbb{R}^n}$ while the identity matrix of order n is denoted by I. We will also denote by $\{e^1, ..., e^n\}$ the canonical basis of $\mathbb{R}^n$.

Let $M \in \mathbb{R}^{n\times n}$ be a real matrix of order n. We denote by $\sigma_{\mathbb{R}}(M)$ the set of real eigenvalues of M.
Let $k \in \mathbb{N}, 1 \le k \le n$. One denotes by $M^{i_1 i_2 \ldots i_k}_{i_1 i_2 \ldots i_k}$ the principal submatrix of M determined by the index set $\{i_1, i_2, \ldots, i_k\}$ and by $\Delta(M)_{i_1 i_2 \ldots i_k} := \det(M^{i_1 i_2 \ldots i_k}_{i_1 i_2 \ldots i_k})$ the corresponding k-by-k principal minor of M.

2. Mathematical Tools

2.1. Mathematical Tools in Convex Analysis

The aim of this Section is to recall some notions and fundamental results in convex analysis which will be used throughout this paper.

[Convex function] The following classes of convex functions will be used throughout this paper:

• We denote by $\Gamma_0(\mathbb{R}^n; \mathbb{R} \cup \{+\infty\})$ the set of proper, convex and lower semicontinuous functions from $\mathbb{R}^n$ to $\mathbb{R} \cup \{+\infty\}$. The domain $D(\Phi)$ of Φ is defined by:

$$D(\Phi) = \{x \in \mathbb{R}^n : \Phi(x) < +\infty\}.$$

• We denote by $\Gamma(\mathbb{R}^n; \mathbb{R} \cup \{+\infty\})$ the set of proper convex and lower semicontinuous functions $\Phi : \mathbb{R}^n \to \mathbb{R} \cup \{+\infty\}$ with closed domain, i.e.:

$$\Gamma(\mathbb{R}^n; \mathbb{R} \cup \{+\infty\}) := \{\Phi \in \Gamma_0(\mathbb{R}^n; \mathbb{R} \cup \{+\infty\}) : D(\Phi) = \overline{D(\Phi)}\}.$$

• We denote by $\Gamma_B(\mathbb{R}^n; \mathbb{R} \cup \{+\infty\})$ the set of proper convex and lower semicontinuous functions $\Phi : \mathbb{R}^n \to \mathbb{R} \cup \{+\infty\}$ with bounded and closed domain, i.e.:

$$\Gamma_B(\mathbb{R}^n; \mathbb{R} \cup \{+\infty\}) := \{\Phi \in \Gamma(\mathbb{R}^n; \mathbb{R} \cup \{+\infty\}) : D(\Phi) \text{ is bounded}\}.$$

• We denote by $P\Gamma(\mathbb{R}^n; \mathbb{R} \cup \{+\infty\})$ the set of functions $\Phi : \mathbb{R}^n \to \mathbb{R} \cup \{+\infty\}$ satisfying the partitionability property:

$$\Phi(x) = \Phi_1(x_1) + \Phi_2(x_2) + \ldots + \Phi_n(x_n),\ \forall x \in \mathbb{R}^n, \tag{3}$$

where, for all $1 \le i \le n$, we have

$$\Phi_i \in \Gamma(\mathbb{R}; \mathbb{R} \cup \{+\infty\}). \tag{4}$$

• We denote by $D\Gamma(\mathbb{R}^n; \mathbb{R} \cup \{+\infty\})$ the set of functions $\Phi : \mathbb{R}^n \to \mathbb{R} \cup \{+\infty\}$ satisfying (3), (4) and the following property:

$$\Phi_i(\lambda x) = \lambda \Phi_i(x),\ \forall \lambda \ge 0,\ \forall x \in D(\Phi_i). \tag{5}$$

• We denote by $SD\Gamma(\mathbb{R}^n; \mathbb{R}\cup\{+\infty\})$ the set of functions $\Phi : \mathbb{R}^n \to \mathbb{R}\cup\{+\infty\}$ satisfying the conditions (3), (4), (5) and such that the functions Φ_i $(1 \leq i \leq n)$ in (3) are strictly convex.

[Convex subdifferential] Let $\Phi \in \Gamma_0(\mathbb{R}^n; \mathbb{R}\cup\{+\infty\})$ be given. The convex subdifferential $\partial\Phi(x)$ (see e.g. [34], [52]) of Φ at x is defined by:

$$\partial\Phi(x) = \{w \in \mathbb{R}^n : \Phi(v) - \Phi(x) \geq \langle w, v - x\rangle, \forall v \in \mathbb{R}^n\}.$$

The set $\partial\Phi(x)$ describes the differential properties of Φ by means of the supporting hyperplanes to the epigraph of Φ at $(x, \Phi(x))$.

[Fenchel transform] Let $\Phi \in \Gamma_0(\mathbb{R}^n; \mathbb{R} \cup \{+\infty\})$ be given. The Fenchel transform Φ^* of Φ is the function defined by:

$$\Phi^*(z) = \sup_{x \in D(\Phi)} \{\langle x, z\rangle - \Phi(x)\}, \ (z \in \mathbb{R}^n).$$

The function $\Phi^* : \mathbb{R}^n \to \mathbb{R} \cup \{+\infty\}$ is proper convex and lower semicontinuous. A well known result in convex analysis (see e.g. [34], [52]) ensures that:

$$z \in \partial\Phi(x) \Longleftrightarrow x \in \partial\Phi^*(z) \Longleftrightarrow \Phi(x) + \Phi^*(z) = \langle x, z\rangle.$$

[Closed convex set] Let $K \subset \mathbb{R}^n$ be a nonempty closed convex set. We denote by Ψ_K the indicator function of K, that is:

$$\Psi_K(x) := \begin{cases} 0 & \text{if } x \in K \\ +\infty & \text{if } x \notin K \end{cases}, \quad (x \in \mathbb{R}^n). \tag{6}$$

Then

$$\partial\Psi_K(x) = \begin{cases} \{w \in \mathbb{R}^n : \langle w, v - x\rangle \leq 0, \forall v \in K\} & \text{if } x \in K \\ \emptyset & \text{if } x \notin K \end{cases}$$

and

$$\Psi_K^*(z) = \sup_{x \in K} \langle x, z\rangle, \quad (z \in \mathbb{R}^n).$$

The dual cone of K is the nonempty closed convex cone K^* defined by

$$K^* := \{w \in \mathbb{R}^n : \langle w, v\rangle \geq 0, \ \forall v \in K\}. \tag{7}$$

[Recession function] Let x_0 be any element in $D(\Phi)$. The recession function of Φ is defined by

$$\Phi_\infty(x) = \lim_{\lambda \to +\infty} \frac{1}{\lambda}\Phi(x_0 + \lambda x) \quad (x \in \mathbb{R}^n).$$

The function $\Phi_\infty : \mathbb{R}^n \to \mathbb{R} \cup \{+\infty\}$ is a proper convex and lower semicontinuous function which describes the asymptotic behavior of Φ.

[Recession cone] Let x_0 be some arbitrary element of K. The recession cone of K is defined by

$$K_\infty = \bigcap_{\lambda > 0} \frac{1}{\lambda}(K - x_0).$$

The set K_∞ is a nonempty closed convex cone that is described in terms of the directions which recede from K.

Let us here recall some important properties of the recession function and recession cone (see Section 1.4 in [28]):

Proposition 1 *a) Let $\Phi : \mathbb{R}^n \to \mathbb{R} \cup \{+\infty\}$ be a proper convex and lower semicontinuous function. Then*

$$\Phi_\infty(\alpha x) = \alpha \Phi_\infty(x), \ (\alpha \geq 0, \ x \in \mathbb{R}^n), \tag{8}$$

$$\Phi_\infty(x) = \liminf_{t \to +\infty, v \to x} \frac{\Phi(tv)}{t}, \ (x \in \mathbb{R}^n), \tag{9}$$

$$(x \in D(\Phi), e \in \mathbb{R}^n) \implies \Phi_\infty(e) \geq \Phi(x + e) - \Phi(x). \tag{10}$$

b) Let $\Phi_1 : \mathbb{R}^n \to \mathbb{R} \cup \{+\infty\}$ and $\Phi_2 : \mathbb{R}^n \to \mathbb{R} \cup \{+\infty\}$ be two proper, convex and lower semicontinuous functions. Then

$$(\Phi_1 + \Phi_2)_\infty(x) \geq (\Phi_1)_\infty(x) + (\Phi_2)_\infty(x), \ (x \in \mathbb{R}^n). \tag{11}$$

c) Let $K \subset \mathbb{R}^n$ be a nonempty, closed and convex set. Then

$$(\Psi_K)_\infty(x) = \Psi_{K_\infty}(x), \ (x \in \mathbb{R}^n), \tag{12}$$

$$(x \in K, \ e \in K_\infty) \implies x + e \in K. \tag{13}$$

d) Let $K \subset \mathbb{R}^n$ be a nonempty closed and convex cone. Then

$$K_\infty = K. \tag{14}$$

e) Let $K \subset \mathbb{R}^n$ be a nonempty compact and convex set. Then

$$K_\infty = \{0\}. \tag{15}$$

[Proximation operator] Let $\Phi \in \Gamma_0(\mathbb{R}^n; \mathbb{R} \cup \{+\infty\})$ be given. It is known that for each $y \in \mathbb{R}^n$, there exists a unique $x \in \mathbb{R}^n$ such that

$$\langle x - y, v - x \rangle + \Phi(v) - \Phi(x) \geq 0, \; \forall v \in \mathbb{R}^n,$$

that is

$$y \in x + \partial\Phi(x).$$

The mapping $P_\Phi : \mathbb{R}^n \to \mathbb{R}^n; y \mapsto P_\Phi(y)$, called the proximation operator (see e.g. [51]), and defined by

$$P_\Phi(y) = (id_{\mathbb{R}^n} + \partial\Phi)^{-1}(y), \; (y \in \mathbb{R}^n) \tag{16}$$

is thus a well-defined singled-valued operator. Moreover, it is easy to check that:

$$y \in x + \partial\Phi(x) \iff x = (id_{\mathbb{R}^n} + \partial\Phi)^{-1}(y) \iff x = \operatorname{argmin}_{v \in \mathbb{R}^n}\{\frac{1}{2}||v - y||^2 + \Phi(v)\}.$$

If K is a nonempty closed convex set, then

$$P_{\Psi_K} \equiv P_K$$

where P_K denotes the projector from $\mathbb{R}^n$ onto K, i.e.

$$P_K(x) = \operatorname{argmin}_{v \in K}\{\frac{1}{2}||v - x||^2\}.$$

2.2. Mathematical Tools in Nonlinear Analysis

Let us here recall some properties of Brouwer topological degree that we will use to prove our next result. Let $D \subset \mathbb{R}^n$ be an open and bounded set. If $f : \bar{D} \to \mathbb{R}^n$ is continuous and $0 \notin f(\partial D)$ then the Brouwer topological degree of f with respect to D and 0 is well-defined (see e.g. [43]) and denoted by $\deg(f, D, 0)$. The Brouwer topological degree $deg(f, D, 0)$ is an integer that estimates the number of solutions of $f(x) = 0$ in D and which is continuous in f and additive in the domain D. More precisely, the following properties hold:

P1. Solution property: If $0 \notin f(\partial D)$ and $\deg(f, D, 0) \neq 0$ then there exists $x \in D$ such that $f(x) = 0$.

P2. Homotopy invariance property: Let $\varphi : [0,1] \times \bar{D} \to \mathbb{R}^n; (\lambda, x) \mapsto \varphi(\lambda, x)$, be continuous such that, for each $\lambda \in [0,1]$, one has $0 \notin \varphi(\lambda, \partial D)$. Then the map $\lambda \mapsto \deg(\varphi(\lambda, .), D, 0)$ is constant on $[0,1]$.

P3. Normalized property: If $p \in D$ then

$$\deg(id_{\mathbb{R}^n} - p, D, 0) = 1.$$

P4. If D is the disjoint union of open sets D_1 and D_2 then

$$deg(f, D, 0) = deg(f, D_1, 0) + deg(f, D_2, 0).$$

If $K \subset \bar{D}$ is closed and $0 \notin f(K)$ then

$$deg(f, D, 0) = deg(f, D \setminus K, 0).$$

3. Variational Inequality of the Second Kind

Let $\Phi : \mathbb{R}^n \to \mathbb{R} \cup \{+\infty\}$ be a proper, convex and lower semicontinuous function with closed domain, i.e. $\Phi \in \Gamma(\mathbb{R}^n; \mathbb{R} \cup \{+\infty\})$. Let $M \in \mathbb{R}^{n\times n}$ be a given matrix and $q \in \mathbb{R}^n$ a given vector. We consider the variational inequality problem:

$\mathbf{VI(M, q, \Phi)}$: Find $u \in \mathbb{R}^n$ such that:

$$\langle Mu + q, v - u\rangle + \Phi(v) - \Phi(u) \geq 0, \ \forall v \in \mathbb{R}^n. \tag{17}$$

The solution set of problem $\mathbf{VI(M, q, \Phi)}$ will be denoted by $SOL(M, q, \Phi)$ and the resolvant set by $\mathcal{R}(M, \Phi)$, that are:

$$SOL(M, q, \Phi) := \{u \in \mathbb{R}^n : u \text{ solution of } (17)\}$$

and

$$\mathcal{R}(M, \Phi) := \{q \in \mathbb{R}^n : SOL(M, q, \Phi) \neq \emptyset\}.$$

Remark 1 *i) The condition $\overline{D(\Phi)} = D(\Phi)$ that is required in assuming that $\Phi \in \Gamma(\mathbb{R}^n; \mathbb{R} \cup \{+\infty\})$ is not really necessary to develop our approach but it is assumed in order to simplify its presentation.*

ii) The variational inequality in (17) is equivalent to the differential inclusion:

$$Mu + q \in -\partial\Phi(u). \tag{18}$$

iii) If $K \subset \mathbb{R}^n$ denotes a nonempty closed convex set then the variational inequality model $\mathbf{VI(M, \Psi_K, q)}$ in (17) reduces to the variational inequality of the first kind:

$$u \in K, \ \ \langle Mu + q, v - u\rangle \geq 0, \ \forall v \in K. \tag{19}$$

If in addition K is a cone then the inequality model in (19) is equivalent to the complementarity problem:

$$u \in K, \ \ Mu + q \in K^*, \ \ \langle Mu + q, u\rangle = 0. \tag{20}$$

iv) If the matrix M is symmetric then u is a solution of problem $\mathbf{VI(M, q, \Phi)}$ if and only if u is a solution of the optimization problem:

$$\min_{x\in\mathbb{R}^n} \{\frac{1}{2}\langle Mx, x\rangle + \langle q, x\rangle + \Phi(x)\}. \tag{21}$$

The structure of the model in (17) can be easily employed to deduce some necessary conditions of solvability.

Proposition 2 *Let $\Phi \in \Gamma(\mathbb{R}^n; \mathbb{R} \cup \{+\infty\})$, $M \in \mathbb{R}^{n\times n}$ and $q \in \mathbb{R}^n$ be given. If $u \in SOL(M, q, \Phi)$ denotes a solution of problem $\mathbf{VI(M, \Phi, q)}$ then the following conditions hold:*

i) There exists $x_0 \in D(\Phi)$ such that:

$$\langle Mx_0 + q, e\rangle + \Phi_\infty(e) \geq 0, \ \forall e \in \mathbb{R}^n. \tag{22}$$

ii) There exists $x_0 \in D(\Phi)$ *such that:*

$$\langle -M^T x_0 + q, e\rangle + \Phi_\infty(e) \geq 0, \ \forall e \in \ker\{M + M^T\}. \tag{23}$$

iii) We have:

$$\langle q, e\rangle + \Phi_\infty(e) \geq 0, \ \forall e \in \ker\{M^T\}. \tag{24}$$

Proof: i) We have $\langle Mu + q, e\rangle + \Phi(u + e) - \Phi(u) \geq 0, \ \forall e \in \mathbb{R}^n$, and then using (10), we get:

$$\langle Mu + q, e\rangle + \Phi_\infty(e) \geq 0, \ \forall e \in \mathbb{R}^n.$$

It results that necessarily there exists $x_0 := u \in D(\Phi)$ satisfying condition (22).

ii) We obtain also:

$$\langle q, e\rangle + \langle u, M^T e\rangle + \Phi_\infty(e) \geq 0, \ \forall e \in \mathbb{R}^n. \tag{25}$$

If $e \in \ker\{M + M^T\}$ then $\langle u, M^T e\rangle = -\langle u, Me\rangle = -\langle M^T u, e\rangle$ and thus:

$$\langle q - M^T u, e\rangle + \Phi_\infty(e) \geq 0, \ \forall e \in \ker\{M + M^T\}.$$

It results that necessarily there exists $x_0 := u \in D(\Phi)$ satisfying condition (23).

iii) If $e \in \ker\{M^T\}$ then the result in (24) is a direct consequence of the relation in (25). □

Let us here also give some property characterizing the possible solutions of Problem $\mathbf{VI}(\mathbf{M}, \mathbf{\Phi}, \mathbf{q})$ which will be helpful later to get uniqueness results.

Proposition 3 *Let* $\Phi \in \Gamma(\mathbb{R}^n; \mathbb{R} \cup \{+\infty\})$, $M \in \mathbb{R}^{n\times n}$ *and* $q \in \mathbb{R}^n$ *be given. If* u_1 *and* u_2 *denote two solutions of Problem* $\mathbf{VI}(\mathbf{M}, \mathbf{\Phi}, \mathbf{q})$ *then*

$$\langle M(u_1 - u_2), u_1 - u_2\rangle \leq 0.$$

Proof: If u_1 and u_2 denote two solutions of $\mathbf{VI}(\mathbf{M}, \mathbf{q}, \mathbf{\Phi})$ then $\langle Mu_1 + q, u_2 - u_1\rangle + \Phi(u_2) - \Phi(u_1) \geq 0$ and $\langle Mu_2 + q, u_1 - u_2\rangle + \Phi(u_1) - \Phi(u_2) \geq 0$ from which we deduce that $\langle M(u_1 - u_2), u_1 - u_2\rangle \leq 0$. □

If Φ is partitionable then we may prove the following more precise result (see also [41]):

Proposition 4 *Let* $\Phi \in P\Gamma(\mathbb{R}^n; \mathbb{R} \cup \{+\infty\})$, $M \in \mathbb{R}^{n\times n}$ *and* $q \in \mathbb{R}^n$ *be given. If* u_1 *and* u_2 *denote two solutions of problem* $\mathbf{VI}(\mathbf{M}, \mathbf{\Phi}, \mathbf{q})$ *then*

$$(\forall i \in \{1, ..., n\}): \ (M(u_1 - u_2))_i (u_1 - u_2)_i \leq 0.$$

Proof: If u denotes a solution of $\mathbf{VI}(\mathbf{M}, \mathbf{q}, \mathbf{\Phi})$ then

$$\sum_{i=1}^{n}(Mu+q)_i(v_i-u_i)+\sum_{i=1}^{n}(\Phi(v_i)-\Phi(u_i))\geq 0,\ \forall v\in\mathbb{R}^n.$$

Let $i\in\{1,...,n\}$ and $h_i\in\mathbb{R}$ be given. Setting $v=(u_1,...,u_{i-1},h_i,u_{i+1},...,u_n)$ we obtain that $(Mu+q)_i(h_i-u_i)+\Phi(h_i)-\Phi(u_i)\geq 0$. Thus for all $i\in\{1,...,n\}$, we have:

$$(Mu+q)_i(h_i-u_i)+\Phi(h_i)-\Phi(u_i)\geq 0,\forall h_i\in\mathbb{R}.$$

If u_1 and u_2 denote two solutions of $\mathbf{VI}(\mathbf{M}, \mathbf{q}, \mathbf{\Phi})$ then for all $i\in\{1,...,n\}$ we see that $(Mu_1+q)_i(u_2-u_1)_i+\Phi((u_2)_i)-\Phi((u_1)_i)\geq 0$ and $(Mu_2+q)_i(u_1-u_2)_i+\Phi((u_1)_i)-\Phi((u_2)_i)\geq 0$ from we wich deduce that $(M(u_1-u_2))_i(u_1-u_2)_i\leq 0$. □

4. Semi-Complementarity Problem

Let $\Phi\in\Gamma(\mathbb{R}^n;\mathbb{R}\cup\{+\infty\})$ and $M\in\mathbb{R}^{n\times n}$ be given. Here $D(\Phi)$ is nonempty, convex and assumed to be closed and the set $D(\Phi)_\infty$ is thus a well-defined nonempty closed and convex cone. Let us now consider the semi-complementarity problem:

$\mathbf{SCP}_\infty(\mathbf{M}, \mathbf{\Phi})$:

$$\begin{cases} z\in D(\Phi)_\infty \\ Mz\in D(\Phi_\infty)^* \\ \langle Mz,z\rangle\leq 0. \end{cases} \tag{26}$$

Recall that $D(\Phi_\infty)^*$ denotes the dual cone of $D(\Phi_\infty)$ and the second relation in (26) reads also:

$$\langle Mz,h\rangle\geq 0,\ \forall h\in D(\Phi_\infty).$$

Note that the first relation in (26) involves the recession cone of the domain of Φ while the second relation in (26) invokes the dual cone of the domain of Φ_∞ and not the dual cone of the recession cone of the domain of Φ.

Let us now first remark that if $D(\Phi)_\infty=D(\Phi_\infty)$ then problem $\mathbf{SCP}_\infty(\mathbf{M}, \mathbf{\Phi})$ reduces to a classical complementarity problem.

Proposition 5 *Let $\Phi:\mathbb{R}^n\to\mathbb{R}$ be a proper convex and lower semicontinuous function with closed domain and let $M\in\mathbb{R}^{n\times n}$ be a given matrix. If $D(\Phi)_\infty=D(\Phi_\infty)$ then $z\in\mathbb{R}^n$ is a solution of problem $\mathbf{SCP}_\infty(\mathbf{M}, \mathbf{\Phi})$ if and only if z is a solution of the complementarity problem $\mathbf{CP}(\mathbf{M}, \mathbf{D}(\mathbf{\Phi})_\infty)$:*

$$\begin{cases} z\in D(\Phi)_\infty \\ Mz\in (D(\Phi)_\infty)^* \\ \langle Mz,z\rangle=0. \end{cases} \tag{27}$$

Proof: Let z be a solution of problem $\mathbf{SCP}_\infty(\mathbf{M}, \mathbf{\Phi})$. Then the second relation in (26) reads here

$$\langle Mz, h\rangle \geq 0, \forall h \in D(\Phi)_\infty$$

from which we deduce in particular that $\langle Mz, z\rangle \geq 0$. This together with the third relation in (26) ensures that $\langle Mz, z\rangle = 0$.

If z is a solution of problem $\mathbf{CP}(\mathbf{M}, \mathbf{D}(\mathbf{\Phi})_\infty)$ then z is clearly a solution of problem $\mathbf{SCP}_\infty(\mathbf{M}, \mathbf{\Phi})$ too. □

Remark 2 *Let $\Phi : \mathbb{R}^n \to \mathbb{R}$ be a proper convex and lower semicontinuous function with closed domain.*

i) The following inclusion holds:

$$D(\Phi_\infty) \subset D(\Phi)_\infty. \tag{28}$$

Indeed, let $e \in D(\Phi_\infty)$ be given. Then, for some $x_0 \in D(\Phi)$, we have

$$\Phi_\infty(e) = \lim_{\lambda\to+\infty} \frac{1}{\lambda}\Phi(x_0 + \lambda e) = c < +\infty.$$

Then, remarking that

$$\lim_{\lambda\to+\infty} \frac{1}{\lambda}\Phi(x_0 + \lambda e) = \lim_{\lambda\to+\infty} \frac{1}{\lambda}(\Phi(x_0 + \lambda e) - \Phi(x_0)) = \sup_{\lambda>0} \frac{1}{\lambda}(\Phi(x_0 + \lambda e) - \Phi(x_0)),$$

we see that $\Phi(x_0 + \lambda e) \leq c\lambda + \Phi(x_0)$, $\forall \lambda > 0$, so that $x_0 + \lambda e \in D(\Phi)$, $\forall \lambda > 0$, and thus $e \in \frac{1}{\lambda}(D(\Phi) - x_0)$, $\forall \lambda > 0$, so that $e \in D(\Phi)_\infty$.

ii) If $\Phi \equiv \Psi_K$ where $K \subset \mathbb{R}^n$ is a nonempty closed convex set, then using (12), we see that

$$D((\Psi_K)_\infty) = D(\Psi_{K_\infty}) = K_\infty = D(\Psi_K)_\infty.$$

In this case, problem $\mathbf{SCP}_\infty(\mathbf{M}, \mathbf{\Psi_K})$ reduces to the complementarity problem:

$$\begin{cases} z \in K_\infty \\ Mz \in (K_\infty)^* \\ \langle Mz, z\rangle = 0. \end{cases} \tag{29}$$

iii) Let $\Phi : \mathbb{R} \to \mathbb{R}$ be the function defined by $\Phi(x) = x^2$, $\forall x \in \mathbb{R}$. Then $\Phi_\infty \equiv \Psi_{\{0\}}$, $D(\Phi_\infty) = \{0\}$, $D(\Phi)_\infty = \mathbb{R}$ and the inclusion in (28) is strict.

Let us now denote by $\mathcal{B}(M, \Phi)$ the solution set of problem $\mathbf{SCP}_\infty(\mathbf{M}, \mathbf{\Phi})$. Remark that problem $\mathbf{SCP}_\infty(\mathbf{M}, \mathbf{\Phi})$ has at least one (trivial) solution since $0 \in \mathcal{B}(M, \Phi)$.

Let us also set:

$$\mathcal{K}(M, \Phi) = \{x \in \mathbb{R}^n : Mx \in D(\Phi_\infty)^*\}, \tag{30}$$

$$\mathcal{N}_0(M) = \{x \in \mathbb{R}^n : \langle Mx, x\rangle = 0\}, \tag{31}$$

$$\mathcal{N}_-(M) = \{x \in \mathbb{R}^n : \langle Mx, x\rangle \leq 0\}, \tag{32}$$

and

$$\mathcal{N}_+(M) = \{x \in \mathbb{R}^n : \langle Mx, x\rangle \geq 0\}. \tag{33}$$

It is clear that:

$$\mathcal{B}(M, \Phi) = D(\Phi)_\infty \cap \mathcal{N}_-(M) \cap \mathcal{K}(M, \Phi). \tag{34}$$

Let us now introduce a concept of generalized eigenvalue for the couple (M, Φ) that we will use in the next section to state a general spectral condition ensuring the solvability of problem $\mathbf{VI}(\mathbf{M}, \mathbf{\Phi}, \mathbf{q})$.

Definition 1 *We say that $\mu \in \mathbb{R}$ is a generalized eigenvalue of the couple (M, Φ) provided that there exists a vector $z \in \mathbb{R}^n$, $z \neq 0$ such that $z \in \mathcal{B}(M - \mu I, \Phi)$, i.e.*

$$\begin{cases} z \in D(\Phi)_\infty, \\ Mz - \mu z \in D(\Phi_\infty)^*, \\ \langle Mz - \mu z, z\rangle \leq 0. \end{cases} \tag{35}$$

Let us now denote by $\sigma_\infty(M, \Phi)$ the set of generalized eigenvalues of (M, Φ), i.e.

$$\sigma_\infty(M, \Phi) := \{\mu \in \mathbb{R} : \exists\, z \in \mathbb{R}^n \backslash \{0\} \text{ solution of } (35)\}.$$

Remark 3 *i) Problem (35) appears as a generalized eigenvalue problem that reduces to the classical one as soon as $\Phi \equiv 0$. Indeed, if $\Phi \equiv 0$ then $D(\Phi)_\infty = \mathbb{R}^n$ and $D(\Phi_\infty)^* = \{0\}$ and Problem (35) consists to find $\mu \in \mathbb{R}$ and $z \in \mathbb{R}^n$, $z \neq 0$ such that $Mz - \mu z = 0$.*

ii) If $D(\Phi)_\infty = D(\Phi_\infty)$ then Problem (35) reduces to the eigenvalue problem for complementarity problem:

$$z \in D(\Phi_\infty), \quad Mz - \mu z \in D(\Phi_\infty)^*, \quad \langle Mz - \mu z, z\rangle = 0.$$

iii) If $D(\Phi)$ is bounded then $D(\Phi)_\infty = \{0\}$ and $\sigma_\infty(M, \Phi) = \emptyset$ while if $D(\Phi_\infty) = \mathbb{R}^n$ then $\sigma_\infty(M, \Phi) = \sigma_{\mathbb{R}}(M)$.

5. A Spectral Condition for Variational Inequalities

Definition 2 *We define by* $\mathbf{AC}_n$ *the set of* $(M, \Phi) \in \mathbb{R}^{n \times n} \times \Gamma(\mathbb{R}^n; \mathbb{R} \cup \{+\infty\})$ *such that*

$$\forall t \in [0,1] \; : \; \mathcal{B}((1-t)I + tM, \Phi) = \{0\}.$$

In other words, we say that the couple (M, Φ) is of class $\mathbf{AC_n}$ provided that, for all $t \in [0,1]$, 0 is the unique solution of problem $\mathbf{SCP}_\infty(\mathbf{(1-t)I + tM, \Phi})$.

Remark 4 *Note that the case* $t = 0$ *in Definition 2 is immaterial since* $\mathcal{B}(I, \Phi) = \{0\}$.

The following result shows that the class $\mathbf{AC_n}$ is nothing else that the class of couples (M, Φ) whose generalized eigenvalues are positive.

Proposition 6 *The couple* (M, Φ) *is of class* $\mathbf{AC_n}$ *if and only if:*

$$\sigma_\infty(M, \Phi) \; \subset \;]0, +\infty[. \tag{36}$$

Proof: a) Let us first prove that if (36) holds then the couple (M, Φ) is of class $\mathbf{AC_n}$. In view of Remark 4 it suffices to check that for all $t \in]0,1]$ we have $\mathcal{B}((1-t)I + tM, \Phi) = \{0\}$. Suppose on the contrary that there exists $t \in]0,1]$ and $z \neq 0$ such that $z \in \mathcal{B}((1-t)I + tM, \Phi)$. Then setting $\mu := -\frac{(1-t)}{t}$ we check easily that $z \in \mathcal{B}(M - \mu I, \Phi)$ and this is a contradiction since $\mu \leq 0$. b) Let us now check that if the couple (M, Φ) is of class $\mathbf{AC_n}$ then (36) holds. Suppose on the contrary that there exists $\mu \leq 0$ and $z \neq 0$ such that $z \in \mathcal{B}(M - \mu I, \Phi)$. We set $t := \frac{1}{1-\mu} = \frac{1}{1+|\mu|}$. It is clear that $t \in]0,1]$ and $z \in \mathcal{B}(M + \frac{(1-t)}{t} I, \Phi)$ and thus $z \in \mathcal{B}((1-t)I + tM, \Phi)$ which is a contradiction. □

Lemma 1 *Let* $M \in \mathbb{R}^{n \times n}$ *and* $\Phi \in \Gamma(\mathbb{R}^n; \mathbb{R} \cup \{+\infty\})$ *be given. We suppose that:*

$$\sigma_\infty(M, \Phi) \; \subset \;]0, +\infty[.$$

Let $q \in \mathbb{R}^n$ *be given. Then for* $R > 0$ *large enough we have:*

$$\deg(id_{\mathbb{R}^n} - P_\Phi(id_{\mathbb{R}^n} - (M. + q)), D_R, 0) = 1,$$

where

$$D_R := \{x \in \mathbb{R}^n : ||x|| < R\}.$$

Proof: Let $q \in \mathbb{R}^n$ be given. From (16) and (18), Problem $\mathbf{VI(M, q, \Phi)}$ is equivalent to the fixed point problem:

$$u = P_\Phi(u - (Mu + q)).$$

Let us now define by $H : [0,1] \times \mathbb{R}^n \to \mathbb{R}^n$ the continuous homotopy denoted as

$$H(t,u) = P_\Phi(tu - t(Mu+q)) = P_\Phi(u - \big(t(Mu+q) + (1-t)u\big)).$$

We claim that there exists $R_0 > 0$ such that for all $R \geq R_0$ and for all $t \in [0,1]$,

$$H(t,u) \neq u, \ \forall u \in \mathbb{R}^n, ||u|| = R. \tag{37}$$

Indeed, if we suppose the contrary then we may find sequences $\{t_i\}_{i\in\mathbb{N}} \subset [0,1]$ and $\{u_i\}_{i\in\mathbb{N}} \subset \mathbb{R}^n$ satisfying $||u_i|| \to +\infty$ and $u_i = H(t_i, u_i)$. Then

$$\langle t_i(Mu_i + q) + (1-t_i)u_i, v - u_i\rangle + \Phi(v) - \Phi(u_i) \geq 0, \forall v \in \mathbb{R}^n. \tag{38}$$

It is clear from (38) that

$$u_i \in D(\Phi) \ (i \in \mathbb{N}).$$

Moreover, for i large enough, $||u_i|| \neq 0$ and we may set:

$$z_i := \frac{u_i}{||u_i||}.$$

There exists subsequences, again denoted by $\{t_i\}$ and $\{z_i\}$, such that $\lim_{i\to+\infty} t_i = t \in [0,1]$ and $\lim_{i\to+\infty} z_i = z$ with $||z|| = 1$.

Let $x_0 \in D(\Phi)$ be any element in the domain of Φ. Let $\lambda > 0$ be given. For i large enough, $\frac{\lambda}{||u_i||} < 1$ and thus

$$\frac{\lambda}{||u_i||}u_i + (1 - \frac{\lambda}{||u_i||})x_0 \in D(\Phi)$$

since the set $D(\Phi)$ is convex.

Recalling that the set $D(\Phi)$ is assumed to be closed and taking the limit as $i \to +\infty$, we get $\lambda z + x_0 \in D(\Phi)$. This result holds for any $\lambda > 0$ and thus

$$z \in \bigcap_{\lambda>0} \frac{1}{\lambda}(D(\Phi) - x_0) = D(\Phi)_\infty. \tag{39}$$

Let $e \in D(\Phi_\infty)$ be given. Then from (28), $e \in D(\Phi)_\infty$ and from (13), $u_i + e \in D(\Phi)$. We may thus set $v = u_i + e$ in (38) to get

$$\langle (1-t_i)u_i + t_i(Mu_i + q), e\rangle + \Phi(u_i + e) - \Phi(u_i) \geq 0$$

and thus using (10), we obtain

$$\langle (1-t_i)u_i, e\rangle + \langle t_i(Mu_i + q), e\rangle + \Phi_\infty(e) \geq 0.$$

Note that $\Phi_\infty(e) < +\infty$ since $e \in D(\Phi_\infty)$ and we may therefore divide this last relation by $||u_i||$ to get:

$$\langle (1-t_i)z_i, e\rangle + \langle t_i M z_i + t_i \frac{q}{||u_i||}, e\rangle + \frac{1}{||u_i||}\Phi_\infty(e) \geq 0.$$

Taking the limit as $i \to +\infty$, we get $\langle (1-t)z + tMz, e\rangle \geq 0$. This holds for any $e \in D(\Phi_\infty)$ and thus

$$(1-t)z + tMz \in D(\Phi_\infty)^*. \tag{40}$$

Setting now $v = x_0$ in (38), we obtain:

$$(1-t_i)||u_i||^2 + t_i\langle Mu_i, u_i\rangle \leq \langle (1-t_i)u_i, x_0\rangle + \langle t_i M u_i, x_0\rangle + \langle t_i q, x_0 - u_i\rangle + \Phi(x_0) - \Phi(u_i). \tag{41}$$

The function Φ is proper, convex and lower semicontinuous, and thus (see e.g. Theorem 1.1.11 in [28]) there exists $a \geq 0$ and $b \in \mathbb{R}$ such that:

$$\Phi(x) \geq -a||x|| + b, \ \forall x \in \mathbb{R}^n.$$

Thus

$$(1-t_i)||u_i||^2 + t_i\langle Mu_i, u_i\rangle \leq a||u_i|| - b + \langle (1-t_i)u_i, x_0\rangle + \langle t_i M u_i, x_0\rangle + \langle t_i q, x_0 - u_i\rangle + \Phi(x_0).$$

Dividing this last relation by $||u_i||^2$, we get:

$$(1-t_i)||z_i||^2 + t_i\langle Mz_i, z_i\rangle \leq \frac{a}{||u_i||} - \frac{b}{||u_i||^2} + \langle t_i M z_i, \frac{x_0}{||u_i||}\rangle +$$

$$+(1-t_i)\langle z_i, \frac{x_0}{||u_i||}\rangle + \langle t_i \frac{q}{||u_i||}, \frac{x_0}{||u_i||} - z_i\rangle + \frac{\Phi(x_0)}{||u_i||^2}.$$

Taking the limit as $i \to +\infty$, we get

$$\langle (1-t)z + tMz, z\rangle \leq 0.$$

This last relation together with (39) and (40) imply that $z \in \mathcal{B}((1-t)I + tM, \Phi)$. Moreover $z \neq 0$ and we obtain a contradiction to our assumption requiring (see Proposition 6) that (M, Φ) is of class $\mathbf{AC}_n$.

Thus, for $R \geq R_0$, (37) holds and the Brouwer degree with respect to the set $D_R = \{x \in \mathbb{R}^n : ||x|| < R\}$ and 0 of the map $u \mapsto u - H(t,u)$ is well-defined for all $t \in [0,1]$. Set $R_1 := P_\Phi(0)$ and let $R > \max\{R_0, R_1\}$ be given. Using the homotopy invariance property as well as the normalized property of Brouwer degree, we obtain:

$$\deg(id_{\mathbb{R}^n} - P_\Phi(id_{\mathbb{R}^n} - (M. + q)), D_R, 0) = \deg(id_{\mathbb{R}^n} - H(1,.), D_R, 0)$$

$$= \deg(id_{\mathbb{R}^n} - H(0,.), D_R, 0) = \deg(id_{\mathbb{R}^n} - P_\Phi(0), D_R, 0) = 1.$$

□

It results from solution property of Brouwer degree the following result:

Theorem 1 *(Spectral Condition) Let $M \in \mathbb{R}^{n\times n}$ and $\Phi \in \Gamma(\mathbb{R}^n; \mathbb{R}\cup\{+\infty\})$. If*

$$\sigma_\infty(M,\Phi) \subset]0,+\infty[$$

then, for each $q \in \mathbb{R}^n$, problem $\mathbf{VI(M,q,\Phi)}$ *has at least one solution.* □

We obtain also the following generalized spectral condition ensuring the solvability of problem $\mathbf{VI(M,q,\Phi)}$:

Theorem 2 *Let $M \in \mathbb{R}^{n\times n}$ and $\Phi \in \Gamma(\mathbb{R}^n; \mathbb{R}\cup\{+\infty\})$ be given. We suppose that*

$$\mathcal{B}(M,\Phi) = \{0\}$$

and assume the existence of $\lambda_0 > 0$ such that:

$$(\forall\lambda \in]0,\lambda_0[) : \sigma_\infty(\lambda I + M,\Phi) \subset]0,+\infty[.$$

Then for each $q \in \mathbb{R}^n$, problem $\mathbf{VI(M,q,\Phi)}$ *has at least one solution.*

Proof: Using Theorem 1, we see that:

$$\mathcal{R}(\lambda I + M,\Phi) = \mathbb{R}^n, \ \forall\, 0 < \lambda < \lambda_0. \tag{42}$$

It results that for all $i \in \mathbb{N}, i > \frac{1}{\lambda_0}$, there exists $u_i \in \mathbb{R}^n$ such that

$$\langle(\frac{1}{i}I + M)u_i + q, v - u_i\rangle + \Phi(v) - \Phi(u_i) \geq 0, \ \forall v \in \mathbb{R}^n. \tag{43}$$

We claim that the sequence $\{u_i\} \equiv \{u_i; i \in \mathbb{N}, i > \frac{1}{\lambda_0}\}$ is bounded. Suppose on the contrary that $||u_i|| \to +\infty$ as $i \to +\infty$. Then, for i large enough, $||u_i|| \neq 0$ and we may set $z_i := \frac{u_i}{||u_i||}$. There exists a subsequence, again denoted by $\{z_i\}$, such that $\lim_{i\to+\infty} z_i = z$ with $||z|| = 1$.

It is clear from (43) that

$$u_i \in D(\Phi) \ (i \in \mathbb{N}, i > \frac{1}{\lambda_0}).$$

Let $x_0 \in D(\Phi)$ be any element in the domain of Φ. Let $\lambda > 0$ be given. For i large enough, $\frac{\lambda}{||u_i||} < 1$ and thus

$$\frac{\lambda}{||u_i||}u_i + (1 - \frac{\lambda}{||u_i||})x_0 \in D(\Phi)$$

since $D(\Phi)$ is convex and $x_0 \in D(\Phi)$. Taking the limit as $i \to +\infty$, we get $\lambda z + x_0 \in D(\Phi)$ since $D(\Phi)$ is assumed to be closed. This result holds for any $\lambda > 0$ and thus

$$z \in \bigcap_{\lambda>0} \frac{1}{\lambda}(D(\Phi) - x_0) = D(\Phi)_\infty. \tag{44}$$

Let $e \in D(\Phi_\infty)$ be given. Then $e \in D(\Phi)_\infty$ and from (13), $u_i + e \in D(\Phi)$. We may set $v = u_i + e$ in (43) to get

$$\langle(\frac{1}{i}I + M)u_i + q, e\rangle + \Phi(u_i + e) - \Phi(u_i) \geq 0$$

and thus using (10), we obtain

$$\langle \frac{1}{i}u_i, e\rangle + \langle Mu_i + q, e\rangle + \Phi_\infty(e) \geq 0.$$

Remarking that $\Phi_\infty(e) < +\infty$ since $e \in D(\Phi_\infty)$, we may divide this last relation by $||u_i||$ to get:

$$\langle \frac{1}{i}z_i, e\rangle + \langle Mz_i + \frac{q}{||u_i||}, e\rangle + \frac{1}{||u_i||}\Phi_\infty(e) \geq 0.$$

Taking the limit as $i \to +\infty$, we get $\langle Mz, e\rangle \geq 0$. This holds for any $e \in D(\Phi_\infty)$ and thus

$$z \in \mathcal{K}(M, \Phi). \tag{45}$$

Setting now $v = x_0$ in (43), we obtain:

$$\frac{1}{i}||u_i||^2 + \langle Mu_i, u_i\rangle \leq \langle(\frac{1}{i}I + M)u_i, x_0\rangle + \langle q, x_0 - u_i\rangle + \Phi(x_0) - \Phi(u_i). \tag{46}$$

The function Φ is proper, convex and lower semicontinuous, and thus (see e.g. Theorem 1.1.11 in [28]) there exists $a \geq 0$ and $b \in \mathbb{R}$ such that:

$$\Phi(x) \geq -a||x|| + b,\ \forall x \in \mathbb{R}^n.$$

Thus

$$\langle Mu_i, u_i\rangle \leq a||u_i|| - b + \langle(\frac{1}{i}I + M)u_i, x_0\rangle + \langle q, x_0 - u_i\rangle + \Phi(x_0).$$

Dividing this last relation by $||u_i||^2$, we get:

$$\langle Mz_i, z_i\rangle \leq \frac{a}{||u_i||} - \frac{b}{||u_i||^2} + \langle(\frac{1}{i}I + M)z_i, \frac{x_0}{||u_i||}\rangle + \langle \frac{q}{||u_i||}, \frac{x_0}{||u_i||} - z_i\rangle + \frac{\Phi(x_0)}{||u_i||^2}.$$

Taking the limit as $i \to +\infty$, we get $\langle Mz, z\rangle \leq 0$. Thus

$$z \in \mathcal{N}_-(M). \tag{47}$$

Using (44), (45) and (47) we obtain that

$$z \in \mathcal{B}(M, \Phi).$$

Finally, we get a contradiction since we have proved that $z \neq 0$ and $z \in \mathcal{B}(M, \Phi)$.

The sequence $\{u_i\}$ is thus bounded and there exists a subsequence, again denoted by $\{u_i\}$ which converges. Let us set

$$u = \lim_{i \to +\infty} u_i.$$

Let $v \in \mathbb{R}^n$ be fixed. We have:

$$\langle(\frac{1}{i}I + M)u_i + q, u_i - v\rangle - \Phi(v) + \Phi(u_i) \leq 0.$$

Taking the limit inferior as $i \to +\infty$, and using the lower semicontinuity of Φ, we obtain:

$$\langle Mu + q, u - v\rangle - \Phi(v) + \Phi(u) \leq 0. \tag{48}$$

The vector v has been chosen arbitrarily in $\mathbb{R}^n$ and thus the result in (48) holds for all $v \in \mathbb{R}^n$. The existence result follows. □

6. Qualitative Analysis of Problem $\mathbf{VI}(\mathbf{M}, \mathbf{q}, \mathbf{\Phi})$

In this Section, we will use our previous spectral condition to study the Problem $\mathbf{VI}(\mathbf{M}, \mathbf{q}, \mathbf{\Phi})$ for various special and important classes of matrices M and functions Φ.

6.1. Variational Inequalities with Bounded Constraints

Let $\Phi \in \Gamma_B(\mathbb{R}^n; \mathbb{R} \cup \{+\infty\})$ be a proper convex and lower semicontinuous function with closed and bounded domain domain and let $M \in \mathbb{R}^{n \times n}$ be a given matrix.
Then $D(\Phi)_\infty = \{0\}$ and thus:

$$\mathcal{B}(M, \Phi) = \{0\}$$

and

$$\sigma_\infty(M, \Phi) = \emptyset.$$

It results from Theorem 1 that:

Corollary 1 *Suppose that $\Phi \in \Gamma_B(\mathbb{R}^n; \mathbb{R} \cup \{+\infty\})$ and $M \in \mathbb{R}^{n \times n}$. Then a) $\sigma_\infty(M, \Phi) = \emptyset$ and b) $\mathcal{R}(M, \Phi) = \mathbb{R}^n$, i.e. for all $q \in \mathbb{R}^n$, problem $\mathbf{VI}(\mathbf{M}, \mathbf{q}, \mathbf{\Phi})$ has at least one solution.* □

Example 1 *Let $M \in \mathbb{R}^{2 \times 2}$ and $\Phi \equiv \Psi_{[0,1] \times [0,1]}$. Here $D(\Phi) = [0, 1] \times [0, 1]$ is compact and consequently we may apply Corollary 1.*

6.2. Variational Inequalities Involving Weakly Positive Definite Matrices

Recall that one says that a matrix $M \in \mathbb{R}^{n \times n}$ is weakly positive definite provided that:

$$\sigma_\mathbb{R}(M) \subset]0, +\infty[.$$

It recovers the class of positive stable matrices defined as the matrices all of whose eigenvalues have positive real parts (see e.g. [36]). Let us now introduce the following related notion:

Definition 3 *We define by $\mathbf{PS_n}$ the set of $(M, \Phi) \in \mathbb{R}^{n \times n} \times \Gamma(\mathbb{R}^n; \mathbb{R} \cup \{+\infty\})$ such that:*

$$D(\Phi_\infty) = \mathbb{R}^n \tag{49}$$

and

$$\sigma_\mathbb{R}(M) \subset]0, +\infty[. \tag{50}$$

If $(M, \Phi) \in \mathbf{PS_n}$ then $D(\Phi_\infty) = \mathbb{R}^n$ and thus

$$\mathcal{B}(M, \Phi) = \ker\{M\}.$$

Indeed, here $D(\Phi_\infty)^* = \{0\}$ and $\mathcal{K}(M, \Phi) = \ker\{M\} \subset \mathcal{N}_-(M)$ so that $\mathcal{B}(M, \Phi) = \ker\{M\}$. Using our approach, we obtain:

Corollary 2 *Suppose that* $(M,\Phi) \in \mathbf{PS_n}$. *Then a)* $\sigma_\infty(M,\Phi) \subset]0,+\infty[$ *and b)* $\mathcal{R}(M,\Phi) = \mathbb{R}^n$, *i.e. for all* $q \in \mathbb{R}^n$, *problem* $\mathbf{VI(M,q,\Phi)}$ *has at least one solution.*

Proof: We first check that $\mathbf{PS_n} \subset \mathbf{AC_n}$. Indeed, let $t \in [0,1]$ and $z \in \mathcal{B}((1-t)I+tM,\Phi)$ be given. We claim that $z=0$. Suppose on the contrary that $z \neq 0$. Here $D(\Phi_\infty)^* = \{0\}$ and the second relation in (26) yields

$$(1-t)z + tMz = 0.$$

If $t=0$ then $z=0$ and a contradiction. If $0<t\leq 1$ then

$$Mz = -\frac{(1-t)}{t}z$$

so that $\nu^* := -\frac{(1-t)}{t} \leq 0$ is a real eigenvalue of M and a contradiction to (50). It results from Proposition 6 that $\sigma_\infty(M,\Phi) \subset]0,+\infty[$. Using then Theorem 1 we may conclude. □

Example 2 *Let us consider the matrix*

$$M = \begin{pmatrix} 1 & 10 \\ 0 & 4 \end{pmatrix},$$

and the function

$$\Phi(x_1,x_2) = |x_1| + |x_2|, \ \forall (x_1,x_2) \in \mathbb{R}^2.$$

Here $D(\Phi_\infty) = \mathbb{R}^2$ *and* $\sigma_{\mathbb{R}}(M) = \{1,4\} \subset]0,+\infty[$ *imply* $(M,\Phi) \in \mathbf{PS_2}$ *and subsequently we can apply Corollary 2.*

6.3. Variational Inequalities Involving Positive Definite Matrices

Recall that one says that a matrix $M \in \mathbb{R}^{n\times n}$ is positive definite provided that:

$$\langle x, Mx\rangle > 0, \ \forall x \in \mathbb{R}^n, x \neq 0,$$

which is equivalent to the following condition (Sylvester criterion) on the leading principal minors on M:

$$\Delta(M+M^T)_{12\ldots k} > 0, \ \forall 1 \leq k \leq n.$$

Let us now introduce the following related notion:

Definition 4 *We define by* $\mathbf{PD_n}$ *the set of* $(M,\Phi) \in \mathbb{R}^{n\times n} \times \Gamma(\mathbb{R}^n;\mathbb{R}\cup\{+\infty\})$ *such that*

$$\langle Mx, x\rangle > 0, \ \forall x \in D(\Phi)_\infty, x \neq 0. \tag{51}$$

Corollary 3 *Suppose that* $(M, \Phi) \in \mathbf{PD_n}$. *Then a)* $\sigma_\infty(M, \Phi) \subset]0, +\infty[$ *and b)* $\mathcal{R}(M, \Phi) = \mathbb{R}^n$, *i.e. for all* $q \in \mathbb{R}^n$, *problem* $\mathbf{VI(M, q, \Phi)}$ *has at least one solution.*

Proof: Indeed, from Proposition 6 it suffices to check that $(M, \Phi) \in \mathbf{AC_n}$. Let $t \in [0, 1]$ and $z \in \mathcal{B}((1-t)I + tM, \Phi)$ be given. Then $z \in D(\Phi)_\infty$ and $\langle (1-t)z + tMz, z\rangle \leq 0$. If $t = 1$ then $\langle Mz, z\rangle \leq 0$ and (51) yields $z = 0$. If $0 \leq t < 1$ then $||z||^2 \leq -\frac{t}{1-t}\langle Mz, z\rangle$ and from (51), we deduce that necessarily $z = 0$. Thus $\sigma_\infty(M, \Phi) \subset]0, +\infty[$ and the result is a direct consequence of Theorem 1. □

Example 3 *Let*

$$M = \begin{pmatrix} -1 & 0 \\ 0 & 1 \end{pmatrix},$$

and

$$\Phi \equiv \Psi_{\{x \in (\mathbb{R}_+)^2 :\, x_2 \geq x_1 + 1\}}.$$

Here $D(\Phi)_\infty = \{x \in (\mathbb{R}_+)^2 : x_2 \geq x_1\}$ *and*

$$(\forall x = (x_1, x_2) \in D(\Phi)_\infty) : \langle Mx, x\rangle = x_2^2 - x_1^2 \geq 0.$$

Therefore, $(M, \Phi) \in \mathbf{PD_2}$ *and subsequently Corollary 3 may be applied.*

As a consequence of Corollary 3 and Proposition 3 we get the following classical existence and uniqueness result:

Corollary 4 *Let* $\Phi \in \Gamma(\mathbb{R}^n; \mathbb{R} \cup \{+\infty\})$ *be given and suppose that* $M \in \mathbb{R}^{n \times n}$ *is a positive definite matrix. Then for all* $q \in \mathbb{R}^n$, *problem* $\mathbf{VI(M, q, \Phi)}$ *has a unique solution.*

Proof: The existence of a solution is a direct consequence of Corollary 3. To prove the uniqueness, suppose by contradiction that u_1 and u_2 are two different solutions. Then $z = u_1 - u_2 \neq 0$ and Proposition 3 ensures that $\langle Mz, z\rangle \leq 0$. This is a contradiction since M is a positive definite matrix. □

Example 4 *Let* $\Phi \in \Gamma(\mathbb{R}^2; \mathbb{R} \cup \{+\infty\})$ *be given and let us consider the matrix*

$$M = \begin{pmatrix} 2 & -1 \\ -1 & 2 \end{pmatrix}.$$

Here M *is positive definite and Corollary 4 gives, for all* $q \in \mathbb{R}^2$, *the existence and the uniqueness of the solution of* $\mathbf{VI(M, q, \Phi)}$.

6.4. Variational Inequalities Involving P-matrices

Recall that one says that a matrix $M \in \mathbb{R}^{n\times n}$ is a P-matrix provided that:

$$(\forall x \in \mathbb{R}^n, x \neq 0)(\exists \alpha \in \{1, ..., n\}) : \; x_\alpha (Mx)_\alpha > 0,$$

which is equivalent to the following condition on the principal minors of M:

$$\Delta(M)_{i_1 i_2 \dots i_k} > 0, \; \forall 1 \leq k \leq n.$$

Let us now introduce the following related notion:

Definition 5 *We define by* $\mathbf{P_n}$ *the set of* $(M, \Phi) \in \mathbb{R}^{n\times n} \times \Gamma(\mathbb{R}^n; \mathbb{R} \cup \{+\infty\})$ *such that*

$$x \in D(\Phi)_\infty \Longrightarrow \langle x, e^j \rangle e^j \in D(\Phi_\infty) \; (1 \leq j \leq n), \tag{52}$$

and

$$(x \in D(\Phi)_\infty, x \neq 0) \Longrightarrow \exists \alpha \in \{1, ..., n\} : x_\alpha (Mx)_\alpha > 0. \tag{53}$$

Remark 5 *i) Condition (52) means that if* $w \in D(\Phi)_\infty$ *then its projection* $\langle w, e^j \rangle e^j = w_j e^j$ *onto the space*

$$X_j := \{x \in \mathbb{R}^n : x_k = 0, \forall k \in \{1, ..., n\}, k \neq j\}$$

belongs to $D(\Phi_\infty)$.

ii) If $\Phi \equiv \Psi_K$ *with* $K = \mathbb{R}^n, \;\; \mathbb{R}^{n-\alpha} \times (\mathbb{R}_+)^\alpha$ $(\alpha \in \{1, ..., n\})$ *or* $(\mathbb{R}_+)^n$ *then condition (52) holds.*

Corollary 5 *Suppose that* $(M, \Phi) \in \mathbf{P_n}$. *Then a)* $\sigma_\infty(M, \Phi) \subset]0, +\infty[$ *and b)* $\mathcal{R}(M, \Phi) = \mathbb{R}^n$, *i.e. for all* $q \in \mathbb{R}^n$, *problem* $\mathbf{VI}(\mathbf{M}, \mathbf{q}, \mathbf{\Phi})$ *has at least one solution.*

Proof: We prove that $\mathbf{P_n} \subset \mathbf{AC_n}$. Let $(M, \Phi) \in \mathbf{P_n}$, $t \in [0, 1]$ and $z \in \mathcal{B}((1-t)I + tM, \Phi)$ be given. We assert that $z = 0$. Suppose on the contrary that $z \neq 0$. We claim that there exists some index $k \in \{1, ..., n\}$ such that $(1-t)z_k^2 + t(Mz)_k z_k > 0$. Indeed, if $0 < t \leq 1$ the result follows from (53) since $z \in D(\Phi)_\infty, z \neq 0$ while if $t = 0$, the result is trivial since $z \neq 0$. We know that

$$\langle (1-t)z + tMz, h \rangle \geq 0, \forall h \in D(\Phi_\infty).$$

Let $j \in \{1, ..., n\}$ be given. Using (52), we may set $h = \langle z, e^j \rangle e^j$ to get $(1-t)z_j^2 + t(Mz)_j z_j \geq 0$. This last relation holds for all $j \in \{1, ..., n\}$ and since $\langle (1-t)z + tMz, z \rangle \leq 0$, we get the contradiction:

$$0 \geq (1-t)z_k^2 + t(Mz)_k z_k + \sum_{j \neq k} (1-t)z_j^2 + t(Mz)_j z_j > 0.$$

Thus $\sigma_\infty(M, \Phi) \subset]0, +\infty[$ and the result is a consequence of Theorem 1. □

Example 5 *Let us consider the matrix:*

$$M = \begin{pmatrix} 1 & 1 \\ -4 & 1 \end{pmatrix}$$

and the function:

$$\Phi_5 \equiv \Psi_{[1,\infty[\times[1,\infty[}.$$

Here $D(\Phi_\infty) = D(\Phi)_\infty = (\mathbb{R}_+)^2$ *and it can be easily checked that (52) holds so that* $(M, \Phi) \in \mathbf{P_2}$. *Therefore we may apply Corollary 5.*

Using now Corollary 5 together with Proposition 4 we get the following existence and uniqueness result which can be applied to study a partitionable variational inequality involving a P-matrix.

Corollary 6 *Let* $\Phi \in P\Gamma(\mathbb{R}^n; \mathbb{R} \cup \{+\infty\})$ *satisfying condition (52) and let* $M \in \mathbb{R}^{n\times n}$ *be a P-matrix . Then for all* $q \in \mathbb{R}^n$, *problem* $\mathbf{VI(M, q, \Phi)}$ *has a unique solution.*

Proof: The existence of a solution is a direct consequence of Corollary 5. To prove the uniqueness, suppose by contradiction that u_1 and u_2 are two different solutions. Then $z = u_1 - u_2 \neq 0$ and Proposition 4 ensures that for all $i \in \{1, ..., n\}$ we have $(Mz)_i z_i \leq 0$. This is a contradiction since M is a P-matrix. □

Example 6 *Let us consider the matrix:*

$$M = \begin{pmatrix} 2 & 1 \\ -10 & 2 \end{pmatrix},$$

and the function:

$$\Phi(x) = |x_1| + |x_2|, \ \forall x = (x_1, x_2) \in \mathbb{R}^2.$$

We can see that M *is a P-matrix and* $\Phi \in P\Gamma(\mathbb{R}^2; \mathbb{R} \cup \{+\infty\})$. *Here* $D(\Phi_\infty) = \mathbb{R}^2$ *and condition (52) is satisfied. Thus we can apply Corollary 6.*

Remark 6 *If* $\Phi \in D\Gamma(\mathbb{R}^n; \mathbb{R} \cup \{+\infty\})$ *then* $\Phi \in D\Gamma(\mathbb{R}^n; \mathbb{R} \cup \{+\infty\})$ *and satisfies condition (52). Indeed, we first remark that here* $D(\Phi) = D(\Phi_1) \times D(\Phi_2) \times ... \times D(\Phi_n)$. *We remark also that each set* $D(\Phi_i)$ *is a nonempty closed convex cone and thus* $D(\Phi)_\infty = D(\Phi)$. *Moreover, the function* Φ *is positively homogeneous and thus* $\Phi_\infty \equiv \Phi$, $D(\Phi_\infty) = D(\Phi)$. *If* $x \in D(\Phi)_\infty = D(\Phi)$ *then for all* $j \in \{1, ..., n\}$, *we see that* $\langle x, e^j \rangle e^j = (0 \dots 0 \ x_j \ 0 \dots 0)^T \in D(\Phi_1) \times ... D(\Phi_{j-1}) \times D(\Phi_j) \times D(\Phi_{j+1}) \times ... \times D(\Phi_n)$ *and thus* $\langle x, e^j \rangle e^j \in D(\Phi) = D(\Phi_\infty)$.

Remark 7 *The famous result on complementarity problems with P-matrices (see e.g. [21]) can be easily deduced from Corollary 6. Indeed, if* $\Phi = \Psi_{(\mathbb{R}_+)^n}$ *then* Φ *can be written as in (3) with* $\Phi_i = \Psi_{\mathbb{R}_+}$ $(1 \leq i \leq n)$ *and if* M *is a P-matrix then Corollary 9 can be applied to ensure that for each* $q \in \mathbb{R}^n$, *there exists a unique* $u \in (\mathbb{R}_+)^n$ *such that* $Mu + q \in (\mathbb{R}_+)^n$ *and* $\langle u, Mu + q \rangle = 0$.

6.5. Variational Inequalities Involving Positive Semidefinite Matrices

One says that a matrix $M \in \mathbb{R}^{n\times n}$ is positive semidefinite provided that:

$$\langle x, Mx\rangle \geq 0,\ \forall x \in \mathbb{R}^n.$$

This last condition can also be expressed equivalently as the following criterion on the principal minors:

$$\Delta(M + M^T)_{i_1 i_2 \dots i_k} \geq 0,\ \forall 1 \leq k \leq n.$$

Definition 6 *We define by* $\mathbf{PD0_n}$ *the set of* $(M, \Phi) \in \mathbb{R}^{n\times n} \times \Gamma(\mathbb{R}^n; \mathbb{R} \cup \{+\infty\})$ *such that*

$$\langle Mx, x\rangle \geq 0,\ \forall x \in D(\Phi)_\infty. \tag{54}$$

Corollary 7 *Suppose that* $(M, \Phi) \in \mathbf{PD0_n}$. *Then a)* $(\forall\lambda > 0)$: $\sigma_\infty(\lambda I + M, \Phi) \subset]0, +\infty[$ *and b) if* $\mathcal{B}(M, \Phi) = \{0\}$ *then* $\mathcal{R}(M, \Phi) = \mathbb{R}^n$, *i.e. for all* $q \in \mathbb{R}^n$, *problem* $\mathbf{VI(M, q, \Phi)}$ *has at least one solution.*

Proof: Here $(M, \Phi) \in \mathbf{PD0_n}$ and condition (54) on M entails that for all $\lambda > 0$ the matrix $\lambda I + M$ satisfies condition (51) and thus for all $\lambda > 0$, $\sigma_\infty(\lambda I + M, \Phi) \subset]0, +\infty[$. We may thus apply Theorem 2 to conclude. □

Note that if $(M, \Phi) \in \mathbf{PD0_n}$ then

$$\mathcal{B}(M, \Phi) = D(\Phi)_\infty \cap \mathcal{N}_0(M) \cap \mathcal{K}(M, \Phi).$$

Indeed, we have $\langle Mx, x\rangle \geq 0, \forall x \in D(\Phi)_\infty$ and thus $D(\Phi)_\infty \cap \mathcal{N}_-(M) = D(\Phi)_\infty \cap \mathcal{N}_0(M)$.

If M is positive semidefinite then $\mathcal{N}_0(M) = \ker\{M + M^T\}$ and thus:

$$\mathcal{B}(M, \Phi) = D(\Phi)_\infty \cap \ker\{M + M^T\} \cap \mathcal{K}(M, \Phi).$$

Example 7 *Let us consider the matrix:*

$$M = \begin{pmatrix} 1 & -1 \\ -2 & 2 \end{pmatrix}$$

and the function:

$$\Phi \equiv \Psi_{\{x \in (\mathbb{R}_+)^2\,:\,x_2 \geq x_1 + 1\}}.$$

Here $D(\Phi)_\infty = \{x \in (\mathbb{R}^2_+) : x_2 \geq x_1\}$ *and* $\langle Mx, x\rangle = (x_1 - x_2)^2 + x_2(x_2 - x_1) \geq 0,\ \forall x = (x_1, x_2) \in D(\Phi)_\infty$. *Therefore* $(M, \Phi) \in \mathbf{PD0_2}$ *and we may apply Corollary 7.*

6.6. Variational Inequalities Involving $\mathbf{P}_0$-matrices

One says that $M \in \mathbb{R}^{n\times n}$ is a P_0-matrix provided that:

$$(\forall x \in \mathbb{R}^n, x \neq 0)(\exists \alpha \in \{1, ..., n\}) : x_\alpha \neq 0 \text{ and } x_\alpha(Mx)_\alpha \geq 0,$$

which is equivalent to the following condition on the principal minors:

$$\Delta(M)_{i_1 i_2 \ldots i_k} \geq 0, \; \forall 1 \leq k \leq n$$

Definition 7 *We define by* $\mathbf{P0_n}$ *the set of* $(M, \Phi) \in \mathbb{R}^{n\times n} \times \Gamma(\mathbb{R}^n; \mathbb{R} \cup \{+\infty\})$ *such that*

$$x \in D(\Phi)_\infty \implies \langle x, e^j\rangle e^j \in D(\Phi_\infty) \; (1 \leq j \leq n), \tag{55}$$

and

$$(x \in D(\Phi)_\infty, x \neq 0) \implies \exists \alpha \in \{1, ..., n\} : x_\alpha \neq 0 \text{ and } x_\alpha(Mx)_\alpha \geq 0. \tag{56}$$

Corollary 8 *Suppose that* $(M, \Phi) \in \mathbf{P0_n}$. *Then a)* $(\forall \lambda > 0) : \; \sigma_\infty(\lambda I + M, \Phi) \subset]0, +\infty[$ *and b) if* $\mathcal{B}(M, \Phi) = \{0\}$ *then* $\mathcal{R}(M, \Phi) = \mathbb{R}^n$, *i.e. for all* $q \in \mathbb{R}^n$, *problem* $\mathbf{VI(M, q, \Phi)}$ *has at least one solution.*

Proof: Let $\lambda > 0$ be given. Here$(M, \Phi) \in \mathbf{P0_n}$ and condition (56) on M entails that for each $x \in D(\Phi)_\infty$, $x \neq 0$, there exists $\alpha \in \{1, ..., n\}$ such that

$$x_\alpha(\lambda x + Mx)_\alpha = \lambda x_\alpha^2 + x_\alpha(Mx)_\alpha > 0$$

and thus $\lambda I + M$ satisfies condition (53). It results that for all $\lambda > 0$, $\sigma_\infty(\lambda I + M, \Phi) \subset]0, +\infty[$. We may thus apply Theorem 2 to conclude. □

Example 8 *Let*

$$M = \begin{pmatrix} 1 & 1 \\ -4 & 0 \end{pmatrix}$$

and

$$\Phi \equiv \Psi_{[1,\infty[\times[1,\infty[}.$$

Here $D(\Phi_\infty) = D(\Phi)_\infty = (\mathbb{R}_+)^2$ *and we can easily check that* $(M, \Phi) \in \mathbf{P0_2}$ *and then apply Corollary 8.*

Note that if $(M, \Phi) \in \mathbf{P0_n}$ then

$$\mathcal{B}(M, \Phi) = D(\Phi)_\infty \cap \mathcal{N}_0(M) \cap \mathcal{K}(M, \Phi).$$

Indeed, let $w \in \mathcal{B}(M, \Phi)$ be given. It suffices to check that $\langle Mw, w\rangle = 0$. We know that: $\langle Mw, h\rangle \geq 0, \forall h \in D(\Phi_\infty)$. Let $j \in \{1, ..., n\}$ be given. We may set $h = \langle w, e^j\rangle e^j$ to get $(Mw)_j w_j \geq 0$. This last relation holds for all $j \in \{1, ..., n\}$ and since $0 \geq \langle w, Mw\rangle = \sum_{j=1}^n (Mw)_j w_j$ we obtain finally that $\langle Mw, w\rangle = 0$.
Let us now state an existence and uniqueness result which can be applied to study a partitionable variational inequality involving a P_0-matrix and a strictly convex function Φ.

Corollary 9 *Suppose that $\Phi \in SD\Gamma(\mathbb{R}^n; \mathbb{R} \cup \{+\infty\})$ and let $M \in \mathbb{R}^{n\times n}$ be a P_0-matrix. If $\mathcal{B}(M, \Phi) = \{0\}$ then for each $q \in \mathbb{R}^n$, problem* $\mathbf{VI(M, q, \Phi)}$ *has a unique solution.*

Proof: The existence result is a direct consequence of Corollary 8 and Remark 6.

To prove the uniqueness, suppose by contradiction that problem $\mathbf{VI(M, q, \Phi)}$ has two different solutions u and U. We set

$$w = Mu + q, \quad W = MU + q.$$

We have

$$\langle w, v - u\rangle + \Phi(v) - \Phi(u) \geq 0, \forall v \in \mathbb{R}^n \tag{57}$$

and

$$\langle W, v - U\rangle + \Phi(v) - \Phi(U) \geq 0, \forall v \in \mathbb{R}^n. \tag{58}$$

We may set $v = u + \langle u, e^j\rangle e^j \in D(\Phi)$ $(1 \leq j \leq n)$ in (57) to get

$$0 \leq w_j u_j + \sum_{k=1}^{n} \Phi_k(u_k + u_j e_k^j) - \sum_{k=1}^{n} \Phi_k(u_k) = w_j u_j + \Phi_j(2u_j) - \Phi_j(u_j).$$

Thus, for all integers $1 \leq j \leq n$:

$$0 \leq w_j u_j + \Phi_j(u_j). \tag{59}$$

Using (58), we check in the same way that, for all integers $1 \leq j \leq n$:

$$0 \leq W_j U_j + \Phi_j(U_j). \tag{60}$$

Let us now set $v = u + \langle U, e^j\rangle e^j \in D(\Phi)$ $(1 \leq j \leq n)$ in (57) to get

$$0 \leq w_j U_j + \Phi_j(u_j + U_j) - \Phi_j(u_j) \tag{61}$$

and $v = U + \langle u, e^j\rangle e^j \in D(\Phi)$ $(1 \leq j \leq n)$ in (58) to get

$$0 \leq W_j u_j + \Phi_j(U_j + u_j) - \Phi_j(U_j). \tag{62}$$

Setting $v = 0$ in (57), we get also

$$0 \geq \langle w, u\rangle + \Phi(u)$$

and thus

$$0 \geq \sum_{j=1}^{n} [w_j u_j + \Phi_j(u_j)]. \tag{63}$$

Using (58) we prove in the same way that

$$0 \geq \sum_{j=1}^{n} [W_j U_j + \Phi_j(U_j)]. \tag{64}$$

Using (59) and (60) together with (63) and (64), we see that, for all integers $1 \leq j \leq n$:

$$w_j u_j + \Phi_j(u_j) = 0, \quad W_j U_j + \Phi_j(U_j) = 0.$$

Then, for all integers $1 \leq j \leq n$:

$$(u-U)_j(M(u-U))_j = (u_j - U_j)(w_j - W_j) = u_j w_j + U_j W_j - u_j W_j - U_j w_j \leq$$

$$\leq -\Phi_j(u_j) - \Phi_j(U_j) + \Phi_j(U_j + u_j) - \Phi_j(U_j) + \Phi_j(u_j + U_j) - \Phi_j(u_j)$$

$$= 2\Phi_j(u_j + U_j) - 2(\Phi_j(u_j) + \Phi_j(U_j)).$$

Using the strict convexity of the functions Φ_j $(1 \leq j \leq n)$, we note that for all integers $\alpha \in \{1, ..., n\}$ such that $u_\alpha \neq U_\alpha$:

$$2\Phi_\alpha(u_\alpha + U_\alpha) = 2\Phi_\alpha(2(\frac{1}{2}u_\alpha + \frac{1}{2}U_\alpha)) = 4\Phi_\alpha(\frac{1}{2}u_\alpha + \frac{1}{2}U_\alpha) < 2(\Phi_\alpha(u_\alpha) + \Phi_\alpha(U_\alpha))$$

and then:

$$(u-U)_\alpha(M(u-U))_\alpha < 0$$

which is a contradiction since M is a P_0-matrix. □

Example 9 *Let*

$$M = \begin{pmatrix} 1 & 1 \\ -4 & 0 \end{pmatrix}$$

and

$$\Phi(x) = |x_1| + |x_2|, \ \forall x = (x_1, x_2) \in \mathbb{R}^2.$$

We see that M is a P_0-matrix and $\Phi \in SD\Gamma(\mathbb{R}^2; \mathbb{R} \cup \{+\infty\})$. Here $D(\Phi_\infty) = D(\Phi)_\infty = \mathbb{R}^2$ and $D(\Phi_\infty)^* = \{0\}$. So $\mathcal{K}(M, \Phi) = \{0\}$ and thus $\mathcal{B}(M, \Phi) = \{0\}$ so that we can apply Corollary 9.

Remark 8 *Corollaries 3, 5, 7 and 8 give conditions on M and Φ ensuring that the mapping $y \mapsto (M + \partial\Phi)^{-1}(y) := \{x \in \mathbb{R}^n : y \in Mx + \partial\Phi(x)\}$ is a well-defined set-valued mapping on $\mathbb{R}^n$, i.e.*

$$(\forall y \in \mathbb{R}^n) : \ (M + \partial\Phi)^{-1}(y) \neq \emptyset.$$

Moreover, if $q \in \mathbb{R}^n$ then any $x \in (M + \partial\Phi)^{-1}(-q)$ is solution of $\mathbf{VI(M, q, \Phi)}$. Corollaries 4, 6 and 9 give conditions on M and Φ ensuring that the mapping $y \mapsto (M + \partial\Phi)^{-1}(y)$ is a well-defined single-valued mapping on $\mathbb{R}^n$. Then if $q \in \mathbb{R}^n$, we may assert that $x = (M + \partial\Phi)^{-1}(-q)$ is the unique solution of $\mathbf{VI(M, q, \Phi)}$.

6.7. Solvability Conditions for Variational Inequalities

The results discussed in the previous section require that $\mathcal{B}(M,\Phi) = \{0\}$. If $\mathcal{B}(M,\Phi) \neq \{0\}$ then in assuming some nonnegativity property on the matrix M, we may determine conditions on q ensuring the solvability of problem $\mathbf{VI(M,q,\Phi)}$. Throughout this Section, we will suppose that $\mathcal{B}(M,\Phi) \neq \{0\}$.

Theorem 3 *Let $\Phi : \mathbb{R}^n \to \mathbb{R} \cup \{+\infty\}$ be a proper, convex and lower semicontinuous function with closed domain and let $M \in \mathbb{R}^{n\times n}$ be a matrix. Suppose in addition that*

$$\langle Mx, x\rangle \geq 0, \forall x \in D(\Phi) \cup D(\Phi)_\infty.$$

If there exists $x_0 \in D(\Phi)$ such that:

$$\langle q - M^T x_0, v\rangle + \Phi_\infty(v) > 0,\ \forall v \in \ \mathcal{B}(M,\Phi),\ v \neq 0, \tag{65}$$

then problem $\mathbf{VI(M,q,\Phi)}$ has at least one solution.

Proof: Let $q \in \mathbb{R}^n$ be given. Here $(M,\Phi) \in \mathbf{PD0_n}$ and using Corollary 7, we see that for all $i \in \mathbb{N}, i \neq 0$, there exists $u_i \in D(\Phi)$ such that

$$\langle (\frac{1}{i}I + M)u_i + q, v - u_i\rangle + \Phi(v) - \Phi(u_i) \geq 0,\ \forall v \in \mathbb{R}^n. \tag{66}$$

We claim that the sequence $\{u_i\} \equiv \{u_i; i \in \mathbb{N}\backslash\{0\}\}$ is bounded. Suppose on the contrary that $||u_i|| \to +\infty$ as $i \to +\infty$. Then, for i large enough, $||u_i|| \neq 0$ and we may set: $z_i := \frac{u_i}{||u_i||}$. There exists a subsequence, again denoted by $\{z_i\}$, such that $\lim_{i\to+\infty} z_i = z$ with $||z|| = 1$.

As in the proof of Theorem 2 we check that $z \in \mathcal{B}(M,\Phi)$. Using now (66) with $v = x_0$, we get also:

$$\langle (\frac{1}{i}I + M)u_i, u_i - x_0\rangle \leq \langle q, x_0 - u_i\rangle + \Phi(x_0) - \Phi(u_i).$$

Here

$$\langle (\frac{1}{i}I + M)u_i, u_i\rangle > 0, \forall i \in \mathbb{N}, i \neq 0,$$

and thus

$$-\langle (\frac{1}{i}I + M)u_i, x_0\rangle + \langle q, u_i - x_0\rangle - \Phi(x_0) + \Phi(u_i) < 0.$$

Dividing this last relation by $||u_i||$, we get:

$$-\langle (\frac{1}{i}I + M)z_i, x_0\rangle + \langle q, z_i - \frac{x_0}{||u_i||}\rangle - \frac{\Phi(x_0)}{||u_i||} + \frac{\Phi(||u_i||z_i)}{||u_i||} < 0.$$

Taking the limit inferior as $i \to +\infty$, we get:

$$\langle q - M^T x_0, z\rangle + \liminf_{i\to+\infty} \frac{\Phi(||u_i||z_i)}{||u_i||} \leq 0$$

and thus, using (9), we obtain:

$$\langle q - M^T x_0, z\rangle + \Phi_\infty(z) \leq 0.$$

This is a contradiction to condition (65) since we have proved above that $z \in \mathcal{B}(M, \Phi)$ and $z \neq 0$.

The sequence $\{u_i\}$ is thus bounded and we may conclude as in the proof of Theorem 2. □

Remark 9 *i) Note that if $\mathcal{B}(M, \Phi) = \{0\}$ then condition (65) is trivially satisfied on the empty set.*
ii) If $0 \in D(\Phi)$ (which is the case for most practical problems) then we may choose $x_0 = 0$ to see (65) in the more legible form

$$\langle q, v\rangle + \Phi_\infty(v) > 0,\ \forall v \in \mathcal{B}(M, \Phi), v \neq 0. \tag{67}$$

Theorem 3 and Proposition 3 may obviously be applied to the class of positive semidefinite matrices to give the following important result.

Corollary 10 *Let $\Phi \in \Gamma(\mathbb{R}^n; \mathbb{R} \cup \{+\infty\})$ and $M \in \mathbb{R}^{n\times n}$ a positive semidefinite matrix. If there exists $x_0 \in D(\Phi)$ such that:*

$$\langle q - M^T x_0, v\rangle + \Phi_\infty(v) > 0,\ \forall v \in\ D(\Phi_\infty) \cap \ker\{M + M^T\} \cap \mathcal{K}(M, \Phi),\ v \neq 0, \tag{68}$$

then problem $\mathbf{VI(M, q, \Phi)}$ has at least one solution.
Moreover if u_1 and u_2 denote two solutions of problem $\mathbf{VI(M, q, \Phi)}$ then

$$u_1 - u_2 \in\ \ker\{M + M^T\}. \tag{69}$$

□

Example 10 *Let*

$$M = \begin{pmatrix} 0 & 2 \\ -1 & 0 \end{pmatrix},\ q = \begin{pmatrix} 1 \\ 1 \end{pmatrix}$$

and

$$\Phi(x) = \Psi_{(\mathbb{R}_+)^2}(x),\ \forall x \in \mathbb{R}^2.$$

We have

$$D(\Phi) = D(\Phi_\infty) = D(\Phi)_\infty = (\mathbb{R}_+)^2$$

and

$$\langle Mx, x \rangle \geq 0, \forall x \in (\mathbb{R}_+)^2.$$

We have also:

$$D(\Phi)_\infty \cap \mathcal{N}_-(M) = (\{0\} \times \mathbb{R}_+) \cup (\mathbb{R}_+ \times \{0\})$$

and

$$\mathcal{K}(M, \Phi) = \mathbb{R}_- \times \mathbb{R}_+.$$

Thus,

$$\mathcal{B}(M, \Phi) = \{0\} \times \mathbb{R}_+.$$

Let us choose $x_0 = \begin{pmatrix} 0 & 1 \end{pmatrix}^T$. *We remark that If* $v \in \mathcal{B}(M, \Phi)$, $v \neq 0$ *then* $v_1 = 0$ *and* $v_2 > 0$ *so that* $\langle q - M^T x_0, v \rangle = v_2 > 0$. *We may then apply Theorem 3 to assert that* $\mathbf{VI}(\mathbf{M}, \mathbf{q}, \mathbf{\Phi})$ *has at least one solution.*

The folowing result (see [10]) concerning variational inequalities involving symmetric positive semidefinite matrices is also of particular interest.

Corollary 11 *Let* $\Phi \in \Gamma(\mathbb{R}^n; \mathbb{R} \cup \{+\infty\})$ *and* $M \in \mathbb{R}^{n \times n}$ *a positive semidefinite and symmetric matrix.*

(a) If

$$\langle q, v \rangle + \Phi_\infty(v) > 0, \ \forall v \in \ D(\Phi)_\infty \cap \ker\{M\} \cap \mathcal{K}(M, \Phi), \ v \neq 0, \tag{70}$$

then problem $\mathbf{VI}(\mathbf{M}, \mathbf{q}, \mathbf{\Phi})$ *has at least one solution.*

(b) If u_1 *and* u_2 *denote two solutions of problem* $\mathbf{VI}(\mathbf{M}, \mathbf{q}, \mathbf{\Phi})$ *then*

$$u_1 - u_2 \in \ \ker\{M\}. \tag{71}$$

and

$$\langle q, u_1 - u_2 \rangle = \Phi(u_2) - \Phi(u_1). \tag{72}$$

(c) If

$$\Phi(x + z) = \Phi(x), \forall x \in D(\Phi), \ z \in \ \ker\{M\}$$

and

$$\langle q, e \rangle \neq 0, \ \forall e \in \ \ker\{M\}, e \neq 0,$$

then problem $\mathbf{VI}(\mathbf{M}, \mathbf{q}, \mathbf{\Phi})$ *has at most one solution.*

(d) If the function Φ *is stricly convex then problem* $\mathbf{VI}(\mathbf{M}, \mathbf{q}, \mathbf{\Phi})$ *has at most one solution.*

□

Example 11 *Let*

$$M = \begin{pmatrix} 1 & 1 \\ 1 & 1 \end{pmatrix}$$

and

$$\Phi(x) = |x_1| + |x_2| \ \forall x = (x_1, x_2) \in \mathbb{R}^2.$$

Here $D(\Phi) = D(\Phi)_\infty = D(\Phi_\infty) = (\mathbb{R})^2$, $\Phi_\infty = \Phi$ *and* $D(\Phi_\infty)^* = \{0\} \times \{0\}$. *Moreover* $\mathcal{K}(M, \Phi) = \ker\{M\} = \{(a, -a) : a \in \mathbb{R}\}$ *and* $\mathcal{B}(M, \Phi) = D(\Phi)_\infty \cap \ker\{M\} \cap \mathcal{K}(M, \Phi) = \{(a, -a) : a \in \mathbb{R}\}$. *Therefore for all* $q \in \mathbb{R}^2$ *such that* $|q_1 - q_2| < 2$ *condition (70) of corollary 11 is satisfied.*

7. Variational Inequalities Method in Electronics

Several researchers have recently shown that mathematical tools from complementarity systems theory and variational inequalities theory may be used to develop rigorous mathematical study of circuits in electronics involving devices like diodes, Zener diodes and varistors that are characterized by set-valued ampere-volt characteristics. Mathematical approaches using tools from mathematical programming and complementarity systems theory have been particularly developed in [17], [18], [33], [42], [45] and the references cited therein while mathematical approaches using tools from set-valued analysis and variational inequalities theory have been studied in [8], [9], [12], [15], [19] and [23].

7.1. Set-valued Ampere-Volt Characteristics in Electronics

Electrical devices like diodes are described in terms of Ampere-Volt characteristics (i, V) that is a graph expressing the difference of potential V across the device as a function of current i through the device.

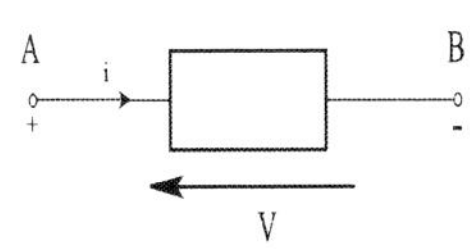

Figure 1. Electrical Device.

The schematic symbol of a circuit element is given in Figure 1. The conventional current flow i will be depicted on the conductor in the direction of the arrow and the potential $V := V_A - V_B$ (V_A (resp. V_B)) potential of point A (resp. B) across the device will be denoted alongside the device. Experimental measures as well as empirical and physical models lead to a variety of monotone graphs that may present vertical branches. The reader can find general descriptions of devices and Ampere-Volt characteristics either in the appropriate electronics literature or in electronics society catalogs (see e.g. [1], [2], [3], [4], [5], [6], [44]).

Let us suppose here that we may write:

$$V \in \mathcal{F}(i), \ (i \in \mathbb{R})$$

for some set-valued function $\mathcal{F} : \mathbb{R} \rightrightarrows \mathbb{R}$. The domain $D(\mathcal{F})$ of $\mathcal{F}$ is defined by:

$$D(\mathcal{F}) = \{x \in \mathbb{R} : \mathcal{F}(x) \neq \emptyset\}.$$

We assume that $\mathcal{F}$ is maximal monotone, i.e.

$$(z_1 - z_2)(x_1 - x_2) \geq 0, \ \forall \, x_1, x_2 \in D(\mathcal{F}), z_1 \in \mathcal{F}(x_1), z_2 \in \mathcal{F}(x_2)$$

and the graph $G(\mathcal{F})$ of $\mathcal{F}$, i.e.

$$G(\mathcal{F}) := \{(x, y) \in \mathbb{R}^n \times \mathbb{R}^n : x \in D(\mathcal{F}), \ y \in \mathcal{F}(x)\}$$

is not properly included in any other monotone subset of $\mathbb{R} \times \mathbb{R}$.

A classical result (see e.g. Proposition 1.3.15 in [28]) ensures that there exists a proper, convex and lower semicontinuous function $\varphi : \mathbb{R} \to \mathbb{R} \cup \{+\infty\}$ such that

$$\mathcal{F}(i) = \partial\varphi(i), \ (i \in \mathbb{R}).$$

Remark 10 *Note that there exists* $-\infty \leq a \leq b \leq +\infty$ *such that* $]a, b[\subset D(\mathcal{F}) \subset [a, b]$ *and* φ *can be determined by the formula:*

$$\varphi(i) = \begin{cases} \int_{i_0}^{i} \beta^0(s)ds \ \text{ if } \ i \in [a, b] \\ \\ +\infty \ \text{ if } \ i \in \mathbb{R}\backslash[a, b] \end{cases} \tag{73}$$

where $i_0 \in]a, b[$ *and* $\beta^0 : D(\mathcal{F}) \to \mathbb{R}$ *denotes the minimal section of* $\mathcal{F}$*, i.e.* $\beta^0(x) \in \mathcal{F}(x)$ *and* $|\beta^0(x)| = \inf\{|w| : w \in \mathcal{F}(x)\}$*. Remark that the function* φ *in (73) is determined by* $\mathcal{F}$ *up to an additive constant.*

Note also that:

$$\partial\varphi(i) = \left[\beta^0(i^-), \beta^0(i^+)\right], \ \forall i \in]a, b[,$$

where

$$\beta^0(i^-) = \lim_{z \to i, z < i} \beta(z)$$

and

$$\beta^0(i^+) = \lim_{z \to i, z > i} \beta(z).$$

We have

$$V \in \partial\varphi(i) \Longleftrightarrow i \in \partial\varphi^*(V) \Longleftrightarrow \varphi(i) + \varphi^*(V) = iV.$$

Remark 11 *(Terminology) We will say that an electrical device is VAM-admissible provided that its ampere-volt characteristic graph* (i, V) *is maximal monotone. Then there exists a proper convex and lower semicontinuous function* $\varphi : \mathbb{R} \to \mathbb{R} \cup \{+\infty\}$ *such that*

$$V \in \partial\varphi(i), \ \forall i \in \mathbb{R}.$$

The function φ *will be called the electrical superpotential (determined up to an additive constant) of the device. Roughly speaking, the electrical superpotential* φ *appears as a "primitive" of* $\mathcal{F}$ *in the sense that the "derivative" (in the generalized sense) of* φ *recovers the set-valued function* $\mathcal{F}$.

7.1.1. Diode Models

Example 12 (IDEAL DIODE MODEL) *The diode is a device that constitutes a rectifier which permits the easy flow of charges in one direction but restrains the flow in the opposite direction. Diodes are used in power electronics applications like rectifier circuits, switching inverter and converter circuits. Figure 2 illustrates the ampere-volt characteristic of an ideal diode.*

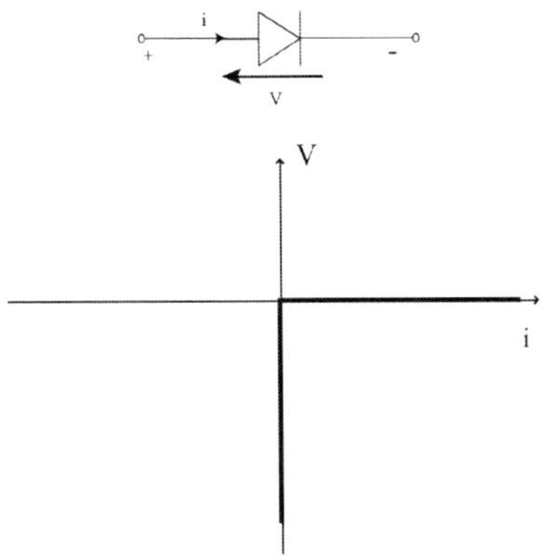

Figure 2. Ideal diode model.

This is a model in which the diode is a simple switch. If $V < 0$ then $i = 0$ and the diode is blocking. If $i > 0$ then $V = 0$ and the diode is conducting. We first see that the ideal diode is described by the complementarity relation

$$V \leq 0, \;\; i \geq 0, \;\; Vi = 0$$

that is also

$$\min\{-V, i\} = 0.$$

The electrical superpotential of the ideal diode is

$$\varphi_D(x) = \Psi_{R_+}(x), \;\; (x \in \mathbb{R}).$$

Then

$$\varphi_D^*(z) = \Psi_{R_-}(z), \;\; (z \in \mathbb{R})$$

and the recession function of the electrical superpotential is:

$$(\varphi_D)_\infty(x) = \varphi_D(x), \; (x \in \mathbb{R}).$$

We have also

$$\partial\varphi_D(x) := \begin{cases} \mathbb{R}_- \;\; \textit{if} \;\; x = 0 \\ \\ 0 \;\; \textit{if} \;\; x > 0 \\ \\ \emptyset \;\; \textit{if} \;\; x < 0 \end{cases}, \;\; (x \in \mathbb{R})$$

and

$$\partial\varphi_D^*(z) := \begin{cases} \mathbb{R}_+ \text{ if } z = 0 \\ 0 \text{ if } z < 0 \\ \emptyset \text{ if } z > 0 \end{cases}, \quad (z \in \mathbb{R}).$$

The complementarity relation can be written as

$$V \in \partial\varphi_D(i) \Longleftrightarrow i \in \partial\varphi_D^*(V) \Longleftrightarrow \varphi_D(i) + \varphi_D^*(V) = iV.$$

Example 13 (PRACTICAL DIODE MODEL) *Figure 3 illustrates the ampere-volt characteristic of a practical diode model.*

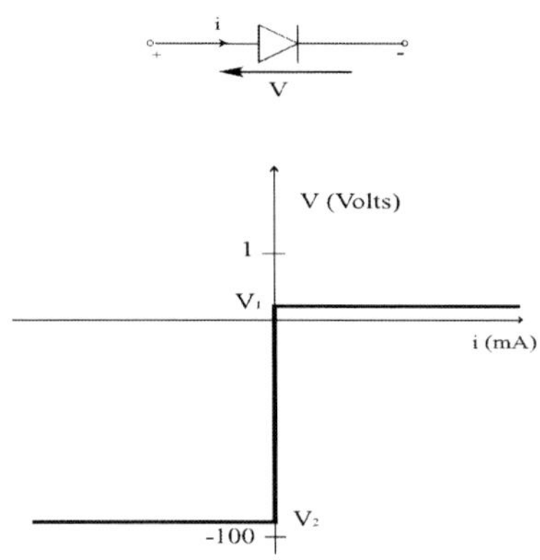

Figure 3. Practical diode model

There is a voltage point, called the knee voltage V_1, at which the diode begins to conduct and a maximum reverse voltage, called the peak reverse voltage V_2, that will not force the diode to conduct. When this voltage is exceeded, the depletion may breakdown and allow the diode to conduct in the reverse direction. Note that usually $\mid V_2 \mid >> \mid V_1 \mid$ and the model is locally ideal.

For general purpose diodes used in low frequency/speed applications, $\mid V_1 \mid \simeq 0.7 - 2.5$ V and $\mid V_2 \mid \simeq 5$ kV; for high voltage rectifier diodes, $\mid V_1 \mid \simeq 10$ V and $\mid V_2 \mid \simeq 30$ kV; for fast diodes used in switched mode power supply and inverter circuits, $\mid V_1 \mid \simeq 0.7 - 1.5$ V and $\mid V_2 \mid \simeq 3$ kV and for Schottky diodes used in high frequency applications, $\mid V_1 \mid \simeq 0.2 - 0.9$ V and $\mid V_2 \mid \simeq 100$ V.

The electrical superpotential of the practical diode is

$$\varphi_{PD}(x) = \begin{cases} V_1 x \text{ if } x \geq 0 \\ V_2 x \text{ if } x < 0 \end{cases}, \quad (x \in \mathbb{R}).$$

Then

$$\varphi_{PD}^*(z) = \Psi_{[V_2,V_1]}(z), \quad (z \in \mathbb{R})$$

and the recession function of the electrical superpotential is given by:

$$(\varphi_{PD})_\infty(x) = \varphi_{PD}(x), \ (x \in \mathbb{R}).$$

We see that

$$\partial\varphi_{PD}(x) = \begin{cases} V_2 \ \text{if} \ x < 0 \\ [V_2, V_1] \ \text{if} \ x = 0 \\ V_1 \ \text{if} \ x > 0 \end{cases}, \quad (x \in \mathbb{R})$$

recovers the ampere-volt characteristic (i, V) *while*

$$\partial\varphi_{PD}^*(z) = \begin{cases} \mathbb{R}_- \ \text{if} \ z = V_2 \\ 0 \ \text{if} \ z \in]V_2, V_1[\\ \mathbb{R}_+ \ \text{if} \ z = V_1 \\ \emptyset \ \text{if} \ z \in \mathbb{R}\backslash [V_2, V_1] \end{cases}, \quad (z \in \mathbb{R}).$$

recovers the volt-ampere characteristic (V, i). *The ampere-volt characteristic of the practical diode can thus be written as*

$$V \in \partial\varphi_{PD}(i) \iff i \in \partial\varphi_{PD}^*(V) \iff \varphi_{PD}(i) + \varphi_{PD}^*(V) = iV.$$

Example 14 (COMPLETE DIODE MODEL) *Figure 4 illustrates a complete diode model which includes the effect of the natural resistance of the diode, called the bulk resistance, the reverse current* I_{R_1}, *the diode capacitance and the diffusion current. This last model is more accurate and represents the true operating characteristics of the diode.*

Note that $| V_4 | << | V_1 |$. *For example, the 10ETS.. rectifier (SAFEIR series [3]) has been designed with* $| V_1 | = 1.1$ *V,* $| V_4 | = 800 - 1600$ *V,* $I_{R1} = 0.05$ *mA and with a bulk resistance equal to* 20 *m*Ω.
Let us use the notation of Figure 4. It is implicitly assumed that

$$I_{R2} < 0 < I_{R1}, \ V_4 < V_2 < 0 < V_1 < V_3.$$

Let us also set:

$$\alpha := \frac{(V_3 - V_1)}{(I_{R3} - I_{R1})}, \ \beta := \frac{(I_{R1}V_3 - I_{R3}V_1)}{(I_{R3} - I_{R1})}, \ \gamma := \frac{I_{R1}(I_{R1}V_3 - I_{R3}V_1)}{2(I_{R3} - I_{R1})}.$$

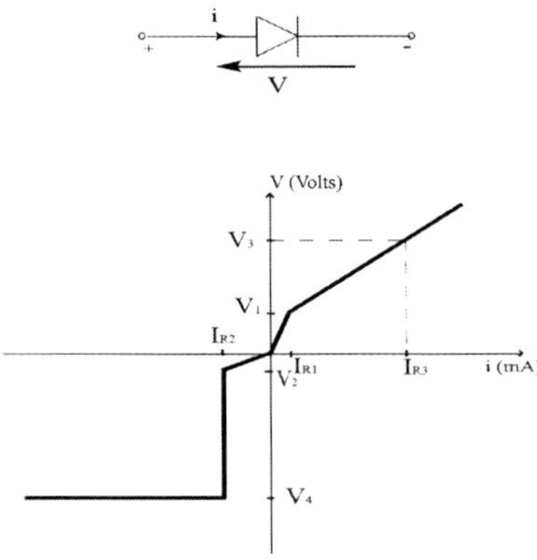

Figure 4. Complete diode model.

The electrical superpotential of the complete diode is

$$\varphi_{CD}(x) = \begin{cases} V_4 x + I_{R2}(\frac{V_2}{2} - V_4) \ \textit{if} \ x \leq I_{R2} \\ \frac{V_2}{2I_{R2}} x^2 \ \textit{if} \ I_{R2} < x \leq 0 \\ \frac{V_1}{2I_{R1}} x^2 \ \textit{if} \ 0 < x \leq I_{R1} \\ \frac{1}{2}\alpha x^2 - \beta x + \gamma \ \textit{if} \ I_{R1} < x \end{cases} , \quad (x \in \mathbb{R})$$

and simple calculations yield:

$$\partial\varphi_{CD}(x) = \begin{cases} V_4 \ \textit{if} \ x < I_{R2} \\ [V_4, V_2] \ \textit{if} \ x = I_{R2} \\ \frac{V_2}{I_{R2}} x \ \textit{if} \ I_{R2} < x \leq 0 \\ \frac{V_1}{I_{R1}} x \ \textit{if} \ 0 < x \leq I_{R1} \\ \alpha x - \beta \ \textit{if} \ I_{R1} < x \end{cases} , \quad (x \in \mathbb{R}).$$

On the other hand, we may compute the conjugate function:

$$\varphi^*_{CD}(z) = \begin{cases} +\infty \ \textit{if} \ z \leq V_4 \\ I_{R2}(z - \frac{V_2}{2}) \ \textit{if} \ V_4 < z \leq V_2 \\ \frac{I_{R2}}{2V_2} z^2 \ \textit{if} \ V_2 < z \leq 0 \\ \frac{I_{R1}}{2V_1} z^2 \ \textit{if} \ 0 < z \leq V_1 \\ \frac{1}{2}\alpha z^2 + (I_{R_1} - \alpha V_1)z + \frac{1}{2}V_1(\alpha V_1 - I_1) \ \textit{if} \ V_1 < z \end{cases}, \quad (z \in \mathbb{R})$$

and

$$\partial\varphi^*_{CD}(z) = \begin{cases} \emptyset \ \textit{if} \ z < V_4 \\ ]-\infty, I_{R2}] \ \textit{if} \ z = V_4 \\ I_{R2} \ \textit{if} \ V_4 < z \leq V_2 \\ \frac{I_{R2}}{V_2} z \ \textit{if} \ V_2 < z \leq 0 \\ \frac{I_{R1}}{V_1} z \ \textit{if} \ 0 < z \leq V_1 \\ \alpha z + (I_{R1} - \alpha V_1) \ \textit{if} \ V_1 < z \end{cases}, \quad (z \in \mathbb{R}).$$

The recession function of the electrical superpotential is here:

$$(\varphi_{CD})_\infty(x) = \begin{cases} V_4 x \ \textit{if} \ x \leq I_{R2} \\ +\infty \ \textit{if} \ x > I_{R2} \end{cases}, \quad (x \in \mathbb{R}).$$

The ampere-volt characteristic of the complete diode can then be written as

$$V \in \partial\varphi_{CD}(i) \Longleftrightarrow i \in \partial\varphi^*_{CD}(V) \Longleftrightarrow \varphi_{CD}(i) + \varphi^*_{CD}(V) = iV.$$

Example 15 (EMPIRICAL DIODE MODEL) *An empirical model used in electronics to describe the ampere-volt characteristic of a diode is:*

$$V(i) = \frac{V_T}{\eta}\ln(\frac{i}{I_S} + 1) \ \ (i > -I_S),$$

where I_S is the saturating reverse current ($10^{-15} \leq I_S \leq 10^{-12}$ A), V_T is the thermodynamic voltage (25 mV) and η is the emission coefficient ($1 \leq \eta \leq 2$). This model is usually *considered in the engineering literature when a rigorous mathematical analysis taking care of the domain of V is not required. It is however possible to proceed to a suitable mathematical treatment as above in defining the set-valued function $\mathcal{V} : \mathbb{R} \rightrightarrows \mathbb{R}$ by*

$$\mathcal{V}(i) = \begin{cases} \frac{V_T}{\eta} \ln(\frac{i}{I_S} + 1) \text{ if } i > -I_S \\ \emptyset \text{ if } i \leq -I_S. \end{cases}$$

Then setting

$$\varphi_{ED}(i) = \begin{cases} \frac{V_T}{\eta} \ln(\frac{i}{I_S} + 1)(i + I_S) - \frac{V_T I_S}{\eta} i \text{ if } i > -I_S \\ \frac{V_T I_S^2}{\eta} \text{ if } i = -I_S \\ +\infty \text{ if } i < -I_S \end{cases}$$

we see that

$$\mathcal{V}(i) = \partial\varphi_{ED}(i) \ (i \in \mathbb{R}).$$

Moreover

$$(\varphi_{ED})_\infty(i) = \Psi_{\{0\}}(i), \ (i \in \mathbb{R}).$$

7.1.2. Zener Diode Models

The Zener diodes are made to permit current to flow in the reverse direction if the voltage is larger than the rated breakdown or "Zener voltage" V_2. For example, for a common Zener diode, $V_1 \simeq 0.7$ volts and $V_2 \simeq -7$ volts.

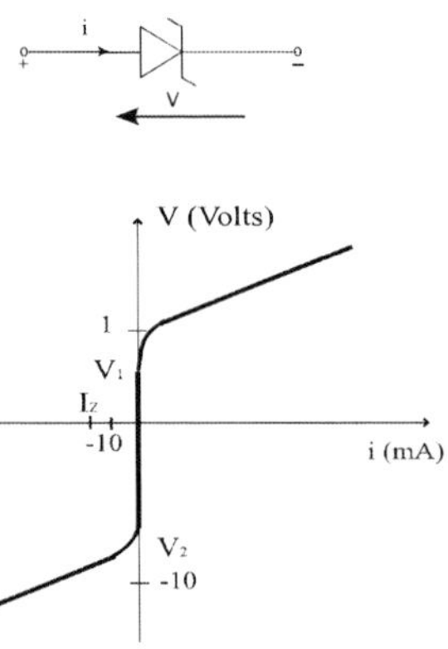

Figure 5. Zener diode model.

The Zener diode (see Figure 5) is a good voltage regulator to maintain a constant voltage regardless of minor variations in load current or input voltage. There is a current point I_Z, called the Zener knee current, which is the minimum value of the Zener current required to maintain voltage regulation and a maximum allowable value of Zener current I_M. Currents above this value will damage or destroy the system.

The graph corresponding to the ampere-volt characteristic (i, V) is maximal monotone and there exists a proper convex and continuous electrical superpotential $\varphi : \mathbb{R} \to \mathbb{R}$ such that

$$V \in \partial\varphi(i), \quad (i \in \mathbb{R}).$$

Example 16 (IDEAL ZENER DIODE MODEL) *The ideal Zener diode model (see Figure 6) is given by the practical diode model (see Figure 3) with the appropriate values for* V_1 *and* V_2*. This means that the voltage across the diode is constant over a wide range of device current values.*

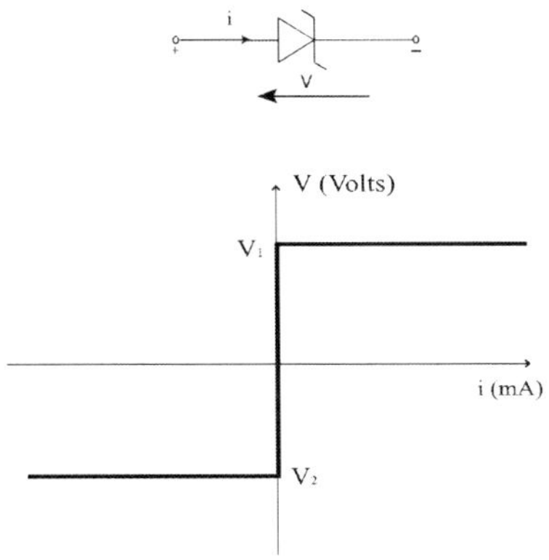

Figure 6. Ideal Zener diode model.

Example 17 (PRACTICAL ZENER DIODE MODEL) *This model (see Figure 7) is a piecewise linear model that includes the effects of the Zener impedance.*

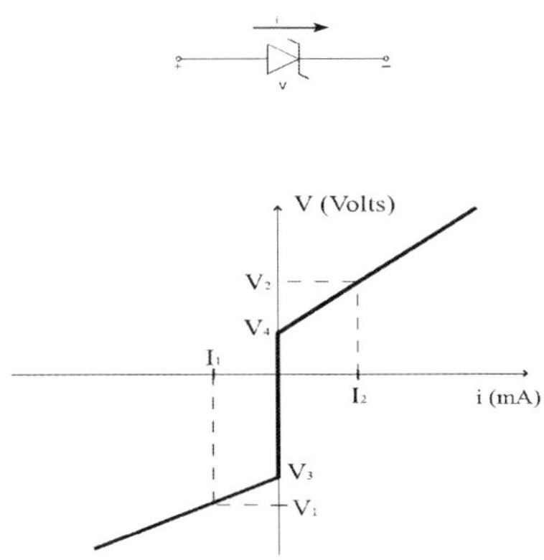

Figure 7. Practical Zener diode model.

Let us use the notation of Figure 7. It is here implicitly assumed that

$$I_1 < 0 < I_2, \quad V_1 < V_3 < 0 < V_4 < V_2.$$

The electrical superpotential of the Zener diode is

$$\varphi_Z(x) = \begin{cases} \frac{(V_1 - V_3)}{2I_1} x^2 + V_3 x \;\; if \;\; x < 0 \\ \\ \frac{(V_2 - V_4)}{2I_2} x^2 + V_4 x \;\; if \;\; x \geq 0 \end{cases}, \quad (x \in \mathbb{R}).$$

Then

$$\varphi_Z^*(z) = \begin{cases} \frac{I_1}{2(V_1 - V_3)}(z^2 - 2V_3 z + V_3^2) \;\; if \;\; z < V_3 \\ \\ 0 \;\; if \;\; V_3 \leq z \leq V_4 \\ \\ \frac{I_2}{2(V_2 - V_4)}(z^2 - 2V_4 z + V_4^2) \;\; if \;\; V_4 < z \end{cases}, \quad (z \in \mathbb{R}).$$

The recession function of the electrical superpotential is given by:

$$(\varphi_Z)_\infty(x) = \Psi_{\{0\}}(x), \; (x \in \mathbb{R}).$$

Moreover

$$\partial\varphi_Z(x) = \begin{cases} \frac{(V_1 - V_3)}{I_1} x + V_3 \;\; if \;\; x < 0 \\ \\ [V_3, V_4] \;\; if \;\; x = 0 \\ \\ \frac{(V_2 - V_4)}{I_2} x + V_4 \;\; if \;\; x > 0 \end{cases}, \quad (x \in \mathbb{R})$$

and

$$\partial\varphi_Z^*(z) = \begin{cases} \frac{I_1}{V_1 - V_3}(z - V_3) \;\; if \;\; z < V_3 \\ \\ 0 \;\; if \;\; V_3 \leq z \leq V_4 \\ \\ \frac{I_2}{V_2 - V_4}(z - V_4) \;\; if \;\; V_4 < z \end{cases}, \quad (z \in \mathbb{R}).$$

The ampere-volt characteristic of the concrete Zener diode can thus be written as

$$V \in \partial\varphi_Z(i) \Longleftrightarrow i \in \partial\varphi_Z^*(V) \Longleftrightarrow \varphi_Z(i) + \varphi_Z^*(V) = iV.$$

Example 18 (VARISTOR) *A varistor is a nonlinear device that has an electrical behavior similar to the Zener diode (with $\mid V_1 \mid = \mid V_2 \mid$). More precisely, the varistor (see Figure 8) is a voltage-dependent resistor with a symmetrical monotone ampere-volt characteristic. It is used connected in parallel with the electronic device or circuit that is to be guarded in order to form a low-resistance shunt when voltage increases and thus prevent any further rise in the overvoltage.*

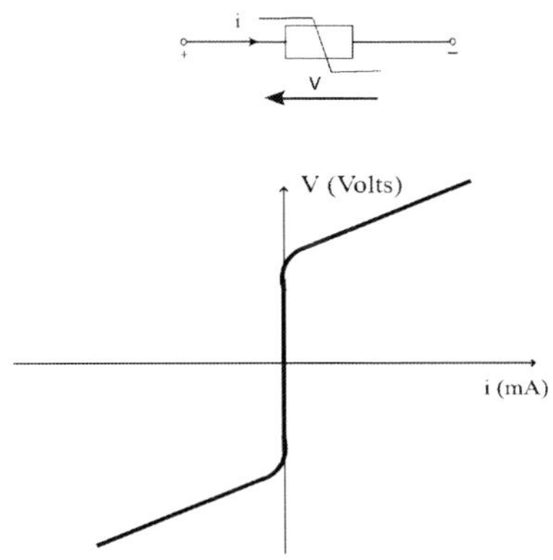

Figure 8. Varistor.

7.1.3. Transistor Models

A junction transistor is a semiconductor triode capable of producing amplification. A P-N-P (resp. N-P-N) transistor consists of a silicon (or germanium) crystal in which a layer of N-type silicon (resp. P-type) is sandwiched between two layers of P-type silicon (resp. N-type). The three portions of transistor are known as emitter, base and collector.

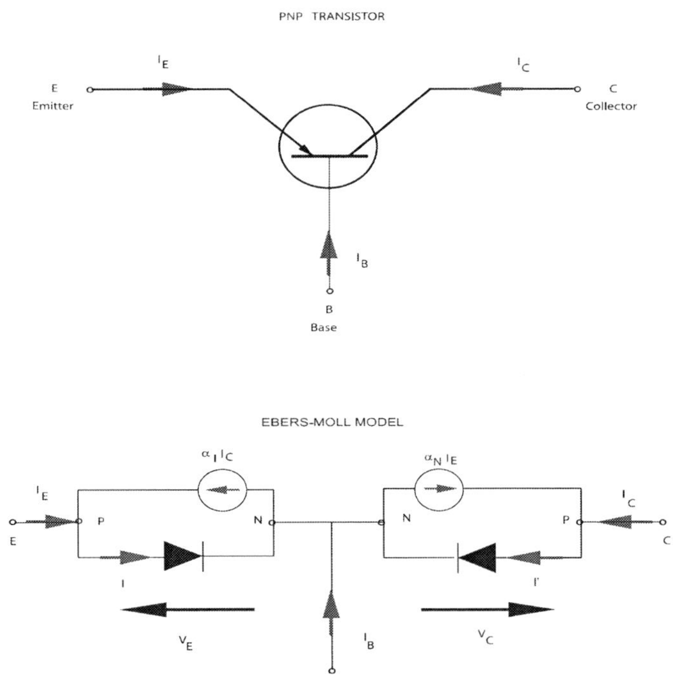

Figure 9. Transistor P-N-P.

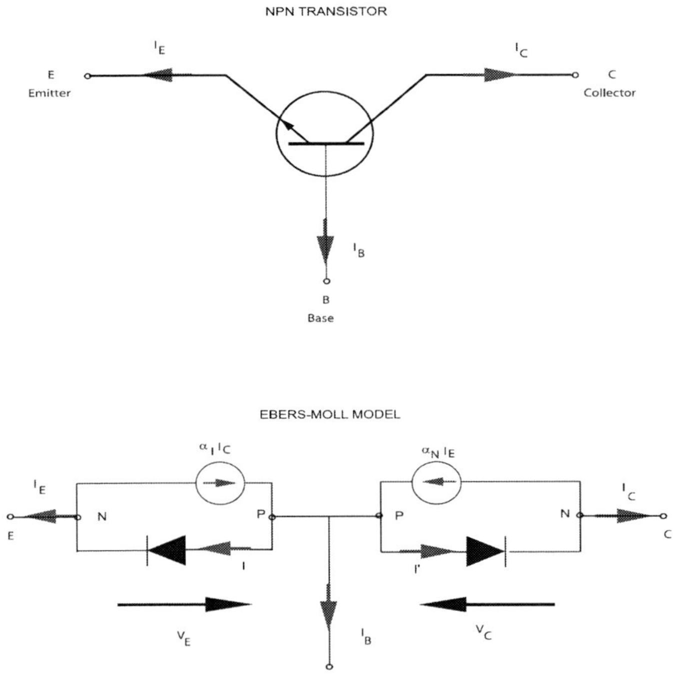

Figure 10. Transistor N-P-N.

The behavior of a transistor can be described by means of the Ebers-Moll model (Ref. [44]) involving two diodes placed back to back and two dependent current-controlled sources $\alpha_I I_C$ and $\alpha_N I_E$ shunting the diodes. Here $\alpha_N \in [0,1[$ is known as the current gain in normal operation and $\alpha_I \in [0,1[$ is known as the inverted common-base gain current. Throughout this section, we will use the notations and conventions of Figures 9 and 10.

Example 19 (IDEAL TRANSISTOR MODEL) *Let us here assume that the two diodes of Ebers-Moll are ideal. That means that each diode acts as a simple switch. If $V_E < 0$ (resp. $V_C < 0$) then $I = 0$ (resp. $I' = 0$) and the diode is blocking. If $I > 0$ (resp. $I' > 0$) then $V_E = 0$ (resp. $V_C = 0$) and the diode is conducting. We may then write:*

$$V_E \leq 0,\ I \geq 0,\ V_E I = 0$$

and

$$V_C \leq 0,\ I' \geq 0,\ V_C I' = 0.$$

That is also:

$$\begin{pmatrix} -V_E \\ -V_C \end{pmatrix} \geq 0, \begin{pmatrix} I \\ I' \end{pmatrix} \geq 0,\ \langle \begin{pmatrix} -V_E \\ -V_C \end{pmatrix}, \begin{pmatrix} I \\ I' \end{pmatrix} \rangle = 0. \tag{74}$$

Moreover:

$$\begin{pmatrix} I \\ I' \end{pmatrix} = \begin{pmatrix} 1 & \alpha_I \\ \alpha_N & 1 \end{pmatrix} \begin{pmatrix} I_E \\ I_C \end{pmatrix} \tag{75}$$

and:

$$I_B = -(I_E + I_C). \tag{76}$$

The relations in (74)-(76) constitutes a handy mathematical model for the transistor.

Example 20 (GENERAL TRANSISTOR MODEL) *A transistor will be called VAM-admissible provided that there exist* $(\varphi_E, \varphi_C) \in \Gamma(\mathbb{R}; \mathbb{R} \cup \{+\infty\})^2$ *such that the ampere-volt characteristics of the two diodes of Ebers-Moll model can be formulated as:*

$$V_E \in \partial\varphi_E(I), \quad V_C \in \partial\varphi_C(I').$$

The function φ_E *will be called the emitter electrical superpotential of the transistor while the function* φ_C *is the collector electrical superpotential. The mathematical model of the transistor reads:*

$$\begin{pmatrix} -V_E \\ -V_C \end{pmatrix} \in \begin{pmatrix} -\partial\varphi_E(I) \\ -\partial\varphi_C(I') \end{pmatrix}, \tag{77}$$

$$\begin{pmatrix} I \\ I' \end{pmatrix} = \begin{pmatrix} 1 & \alpha_I \\ \alpha_N & 1 \end{pmatrix} \begin{pmatrix} I_E \\ I_C \end{pmatrix} \tag{78}$$

and

$$I_B = -(I_E + I_C). \tag{79}$$

The different models of diodes that have been discussed in the previous Section can be here used so as to define corresponding models of transistors.

7.2. Mathematical Formulation of a Class of Non-regular Models in Electronics

A circuit in electronics is formed by the interconnection of electrical devices like generators, resistors, capacitors, inductors, transistors, diodes and various others. The behaviour of a circuit is usually described in terms of currents and voltages that can be specified through each involved electrical device. The approach to state a mathematical model that can be used to determine these currents and voltages consists to formulate the ampere-volt characteristic of each electrical device, to write the Kirchhoff's voltage law expressing that the algebraic sum of the voltages between successive nodes in all mesches in the circuit are zero and to write the Kirchhoff's current law stating that the algebraic sum of the currents in all branches which converge to a common node equal zero.

The practice (see [9] and [14]) shows that a large class of circuits can be studied via the following general mathematical formalism.

Let $A \in \mathbb{R}^{n\times n}$, $B \in \mathbb{R}^{n\times m}$, $C \in \mathbb{R}^{m\times n}$ and $D \in \mathbb{R}^{n\times p}$ be given matrices. Let $\Xi : \mathbb{R}^m \to \mathbb{R} \cup \{+\infty\}$ be a given proper, convex and lower semicontinuous function. Let $u \in \mathbb{R}^p$ be given, we consider the problem:

$\mathbf{NRM(A, B, C, D, u, \Xi)}$: Find $(x, y_L) \in \mathbb{R}^n \times \mathbb{R}^m$ such that

$$Ax - By_L + Du = 0, \tag{80}$$

$$y = Cx, \tag{81}$$

and

$$y_L \in \partial \Xi(y). \tag{82}$$

The matrices A, B, C and D in (80) are structural matrices used to state Kirchhoff's voltage laws and Kirchhoff's current laws in matrix form. The matrix A depends of electrical parameters like resistances, capacitances and inductances. Usually u is a control vector that drives the system, x denotes a current vector and y_L is a voltage vector corresponding to electrical devices like diodes whose (possibly set-valued) ampere-volt characteristics can be described as in (82).

It is noteworthy that (80)–(82) may represent the equations of a static circuit, but also the generalized equation that is to be satisfied by the equilibrium points of a dynamical circuit, or more generally of a class of differential inclusions (see [14] for applications in the absolute stability problem).

Let us now make the following three assumptions:

Assumption $(H1)$: $\Xi \in \Gamma(\mathbb{R}^m; \mathbb{R} \cup \{+\infty\})$.

Assumption $(H2)$: *There exists* $\bar{x}_0 \in \mathbb{R}^n$ *such that* Ξ *is finite and continuous at* $\bar{y}_0 = C\bar{x}_0$.

Assumption $(H3)$: *There exists an invertible matrix* $P \in \mathbb{R}^{n \times n}$ *such that*

$$PB = C^T$$

We set

$$\Phi(x) = \Xi(Cx), \ (\forall x \in \mathbb{R}^n). \tag{83}$$

Then

$$D(\Phi) = \{x \in \mathbb{R}^n : Cx \in D(\Xi)\}. \tag{84}$$

Assumption $(H2)$ entails that $D(\Phi) \neq \emptyset$ and it is clear that $\Phi : \mathbb{R}^n \to \mathbb{R} \cup \{+\infty\}$ is proper convex and lower semicontinuous. Moreover $D(\Phi)$ is convex since $D(\Xi)$ is convex and closed since $D(\Xi)$ is supposed to be closed in $(H1)$.

Proposition 7 *Suppose that assumptions* $(H1)-(H3)$ *are satisfied and let* Φ *be defined as in (83).*

i) If (x, y_L) *is a solution of Problem* $\mathbf{NRM}(\mathbf{A}, \mathbf{B}, \mathbf{C}, \mathbf{D}, \mathbf{u}, \Xi)$ *then* x *is a solution of problem* $\mathbf{VI}(-\mathbf{PA}, -\mathbf{PDu}, \mathbf{\Phi})$, *i.e.*

$$\langle -PAx - PDu, v - x \rangle + \Phi(v) - \Phi(x) \geq 0, \forall v \in \mathbb{R}^n. \tag{85}$$

ii) If x *is a solution of problem* $\mathbf{VI}(-\mathbf{PA}, -\mathbf{PDu}, \mathbf{\Phi})$ *then there exists* $y_L \in \mathbb{R}^m$ *such that* (x, y_L) *is a solution of Problem* $\mathbf{NRM}(\mathbf{A}, \mathbf{B}, \mathbf{C}, \mathbf{D}, \mathbf{u}, \Xi)$.

Proof: Let (x, y_L) be a solution of Problem (80)-(82). Then $0 \in Ax - B\partial\Xi(Cx) + Du$ which is equivalent to $0 \in PAx - PB\partial\Xi(Cx) + PDu$ since P is invertible. Thus $0 \in PAx - C^T\partial\Xi(Cx) + PDu$. The existence of a vector $\bar{y}_0 = C\bar{x}_0$ at which Ξ is finite and continuous ensures that (see e.g. Proposition 1.3.11 in [28]): $C^T\partial\Xi(Cz) = \partial\Phi(z), \ (z \in \mathbb{R}^n)$. Thus $0 \in PAx + PDu - \partial\Phi(x)$, that is $\langle -PAx - PDu, v - x\rangle + \Phi(v) - \Phi(x) \geq 0, \forall v \in \mathbb{R}^n$. Suppose now that x is solution of Problem (85). We see as above that: $0 \in Ax - B\partial\Xi(Cx) + Du$. It results that there exists $y_L \in \partial\Xi(Cx)$ such that: $0 = Ax - By_L + Du$. Then we obtain the relations in (80)-(82) by setting $y = Cx$. □

Proposition 7 means that if assumptions $(H1) - (H3)$ hold then problem (80)-(82) can be studied via the variational inequality $\mathbf{VI}(\mathbf{M}, \mathbf{q}, \mathbf{\Phi})$ with

$$M = -PA, \quad q = -PDu, \quad \Phi \equiv \Xi \circ C. \tag{86}$$

The results discussed in Section 3 can thus be used here too. The following result is then of particular interest to calculate the recession tools involved in $\mathcal{B}(M, q, \Phi)$, that are here $D(\Phi)_\infty$, Φ_∞ and $D(\Phi_\infty)$ with $\Phi \equiv \Xi \circ C$.

Proposition 8 *Suppose that assumptions $(H1)$ and $(H3)$ are satisfied and let Φ be defined as in (83). Then*

$$D(\Phi)_\infty = \{x \in \mathbb{R}^n : Cx \in D(\Xi)_\infty\}, \tag{87}$$

$$\Phi_\infty(x) = \Xi_\infty(Cx), \ \forall x \in \mathbb{R}^n \tag{88}$$

and

$$D(\Phi_\infty) = \{x \in \mathbb{R}^n : Cx \in D(\Xi_\infty)\}. \tag{89}$$

Proof: i) Let us set

$$\mathcal{D}_\infty(C, \Xi) := \{x \in \mathbb{R}^n : Cx \in D(\Xi)_\infty\}.$$

It is easy to see that

$$D(\Phi)_\infty = \mathcal{D}_\infty(C, \Xi). \tag{90}$$

Indeed, if $e \in D(\Phi)_\infty$ then $\lambda e + \bar{x}_0 \in D(\Phi), \forall\lambda > 0$. Thus $C(\lambda e + \bar{x}_0) \in D(\Xi), \forall\lambda > 0$. Here $\bar{y}_0 = C\bar{x}_0 \in D(\Xi)$ and

$$Ce \in \bigcap_{\lambda > 0} \frac{1}{\lambda}(D(\Xi) - \bar{y}_0) = D(\Xi)_\infty.$$

Thus $e \in \mathcal{D}_\infty(C, \Xi)$.

Reciprocally, if $e \in \mathcal{D}_\infty(C, \Xi)$ then $Ce \in D(\Xi)_\infty$. Thus $\lambda Ce + \bar{y}_0 \in D(\Xi), \forall \lambda > 0$. Thus $\lambda e + \bar{x}_0 \in D(\Phi), \forall \lambda > 0$ and then

$$e \in \bigcap_{\lambda > 0} \frac{1}{\lambda}(D(\Phi) - \bar{x}_0) = D(\Phi)_\infty.$$

ii) We have:

$$\Phi_\infty(x) = \lim_{\lambda \to +\infty} \frac{1}{\lambda}\Xi(C(\bar{x}_0 + \lambda x)) = \lim_{\lambda \to +\infty} \frac{1}{\lambda}\Xi(\bar{y}_0 + \lambda Cx) = \Xi_\infty(Cx) \ \ (x \in \mathbb{R}^n).$$

iii) The relation in (89) is a direct consequence of (88). □

8. Diode Circuits

We are now in position to study diode circuits like amplitude selectors that are used to transmit the part of a given waveform which lies above or below some given reference level, double-diode clippers that are used to limit the input amplitude at two independent levels, sampling gates which are transmission circuits in which the output is a reproduction of an input waveform during a selected time interval and is zero otherwise and other circuits involving both diodes and operational amplifiers.

We will present some electrical systems involving ideal diodes that can be studied by means of classical tools from complementarity theory and then discuss some extensions to systems involving practical components.

Let us first start with two simple clipping circuits to illustrate the essence of our methodology.

Example 21 (Clipping circuit 1 / Ideal diode) *Clipping circuits are used to select for transmission a part of a waveform which located above or below some reference level. It works by the switching action of a diode when it is used with a source voltage that changes polarity.*

Let us consider the circuit of figure 11 involving a load resistance $R > 0$, an input-signal source u and corresponding instantaneous current i, an ideal diode as a shunt element and a supply voltage E.

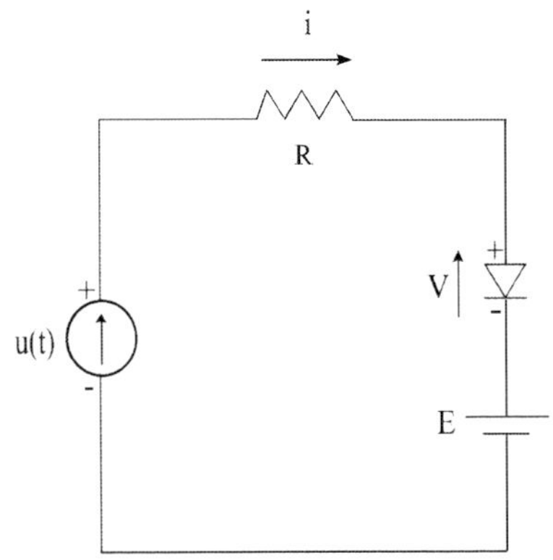

Figure 11. Clipping circuit 1: Diode as shunt element.

Kirchhoff's voltage law gives

$$u = U_R + V + E$$

where $U_R = Ri$ *denotes the difference of potential across the resistor and* $V \in \partial\Psi_{\mathbb{R}_+}(i)$ *is the difference of potential across diode. Thus*

$$E + Ri - u \in -\partial\Psi_{\mathbb{R}_+}(i) \tag{91}$$

which is equivalent to $\mathbf{VI}(\mathbf{R}, \mathbf{E} - \mathbf{u}, \mathbf{\Psi}_{\mathbb{R}_+})$, *i.e.*

$$i \in \mathbb{R}_+ : (Ri + E - u)(v - i) \geq 0, \forall v \in \mathbb{R}_+.$$

Here $R > 0$ *and for each* $E, u \in \mathbb{R}$, *we may apply Corollary 4 to assert that (91) has a unique solution.*

Moreover:

$$(91) \Longleftrightarrow \frac{E}{R} + i - \frac{u}{R} \in -\partial\Psi_{\mathbb{R}_+}(i) \Longleftrightarrow -\frac{E}{R} + \frac{u}{R} \in i + \partial\Psi_{\mathbb{R}_+}(i)$$

$$\Longleftrightarrow i = (id_{\mathbb{R}} + \partial\Psi_{\mathbb{R}_+})^{-1}(\frac{u - E}{R}) = \frac{1}{R}\max\{0, u - E\}.$$

If $u \leq E$ *then* $i = 0$ *and the diode is blocking while if* $u > E$ *then* $i = \frac{u-E}{R}$ *and the diode is conducting.*

Let us now consider a driven time depending input $t \mapsto u(t)$ *and define the output-signal* $t \mapsto V_o(t)$ *as*

$$V_o(t) = E + V(t).$$

The time depending current $t \mapsto i(t)$ *is given by*

$$i(t) = \frac{1}{R}\max\{0, u(t) - E\} \tag{92}$$

and thus

$$V_o(t) = V(t) + E = u(t) - Ri(t) = u(t) + \min\{0, E - u(t)\} = \min\{u(t), E\}. \tag{93}$$

This shows that the circuit in figure 11 can be used to transmit the part of a given input-signal u which lies below some given reference level E.

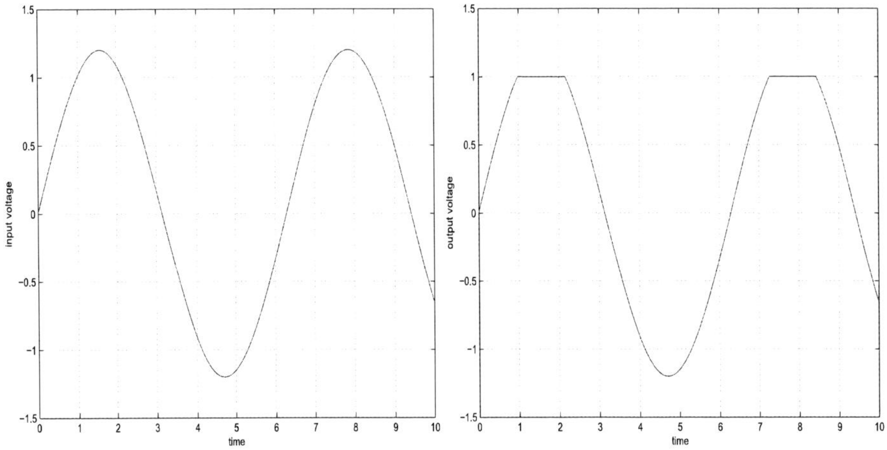

Figure 12. Clipping circuit 1: Ideal diode as shunt element, $E = 1$.

Example 22 (Clipping circuit 1 / General diode) *In the case of a diode with electrical superpotential φ, we may follow the same steps as above to get:*

$$i(t) = (id_{\mathbb{R}} + \frac{1}{R}\partial\varphi)^{-1}(\frac{u(t) - E}{R}) = \text{argmin}_{x\in\mathbb{R}}\{\frac{1}{2}|x - (\frac{u(t) - E}{R})|^2 + \frac{1}{R}\varphi(x)\}. \quad (94)$$

and

$$V_o(t) = u(t) - Ri(t). \quad (95)$$

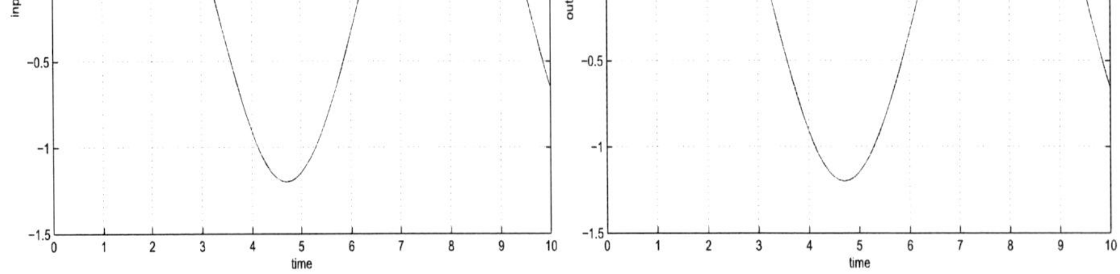

Figure 13. Clipping circuit 1: General diode as shunt element using, $V_1 = 0.1, V_2 = -90$, $E = 1$.

Example 23 (Clipping circuit 2 / Ideal diode) *Let us consider the circuit of figure 14.*

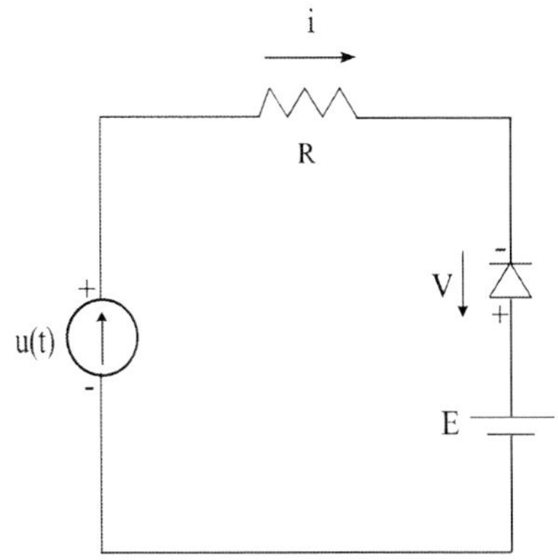

Figure 14. Clipping circuit 2: Diode as shunt element.

Here we have

$$u = U_R - V + E$$

with $U_R = Ri$ *and* $V \in \partial\Psi_{\mathbb{R}_+}(-i)$. *Thus*

$$E + Ri - u \in \partial\Psi_{\mathbb{R}_+}(-i) = -\partial\Psi_{\mathbb{R}_-}(i) \tag{96}$$

which is equivalent to $\mathbf{VI}(\mathbf{R}, \mathbf{E} - \mathbf{u}, \mathbf{\Psi}_{\mathbb{R}_-})$, *i.e.*

$$i \in \mathbb{R}_- : (Ri + E - u)(v - i) \geq 0, \forall v \in \mathbb{R}_-.$$

Here $R > 0$ *and for each* $E, u \in \mathbb{R}$, *we may apply Corollary 4 to assert that (96) has a unique solution. Moreover:*

$$(96) \Longleftrightarrow \frac{E}{R} + i - \frac{u}{R} \in -\partial\Psi_{\mathbb{R}_-}(i) \Longleftrightarrow -\frac{E}{R} + \frac{u}{R} \in i + \partial\Psi_{\mathbb{R}_-}(i)$$

$$\Longleftrightarrow i = (id_{\mathbb{R}} + \partial\Psi_{\mathbb{R}_-})^{-1}(\frac{u - E}{R}) = \frac{1}{R}\min\{0, u - E\}.$$

If $u < E$ *then* $i = \frac{u-E}{R}$ *and the diode is conducting while if* $u \geq E$ *then* $i = 0$ *and the diode is blocking. For a driven time depending input* $t \mapsto u(t)$ *the time depending current* $t \mapsto i(t)$ *is given by*

$$i(t) = \frac{1}{R}\min\{0, u(t) - E\} \tag{97}$$

and the output-signal $t \mapsto V_o(t) = V(t) + E$ *is*

$$V_o(t) = \max\{u(t), E\} \tag{98}$$

and thus the circuit in figure 14 can be used to transmit the part of a given input-signal u *which lies above some given reference level* E.

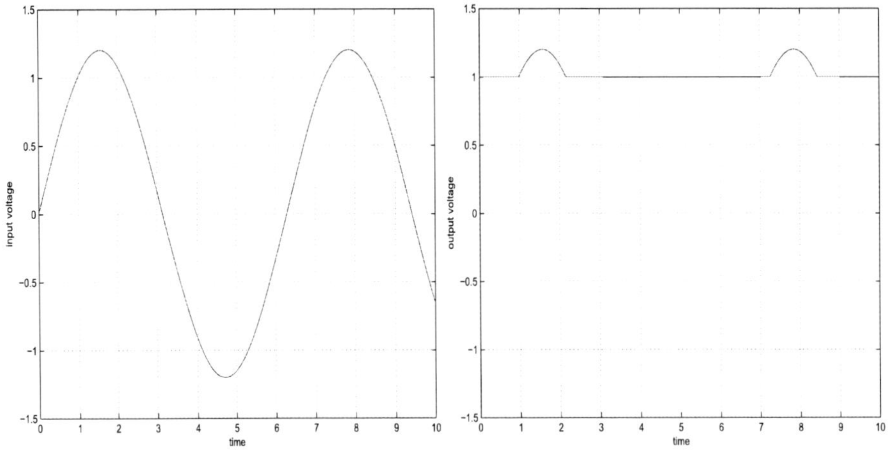

Figure 15. Clipping circuit 2: Ideal diode as shunt element.

Example 24 (Clipping circuit 2 / General diode) *In the case of a general diode with superpotential* φ, *we set:*

$$\bar{\varphi}(x) = \varphi(-x), \quad (x \in \mathbb{R}).$$

Suppose there exists a point $y_0 = -x_0$ *at which* φ *is finite and continuous. Note that this is the case for each model discussed in Section 4. Then:*

$$\partial\bar{\varphi}(x) = -\partial\varphi(-x), \quad (x \in \mathbb{R}).$$

Thus

$$E + Ri - u \in \partial\varphi(-i) = -\partial\bar{\varphi}(i) \Longleftrightarrow -\frac{E}{R} + \frac{u}{R} \in i + \frac{1}{R}\partial\bar{\varphi}(i)$$

$$\Longleftrightarrow i = (id_{\mathbb{R}} + \frac{1}{R}\partial\bar{\varphi})^{-1}\left(\frac{u-E}{R}\right) = \text{argmin}_{x\in\mathbb{R}}\{\frac{1}{2}|x - \left(\frac{u-E}{R}\right)|^2 + \frac{1}{R}\bar{\varphi}(x)\}$$

$$= \text{argmin}_{x\in\mathbb{R}}\{\frac{1}{2}|x - \left(\frac{u-E}{R}\right)|^2 + \frac{1}{R}\varphi(-x)\}.$$

For a driven time depending input $t \mapsto u(t)$ *the time depending current* $t \mapsto i(t)$ *is given by*

$$i(t) = \text{argmin}_{x\in\mathbb{R}}\{\frac{1}{2}|x - \left(\frac{u(t)-E}{R}\right)|^2 + \frac{1}{R}\varphi(-x)\}. \tag{99}$$

and the output-signal $t \mapsto V_o(t) = V(t) + E$ *can be computed by the formula*

$$V_o(t) = u(t) - Ri(t). \tag{100}$$

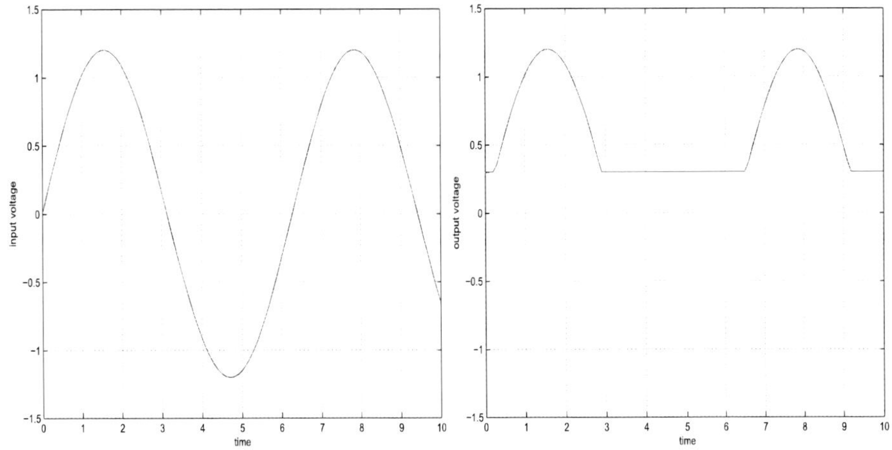

Figure 16. Clipping circuit 2: practical diode as shunt element, $V_1 = 0.2, V_2 = -90$, $E = 0.5$.

Let us now discuss two circuits, a double-diode clipper and a sampling gate, to illustrate the results established in Proposition 7 and Corollary 11.

Example 25 (Double-diode clipper / Ideal diode) *Let us consider the circuit in Figure 17 involving a load resistance $R > 0$, two ideal diodes, an input-signal source and two supply voltages E_1 and E_2. It is assumed that $E_1 < E_2$. We denote by i the current through the resistor R and we set $i = i_1 + i_2$ where $-i_1$ denotes the current through diode D_1 and i_2 is the current through diode D_2.*

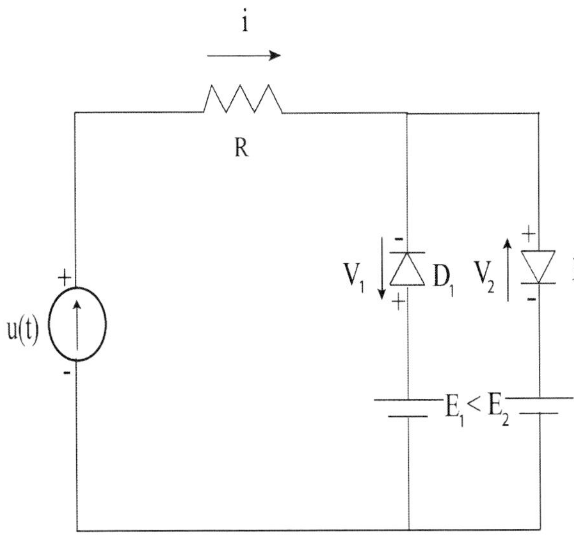

Figure 17. Double-diode clipper.

Using Kirchhoff's voltage laws, we get the system:

$$\begin{cases} E_1 + R(i_1 + i_2) - u = +V_1 \\ \\ E_2 + R(i_1 + i_2) - u = -V_2 \end{cases} \tag{101}$$

where $V_1 \in \partial\Psi_{\mathbb{R}_+}(-i_1) = -\Psi_{\mathbb{R}_-}(i_1)$ *is the difference of potential across diode* D_1 *and* $V_2 \in \partial\Psi_{\mathbb{R}_+}(i_2)$ *is the difference of potential across diode* D_2.

Setting $\Phi \equiv \Psi_{\mathbb{R}_- \times \mathbb{R}_+}$ *and*

$$M = \begin{pmatrix} R & R \\ \\ R & R \end{pmatrix}, \quad q = \begin{pmatrix} E_1 - u \\ \\ E_2 - u \end{pmatrix}, \quad \Upsilon = \begin{pmatrix} i_1 \\ \\ i_2 \end{pmatrix}, \tag{102}$$

we see that the system in (101) is equivalent to the variational inequality $\mathbf{VI}(\mathbf{M}, \mathbf{q}, \mathbf{\Phi})$, *i.e.*

$$\Upsilon \in \mathbb{R}^2 : \langle M\Upsilon + q, v - \Upsilon \rangle + \Phi(v) - \Phi(\Upsilon) \geq 0, \forall v \in \mathbb{R}^2. \tag{103}$$

Here the matrix M *is positive semidefinite and symmetric,* $D(\Phi)_\infty = \mathbb{R}_- \times \mathbb{R}_+$, $\ker\{M\} = \{v \in \mathbb{R}^2 : v_2 = -v_1\}$ *and* $\mathcal{K}(M, \Phi) = \{v \in \mathbb{R}^2 : v_2 = -v_1\}$. *Thus*

$$D(\Phi)_\infty \cap \ker\{M\} \cap \mathcal{K}(M, \Phi) = \{v \in \mathbb{R}^2 : v_1 \leq 0, v_2 = -v_1\}.$$

Then, for all $v \in D(\Phi)_\infty \cap \ker\{M\} \cap \mathcal{K}(M, \Phi)$, $v \neq 0$, *we have* $-v_1 = v_2 > 0$ *and thus*

$$\langle q, v \rangle = (E_1 - u)v_1 + (E_2 - u)v_2 = v_2(E_2 - E_1) > 0. \tag{104}$$

We may apply Corollary 11 which ensures that the system in (103) has at least one solution.

Using the first relation in part (b) of Corollary 11 we first remark that if Υ *and* $\bar{\Upsilon}$ *denote two solutions of (101) then* $\Upsilon - \bar{\Upsilon} \in$ *ker*$\{M\}$ *and thus* $i_1 - \bar{i}_1 = -(i_2 - \bar{i}_2)$, *i.e.*

$$i_1 + i_2 = \bar{i}_1 + \bar{i}_2. \tag{105}$$

It results that if $\Upsilon^* = (i_1^* \; i_2^*)^T$ *denotes a solution of the system in (103) then the current through the resistor* R, *i.e.* $i^* = i_1^* + i_2^*$ *is uniquely determined.*

Using now the second relation of part (b) of Corollary 11, we obtain also that:

$$(E_1 - u)i_1 + (E_2 - u)i_2 = (E_1 - u)\bar{i}_1 + (E_2 - u)\bar{i}_2. \tag{106}$$

Here $E_2 \neq E_1$ *and thus the system in (105)-(106) yields* $i_1 = \bar{i}_1$ *and* $i_2 = \bar{i}_2$. *The solution of (103) is thus unique.*

Using the relations in (101) we see that the unique solution of (103) satisfies:

$$i_1^* + i_2^* = \min\{i_2^*, \frac{u - E_1}{R}\} = \max\{i_1^*, \frac{u - E_2}{R}\}$$

from which we deduce, after elementary calculations, that:

$$i^* = \begin{cases} \frac{u-E_1}{R} \text{ if } u < E_1 \\ 0 \text{ if } E_1 \leq u \leq E_2 \\ \frac{u-E_2}{R} \text{ if } u > E_2 \end{cases} .$$

So, for a driven time depending input $t \mapsto u(t)$ *the time depending current* $t \mapsto i^*(t)$ *through the resistor* R *is given by*

$$i^*(t) = \begin{cases} \frac{u(t)-E_1}{R} \text{ if } u(t) < E_1 \\ 0 \text{ if } E_1 \leq u(t) \leq E_2 \\ \frac{u(t)-E_2}{R} \text{ if } u(t) > E_2 \end{cases} \tag{107}$$

and the output-signal $t \mapsto V_o(t)$ *defined by*

$$V_o(t) = V_2(t) + E_2 = u(t) - Ri^*(t)$$

is then given by the expression:

$$V_o(t) = \begin{cases} E_1 \text{ if } u(t) < E_1 \\ u(t) \text{ if } E_1 \leq u(t) \leq E_2 \\ E_2 \text{ if } u(t) > E_2 \end{cases} . \tag{108}$$

This shows that the circuit can be used to transmit the part of a given input-signal u *that lies above some level* E_1 *and below some level* E_2*.*

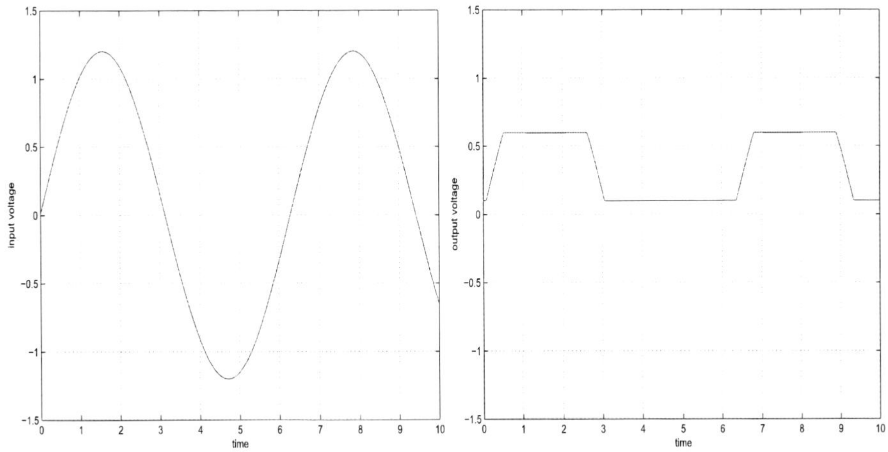

Figure 18. Double-diode clipper: ideal diode, $E_1 = 0.1, E_2 = 0.6$.

Example 26 (Double-diode clipper / Practical diode) *Let us again consider the circuit in Figure 17 and suppose that the electrical superpotential of each diodes D_1 and D_2 is given by (practical diode model):*

$$\varphi_{PD}(x) = \begin{cases} \nu_1 x & \text{if } x \geq 0 \\ \nu_2 x & \text{if } x < 0 \end{cases}, \quad (x \in \mathbb{R})$$

where $\nu_2 < 0 < \nu_1$. We suppose also that

$$|\nu_2| > \frac{E_2 - E_1}{2}. \tag{109}$$

We set

$$\bar{\varphi}_{PD}(x) = \varphi_{PD}(-x), \quad \forall\, x \in \mathbb{R}$$

and

$$\Phi(x) = \bar{\varphi}_{PD}(x_1) + \varphi_{PD}(x_2), \quad \forall\, (x_1, x_2) \in \mathbb{R}^2. \tag{110}$$

Kirchoff's laws yield the system

$$\begin{cases} E_1 + R(i_1 + i_2) - u = +V_1 \in -\partial\bar{\varphi}_{PD}(i_1) \\ E_2 + R(i_1 + i_2) - u = -V_2 \in -\partial\varphi_{PD}(i_2) \end{cases} \tag{111}$$

which is equivalent to the variational inequality $\mathbf{VI}(\mathbf{M}, \mathbf{q}, \mathbf{\Phi})$, *i.e.*

$$\Upsilon \in \mathbb{R}^2 : \langle M\Upsilon + q, v - \Upsilon\rangle + \Phi(v) - \Phi(\Upsilon) \geq 0, \forall v \in \mathbb{R}^2, \tag{112}$$

with M and q as in (102) and Φ as in (110). Here $D(\Phi)_\infty = \mathbb{R}^2$, $\Phi_\infty \equiv \Phi$, $\mathcal{K}(M, \Phi) = \ker\{M\}$ and thus

$$D(\Phi)_\infty \cap \ker\{M\} \cap \mathcal{K}(M, \Phi) = \ker\{M\} = \{v \in \mathbb{R}^2 : v_2 = -v_1\}.$$

Let $v \in \ker\{M\}$, $v \neq 0$, be given. Then:

$$\langle q, v\rangle + \Phi_\infty(v) = v_2(E_2 - E_1) + \varphi_{PD}(-v_1) + \varphi_{PD}(v_2) = v_2(E_2 - E_1) + 2\varphi_{PD}(v_2).$$

It results that if $v_2 > 0$ then

$$\langle q, v\rangle + \Phi_\infty(v) = v_2(E_2 - E_1) + 2\nu_1 v_2 > 0$$

while if $v_2 < 0$ then

$$\langle q, v\rangle + \Phi_\infty(v) = -v_2(2\,|\nu_2| - (E_2 - E_1)) > 0.$$

We may then apply Corollary 11 which ensures that the system in (112) has at least one solution.

If $\Upsilon^* = (i_1^*\ i_2^*)^T$ denotes a solution of the system in (112) then from the first relation in part (b) of Corollary 11, we deduce that the current $i^* = i_1^* + i_2^*$ through the resistor R is uniquely determined.

Moreover, the function Φ is strictly convex and part (d) of Corollary 11 ensures that the solution Υ^* is unique and given by:

$$\Upsilon^* = \operatorname{argmin}_{x\in\mathbb{R}^2}\{\frac{1}{2}\langle Mx, x\rangle + \langle q, x\rangle + \Phi(x)\}. \tag{113}$$

So, for a driven time depending input $t \mapsto u(t)$ the time depending current $t \mapsto i^*(t)$ through the resistor R is given by

$$i^*(t) = i_1^*(t) + i_2^*(t) \tag{114}$$

where

$$\left(\begin{array}{cc} i_1^*(t) & i_2^*(t) \end{array} \right)^T = \operatorname{argmin}_{x\in\mathbb{R}^2}\{\frac{1}{2}\langle Mx, x\rangle + (E_1 - u(t))x_1 + (E_2 - u(t))x_2 + \Phi(x)\} \tag{115}$$

and the output-signal V_o can then be determined by the formula:

$$V_o(t) = u(t) - Ri^*(t).$$

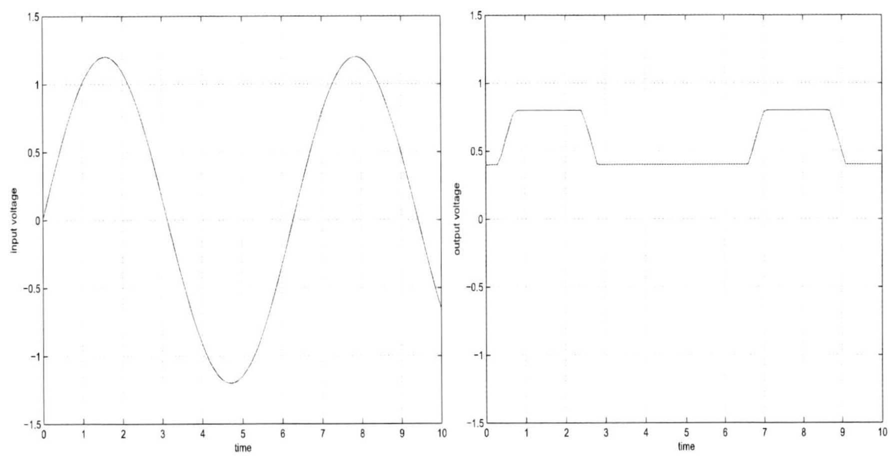

Figure 19. Double-diode clipper: practical diode.

Example 27 (Sampling gate) *A sampling gate is a circuit in which the output is a reproduction of the input waveform during a selected time interval and is zero otherwise. The time interval is selected by the gate signal V_c. The circuit in Figure 20 is a sampling gate involving a bridge of four diodes D_1, D_2, D_3, D_4 and symmetrically controlled by gate voltages $+V_c$ and $-V_c$ through the control resistors $R_c > 0$. The input-signal is given by V_i and the output signal is defined by the voltage V_o through the load resistor $R_L > 0$. Usually, V_i is sinusoidal while V_c is rectangular shaped.*

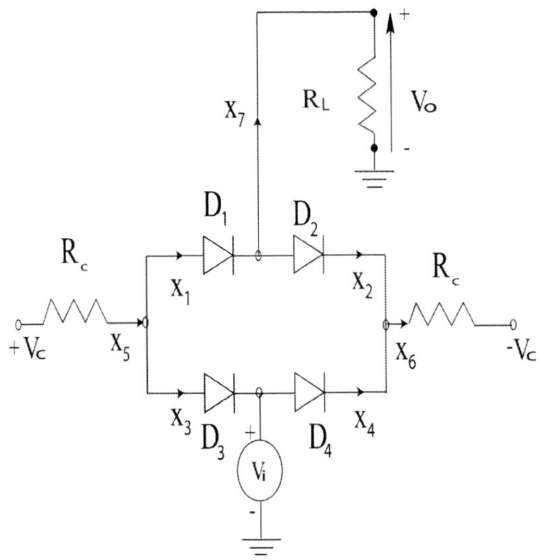

Figure 20. four-diode-bridge sampling gate.

We denote by V_j the voltage of the diode D_j and by x_j the current across the diode D_j $(1 \leq j \leq 4)$. Moreover, x_5 denotes the current through the left resistor R_c, x_6 is the current through the right resistor R_c and x_7 denotes the current trough resistor R_L. Kirchhoff's laws yield

$$\begin{cases} V_i - V_4 + V_2 - R_L x_7 = 0 \\ 2V_c - V_3 - V_4 - R_c x_6 - R_c x_5 = 0 \\ -V_1 - V_2 + V_3 + V_4 = 0. \end{cases}$$

Moreover $x_1 + x_3 = x_2 + x_4 = x_5 = x_6$ *and* $x_1 - x_2 = x_4 - x_3 = x_7$. *Thus*

$$\overbrace{\begin{pmatrix} -R_L & 0 & 0 \\ 0 & -2R_c & 0 \\ 0 & 0 & 0 \end{pmatrix}}^{A} \overbrace{\begin{pmatrix} x_7 \\ x_6 \\ x_1 \end{pmatrix}}^{x} - \overbrace{\begin{pmatrix} 0 & -1 & 0 & 1 \\ 0 & 0 & 1 & 1 \\ 1 & 1 & -1 & -1 \end{pmatrix}}^{B} \overbrace{\begin{pmatrix} V_1 \\ V_2 \\ V_3 \\ V_4 \end{pmatrix}}^{V} + \overbrace{\begin{pmatrix} 1 & 0 \\ 0 & 1 \\ 0 & 0 \end{pmatrix}}^{D} \overbrace{\begin{pmatrix} V_i \\ 2V_c \end{pmatrix}}^{u} = 0,$$

and

$$\begin{cases} V_1 \in \partial\psi_{\mathbb{R}_+}(x_1) \\ V_2 \in \partial\psi_{\mathbb{R}_+}(x_2) = \partial\psi_{\mathbb{R}_+}(x_1 - x_7) \\ V_3 \in \partial\psi_{\mathbb{R}_+}(x_3) = \partial\psi_{\mathbb{R}_+}(x_6 - x_1) \\ V_4 \in \partial\psi_{\mathbb{R}_+}(x_4) = \partial\psi_{\mathbb{R}_+}(x_7 + x_6 - x_1). \end{cases} \tag{116}$$

Setting

$$y = \overbrace{\begin{pmatrix} 0 & 0 & 1 \\ -1 & 0 & 1 \\ 0 & 1 & -1 \\ 1 & 1 & -1 \end{pmatrix}}^{C} \begin{pmatrix} x_7 \\ x_6 \\ x_1 \end{pmatrix}$$

and defining the function $\Xi \equiv \Psi_{(\mathbb{R}_+)^4}$, *we may write the relations in (116) equivalently as:*

$$V \in \partial\Xi(y)$$

and we may then consider problem $\mathbf{NRM}(\mathbf{A}, \mathbf{B}, \mathbf{C}, \mathbf{D}, \mathbf{u}, \Xi)$.

Let us first check that the assumptions of Proposition 7 are satisfied. Indeed, it is clear that Assumption $(H1)$ *holds. Moreover* $\begin{pmatrix} 2 & 1 & 1 & 2 \end{pmatrix}^T = C \begin{pmatrix} 1 & 3 & 2 \end{pmatrix}^T \in \operatorname{int}\{K\}$ *is a point at which* Ξ *is finite and continuous and Assumption* $(H2)$ *is thus satisfied. Finally, we remark that* $C^T = B$ *and thus Assumption* $(H3)$ *holds with* $P = I$.

As a consequence of Proposition 7, problem $\mathbf{NRM}(\mathbf{A}, \mathbf{B}, \mathbf{C}, \mathbf{D}, \mathbf{u}, \Xi)$ *can be studied via the problem* $\mathbf{VI}(-\mathbf{A}, -\mathbf{Du}, \mathbf{\Phi})$ *where* $\Phi = \Xi \circ C$. *Here* $-A$ *is symmetric and positive semi-definite. Moreover* $D(\Phi)_\infty = D(\Phi_\infty) = \{x \in \mathbb{R}^3 : Cx \in (\mathbb{R}_+)^4\}$ *and* $\ker\{-A\} = \{\begin{pmatrix} 0 & 0 & \alpha \end{pmatrix}^T ; \alpha \in \mathbb{R}\}$. *It is then easy to see that*

$$\ker\{-A\} \cap D(\Phi)_\infty \cap \mathcal{K}(-A, \Phi)\} = \{0\}.$$

It results from Corollary 11 that the system in $\mathbf{NRM}(\mathbf{A}, \mathbf{B}, \mathbf{C}, \mathbf{D}, \mathbf{u}, \Xi)$ *has at least one solution* x^*. *The matrix A is symmetric and thus:*

$$-\frac{1}{2}\langle Ax^*, x^*\rangle - \langle Du, x^*\rangle \leq -\frac{1}{2}\langle Ax, x\rangle - \langle Du, x\rangle, \ \forall x \in \mathbb{R}^3 : Cx \in (\mathbb{R}_+)^4. \quad (117)$$

Moreover, if $\bar{x}$ *denotes another solution, then using part (b) in Corollary 11:*

$$\begin{pmatrix} x_7^* - \bar{x}_7 \\ x_6^* - \bar{x}_6 \\ x_1^* - \bar{x}_1 \end{pmatrix} = \begin{pmatrix} 0 \\ 0 \\ \alpha \end{pmatrix}$$

for some $\alpha \in \mathbb{R}$. *It results that* $x_7^* = \bar{x}_7$ *and* $x_6^* = \bar{x}_6$, *i.e. the current through the resistor* R *and the current through the resistor* R_c *are uniquely determined.*

So, for a driven time depending input $t \mapsto V_i(t)$ *and control gate signals* $t \mapsto V_c(t)$ *and* $t \mapsto -V_c(t)$, *the output time depending voltage* $t \mapsto V_o(t)$ *through the resistor* R_L *is uniquely determined by:*

$$V_o(t) = R_L x_7(t)$$

where the current function $t \mapsto x_7(t)$ *is uniquely determined in solving the variational inequality* $\mathbf{VI}(-\mathbf{A}, -\mathbf{Du}(\mathbf{t}), \mathbf{\Phi})$.

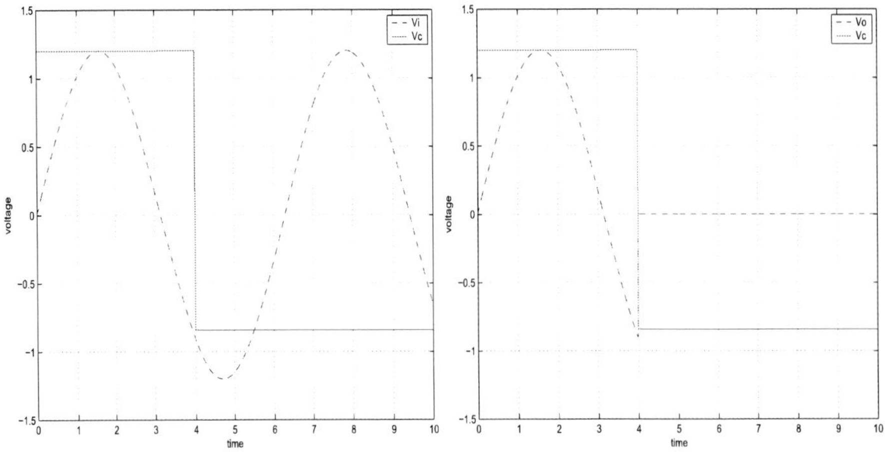

Figure 21. Four-diode-bridge sampling gate.

The previous examples show that the place filled by the class of symmetric and positive semidefinite matrices is as important in Electronics as it is in Mechanics. However, the mathematical formulation of circuits involving devices like transistors and operational amplifiers involves matrices that are neither symmetric nor positive semidefinite. The following example constitutes a nice illustration of Corollary 6.

Example 28 (Ideal transistor circuit) *Let us consider the circuit of Figure 22 involving an ideal transistor PNP and two resistors* $R_L > 0$ *and* $R_B > 0$.

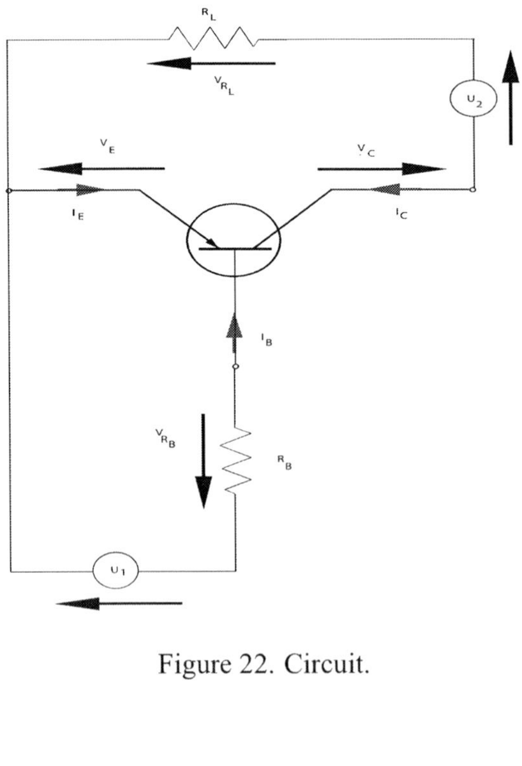

Figure 22. Circuit.

Using Kirchhoff's laws, we obtain:

$$V_{R_B} + U_1 - V_E = 0$$

and

$$V_C + U_2 + V_{R_L} - V_E = 0.$$

Then

$$-V_E = -V_{R_B} - U_1 = -R_B I_B - U_1 = R_B I_E + R_B I_C - U_1$$

and

$$\begin{aligned} -V_C = U_2 + V_{R_L} - V_E &= U_2 + R_L I_C + [R_B I_E + R_B I_C - U_1] \\ &= R_B I_E + (R_B + R_L) I_C + U_2 - U_1. \end{aligned}$$

Thus

$$\begin{pmatrix} -V_E \\ -V_C \end{pmatrix} = \begin{pmatrix} R_B & R_B \\ R_B & R_B + R_L \end{pmatrix} \begin{pmatrix} I_E \\ I_C \end{pmatrix} + \begin{pmatrix} -U_1 \\ U_2 - U_1 \end{pmatrix}$$

$$= \begin{pmatrix} R_B & R_B \\ R_B & R_B + R_L \end{pmatrix} \begin{pmatrix} \frac{1}{1-\alpha_I\alpha_N} & -\frac{\alpha_I}{1-\alpha_I\alpha_N} \\ -\frac{\alpha_N}{1-\alpha_I\alpha_N} & \frac{1}{1-\alpha_I\alpha_N} \end{pmatrix} \begin{pmatrix} I \\ I' \end{pmatrix} + \begin{pmatrix} -U_1 \\ U_2 - U_1 \end{pmatrix}$$

$$= \overbrace{\begin{pmatrix} \frac{(1-\alpha_N)R_B}{1-\alpha_I\alpha_N} & \frac{(1-\alpha_I)R_B}{1-\alpha_I\alpha_N} \\ \frac{(1-\alpha_N)R_B - \alpha_N R_L}{1-\alpha_I\alpha_N} & \frac{(1-\alpha_I)R_B + R_L}{1-\alpha_I\alpha_N} \end{pmatrix}}^{M} \overbrace{\begin{pmatrix} I \\ I' \end{pmatrix}}^{x} + \overbrace{\begin{pmatrix} -U_1 \\ U_2 - U_1 \end{pmatrix}}^{q}.$$

Using now the model of (ideal) transistor, we get the complementarity problem:

$$x \geq 0, \; Mx + q \geq 0, \langle x, Mx + q \rangle = 0. \tag{118}$$

It is clear that this last complementarity problem is equivalent to the inequality problem $\mathbf{VI}(\mathbf{M}, \mathbf{q}, \Psi_{\mathbb{R}_+ \times \mathbb{R}_+})$ *where* $\Psi_{\mathbb{R}_+ \times \mathbb{R}_+}$ *denotes the indicator function of* $\mathbb{R}_+ \times \mathbb{R}_+$. *The principal minors of the matrix* M *are:*

$$\Delta_1(M) = M_{11} = ((1 - \alpha_N)R_B)/(1 - \alpha_I\alpha_N) > 0,$$

$$\Delta_2(M) = M_{22} = ((1 - \alpha_I)R_B + R_L)/(1 - \alpha_I\alpha_N) > 0$$

and

$$\Delta_{12}(M) = \det(M) = (R_B R_L)/(1 - \alpha_I\alpha_N) > 0.$$

The matrix M *is a P-matrix and thus, for each* $q \in \mathbb{R}^n$, *we may apply Corollary 6 to conclude that problem (118) has a unique solution.*

Remark 12 *Note that the matrix* M *is generally not positive definite. For example, if* $\alpha_N = 9/10, \alpha_I = 1/10, R_B = 1, R_L = 1/9$ *then* M *is P-matrix which is not positive definite.*

Example 29 (General transistor circuit) *Let us again consider the circuit of Figure 22 and suppose now that the emitter electrical potential (resp. the collector electrical potential) of the (practical) transistor is given by* $\varphi_E \in \Gamma(\mathbb{R}; \mathbb{R} \cup \{+\infty\})$ *(resp.* $\varphi_C \in \Gamma(\mathbb{R}; \mathbb{R} \cup \{+\infty\})$*). Our model reads here:*

$$\overbrace{\begin{pmatrix} \frac{(1-\alpha_N)R_B}{1-\alpha_I\alpha_N} & \frac{(1-\alpha_I)R_B}{1-\alpha_I\alpha_N} \\ \frac{(1-\alpha_N)R_B-\alpha_N R_L}{1-\alpha_I\alpha_N} & \frac{(1-\alpha_I)R_B+R_L}{1-\alpha_I\alpha_N} \end{pmatrix}}^{M} \overbrace{\begin{pmatrix} I \\ I' \end{pmatrix}}^{x} + \overbrace{\begin{pmatrix} -U_1 \\ U_2 - U_1 \end{pmatrix}}^{q} \in \begin{pmatrix} -\partial\varphi_E(I) \\ -\partial\varphi_C(I') \end{pmatrix}.$$

We set

$$\Phi(z) = \varphi_E(z_1) + \varphi_C(z_2), \ \forall z = (z_1, z_2) \in \mathbb{R}^2.$$

Our model reads now:

$$x \in \mathbb{R}^2 : \langle Mx + q, v - x \rangle + \Phi(v) - \Phi(x) \geq 0, \ \forall v \in \mathbb{R}^2. \tag{119}$$

It is clear that $\Phi \in D\Gamma(\mathbb{R}^n; \mathbb{R} \cup \{+\infty\})$ *and we may apply Corollary 6 to ensure the existence of a unique solution of (128).*

Example 30 (Common emitter amplifier circuit) *Let us consider the common emitter amplifier circuit of Figure 23 involving an ideal transistor NPN. The common emitter configuration lends itself to voltage amplification and is the most common configuration for transistor amplifiers.*

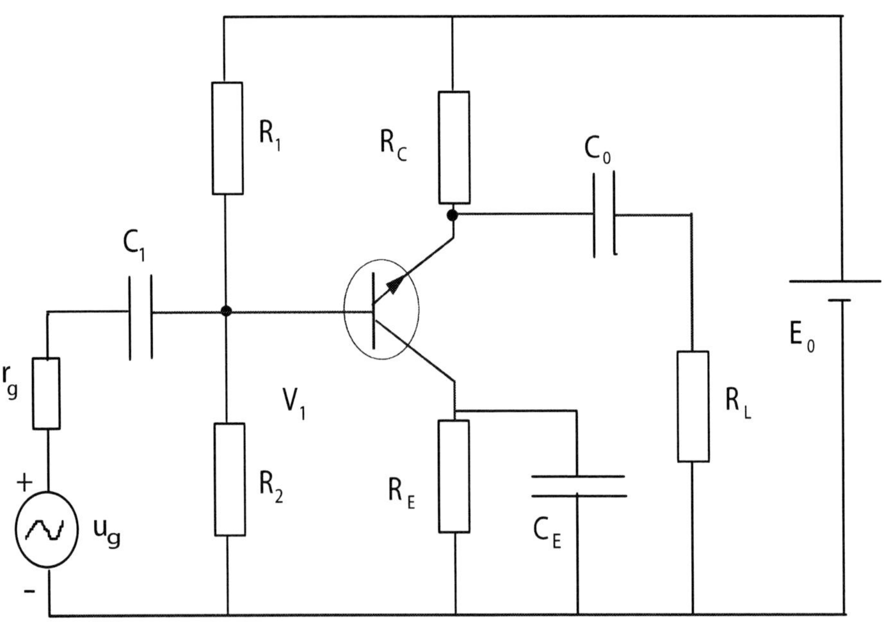

Figure 23. Common emitter amplifier circuit.

In electronics, it is useful to work with equivalent circuits under appropriate conditions. Indeed, in the case of low frequencies the capacitors of the circuit in Figure 23 play a short-circuit role and the circuit reduces to the one of Figure 24.

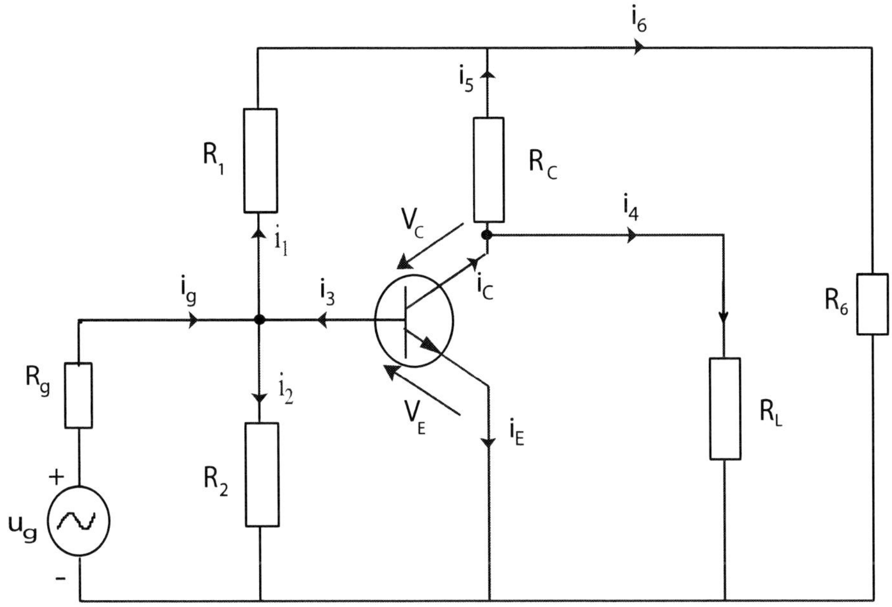

Figure 24. Equivalent common emitter amplifier circuit: Low frequencies.

Using Kirchhoff's law we obtain the following model:

$$\begin{cases} R_g i_g + R_2 i_2 & = & u_g, \\ V_E - R_2 i_2 & = & 0, \\ V_C + R_C i_5 - R_1 i_1 & = & 0, \\ R_L i_4 + V_C - V_E & = & 0, \\ R_L i_4 - R_C i_5 - R_6 i_6 & = & 0, \\ i_g + i_3 & = & i_1 + i_2, \\ -i_C - i_E & = & i_3, \\ i_1 + i_5 & = & i_6, \\ i_4 + i_5 & = & i_C, \end{cases} \tag{120}$$

where $R_i > 0$ $(i = 1, .., 6)$, R_g , R_C and R_L are resistors V_C, V_E are voltages, i_C, i_E and i_i $(i = 1, .., 6)$ are the currents of the circuit as it is depicted in Fig. 24.
By reducing system (120) we obtain:

$$\begin{cases} (R_g K + N)V_E - R_g K V_C & = & -R_g i_E + R_g S i_C + u_g \\ \\ -(L + R_1 K)V_E + (1 + L + R_1 K)V_C & = & -T i_C \end{cases} \tag{121}$$

where $K = (R_L + R_C + R_6)/(R_6 R_L) > 0$, $L = R_C/R_L > 0$, $N = (R_g + R_2)/R_2 > 0$, $S = R_C/R_6 > 0$ *and* $T = R_C + R_1(R_C + R_6)/R_6 > 0$.

The system in (121) is equivalent to:

$$\overbrace{\begin{pmatrix} -(R_g K + N) & R_g K \\ L + R_1 K & -(1 + L + R_1 K) \end{pmatrix}}^{A} \begin{pmatrix} -V_E \\ -V_C \end{pmatrix} =$$

$$= \overbrace{\begin{pmatrix} -R_g & R_g S \\ 0 & -T \end{pmatrix}}^{B} \begin{pmatrix} i_E \\ i_C \end{pmatrix} + \begin{pmatrix} u_g \\ 0 \end{pmatrix}. \tag{122}$$

As the matrix A is invertible it follows:

$$\begin{pmatrix} -V_E \\ -V_C \end{pmatrix} = A^{-1}B \begin{pmatrix} i_E \\ i_C \end{pmatrix} + A^{-1} \begin{pmatrix} u_g \\ 0 \end{pmatrix} \tag{123}$$

where

$$A^{-1}B = \frac{1}{\mu_0} \begin{pmatrix} R_g(1 + L + R_1 K) & -R_g S(1 + L + R_1 K) + R_g K T \\ R_g(L + R_1 K) & -R_g(L + R_1 K)S + T(R_g K + N) \end{pmatrix} \tag{124}$$

with:

$$\mu_0 = R_g K + N(1 + L + R_1 K).$$

Using

$$\begin{pmatrix} i_E \\ i_C \end{pmatrix} = \overbrace{\frac{1}{1 - \alpha_I \alpha_N} \begin{pmatrix} 1 & -\alpha_I \\ -\alpha_N & 1 \end{pmatrix}}^{C} \begin{pmatrix} I \\ I' \end{pmatrix} \tag{125}$$

we obtain

$$\begin{pmatrix} -V_E \\ -V_C \end{pmatrix} = M \overbrace{\begin{pmatrix} I \\ I' \end{pmatrix}}^{x} + \overbrace{A^{-1} \begin{pmatrix} u_g \\ 0 \end{pmatrix}}^{q} \tag{126}$$

where

$$\begin{aligned} M &= A^{-1}BC \\ &= \tfrac{1}{\mu} \times \begin{pmatrix} M_{11} & M_{12} \\ M_{21} & M_{22} \end{pmatrix} \end{aligned} \tag{127}$$

where $\mu = \mu_0(1 - \alpha_I \alpha_N)$, $M_{11} = R_g(1 + L + R_1 K) + \alpha_N R_g S(1 + L + R_1 K) - \alpha_N R_g K T$, $M_{12} = -\alpha_I R_g(1 + L + R_1 K) - R_g S(1 + L + R_1 K) + R_g K T$, $M_{21} = R_g(L + R_1 K) +$

$\alpha_N(R_g(L+R_1K)S-T(R_gK+N))$ *and* $M_{22}=-\alpha_I R_g(L+R_1K)-R_g(L+R_1K)S+T(R_gK+N)$.

Suppose now that the emitter electrical potential (resp. the collector electrical potential) of the (practical) transistor is given by $\varphi_E \in \Gamma(\mathbb{R};\mathbb{R}\cup\{+\infty\})$ *(resp.* $\varphi_C \in \Gamma(\mathbb{R};\mathbb{R}\cup\{+\infty\})$*). We set*

$$\Phi(z)=\varphi_E(z_1)+\varphi_C(z_2),\ \forall z=(z_1,z_2)\in\mathbb{R}^2.$$

Our model reads now:

$$x\in\mathbb{R}^2:\langle Mx+q,v-x\rangle+\Phi(v)-\Phi(x)\geq 0,\ \forall v\in\mathbb{R}^2. \tag{128}$$

It is clear that $\Phi \in D\Gamma(\mathbb{R}^n;\mathbb{R}\cup\{+\infty\})$ *Moreover, the principal minors of the matrix* M *are:*

$$\Delta_1(M)=\frac{1}{\mu}M_{11},\ \ \Delta_2(M)=\frac{1}{\mu}M_{22},\ \ \Delta_{12}(M)=\det(M).$$

We can easily check that $M_{11}>0$ *and* $M_{22}>0$*. Indeed, by replacing L, N, S and T in* M_{11} *and* M_{22} *and after some calculations we obtain*

$$M_{11} \ \ = R_g+(1-\alpha_N)(R_gR_C/R_L+R_1R_gK)>0 \tag{129}$$

and

$$M_{22} \ \ = NT+R_gR_C/R_6+(1-\alpha_I)(R_gL+R_1R_gK)>0. \tag{130}$$

Moreover, as $\det(A)=R_gK+N(1+L+R_K)>0$, $\det(B)=R_gT>0$ *and* $\det(C)>0$, *we can claim that* $\det(M)>0$*. The matrix* M *is then a P-matrix and thus using Corollary 6 we conclude that for each* $q\in\mathbb{R}^n$ *problem* $\mathbf{VI(M,q,\Phi)}$ *has a unique solution.*

Example 31 (Operational amplifier) *An operational amplifier is a great versatile circuit element based on transistors which is used to design audio equipments, oscillators and waveform generators, filters and many other applications. Let us consider the system of Figure 25 involving an ideal Zener diode D and a non-inverting amplifier circuit with resistors* $R_1>0$ *and* $R_2>0$*. We denote by* $R_i>0$ *the input resistor,* R_o *the output resistor and* $\gamma>0$ *the differential gain of the operational amplifier. Note that* R_i *is very large (a few M*Ω*) while* R_o *is very small (a few* Ω*). The differential gain* γ *is very large (*$\simeq 100000$*).*

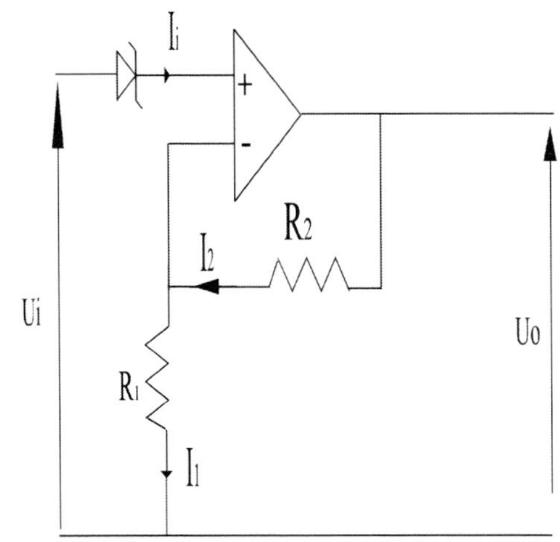

Figure 25. Zener diode + Operational amplifier.

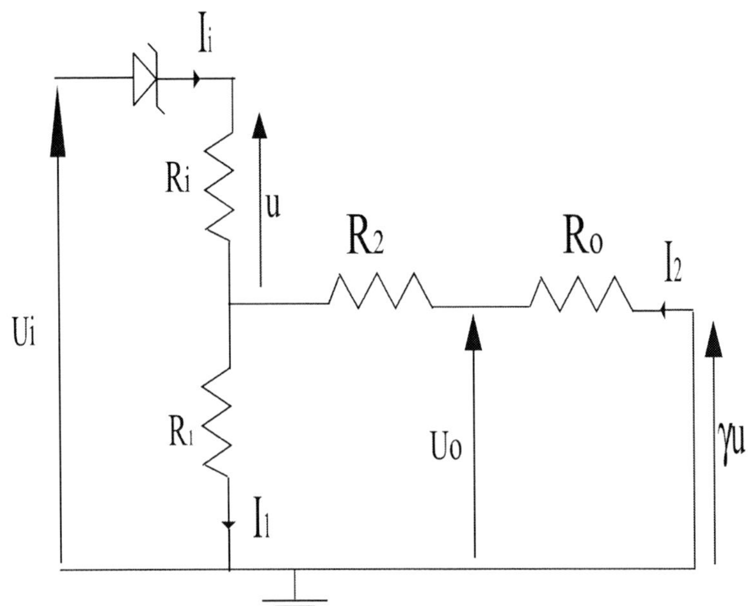

Figure 26. Small signals model.

Considering as in Figure 26 a small signals model for the operational amplifier, Kirchhoff's laws give:

$$\begin{cases} I_i - I_1 + I_2 = 0, \\ R_i I_i + R_1 I_1 - U_i + V_D = 0, \\ -\gamma R_i I_i + R_1 I_1 + (R_2 + R_0) I_2 = 0. \end{cases}$$

with currents I_i, I_1, I_2 *and voltages* V_D, U_i *as defined in Figure 26. Therefore:*

$$\overbrace{\begin{pmatrix} 1 & -1 & 1 \\ R_i & R_1 & 0 \\ -\gamma R_i & R_1 & (R_2 + R_o) \end{pmatrix}}^{A} \overbrace{\begin{pmatrix} I_i \\ I_1 \\ I_2 \end{pmatrix}}^{\Upsilon} - \overbrace{\begin{pmatrix} 0 \\ -1 \\ 0 \end{pmatrix}}^{B} V_D + \overbrace{\begin{pmatrix} 0 \\ -1 \\ 0 \end{pmatrix}}^{D} U_i = 0,$$

and

$$V_D \in \partial\varphi_{ZD}(I_i) \tag{131}$$

where φ_{ZD} *denotes the electrical superpotential of the Zener diode, i.e.*

$$\varphi_{ZD}(x) = \begin{cases} V_1 x & \text{if } x \geq 0 \\ V_2 x & \text{if } x < 0 \end{cases}, \quad (x \in \mathbb{R}),$$

with $V_2 < 0 < V_1$. *Setting*

$$y = \overbrace{\begin{pmatrix} 1 & 0 & 0 \end{pmatrix}}^{C} \begin{pmatrix} I_i \\ I_1 \\ I_2 \end{pmatrix},$$

we may write the relation in (131) equivalently as:

$$V \in \partial\varphi_{ZD}(y)$$

and we may consider the problem $\mathbf{NRM}(\mathbf{A}, \mathbf{B}, \mathbf{C}, \mathbf{D}, \mathbf{U_i}, \varphi_{\mathbf{ZD}})$.

We check that the assumptions of Proposition 7 are satisfied. Indeed, Assumptions $(H1)$ *and* $(H2)$ *are satisfied since* φ_{ZD} *is convex and continuous on* $\mathbb{R}^3$. *Assumption* $(H3)$ *holds with*

$$P = \begin{pmatrix} 0 & -1 & 0 \\ 0 & 0 & -1 \\ -1 & 0 & 0 \end{pmatrix}.$$

As a consequence of Proposition 7, problem $\mathbf{NRM}(\mathbf{A}, \mathbf{B}, \mathbf{C}, \mathbf{D}, \mathbf{U_i}, \varphi_{\mathbf{ZD}})$ *can be studied via the problem* $\mathbf{VI}(-\mathbf{PA}, -\mathbf{PDU_i}, \boldsymbol{\Phi})$ *where*

$$\Phi(x) = \varphi_{ZD}(Cx) = \varphi_{ZD}(x_1), \; \forall x \in \mathbb{R}^3.$$

Here we have:

$$-PA = \begin{pmatrix} R_i & R_1 & 0 \\ -\gamma R_i & R_i & (R_2 + R_0) \\ 1 & -1 & 1 \end{pmatrix}.$$

The matrix $-PA$ is a P-matrix and $\Phi \in D\Gamma(\mathbb{R}^3; \mathbb{R} \cup \{+\infty\}) \subset P\Gamma(\mathbb{R}^3; \mathbb{R} \cup \{+\infty\})$ (with $\Phi_1 \equiv \varphi_{ZD}, \Phi_2 \equiv \Phi_3 \equiv 0$). We may thus apply Corollary 6 to ensure that the system in $\mathbf{NRM(A, B, C, D, U_i, \varphi_{ZD})}$ has a unique solution.

Thus, for a driven time depending input $t \mapsto U_i(t)$ the output time depending voltage $t \mapsto U_o(t)$ defined by (see Figure 25):

$$U_o(t) = \gamma R_i I_i(t) - R_o I_2(t)$$

is uniquely defined with the current functions $t \mapsto I_i(t)$ and $t \mapsto I_2(t)$ that are uniquely determined in solving $\mathbf{VI(-PA, -PDU_i(t), \Phi)}$.

In the next example we analyse the stationary solutions of a bridge rectifier. This application constitutes an example that does not fit within this class of generalized equations in (80)–(82). Here the unknown in (17) is not x but y_L (see (80)–(82)).

Example 32 (Four-diodes bridge full-wave rectifier) *Let us consider the circuit in figure 27 involving four diodes D_1, D_2, D_3 and D_4, a resistor $R > 0$ and a capacitor $C > 0$. We denote by $\varphi_{D_i} \in \Gamma(\mathbb{R}; \mathbb{R} \cup \{+\infty\})$ $(1 \leq i \leq 4)$ the electrical superpotential of diode D_i.*

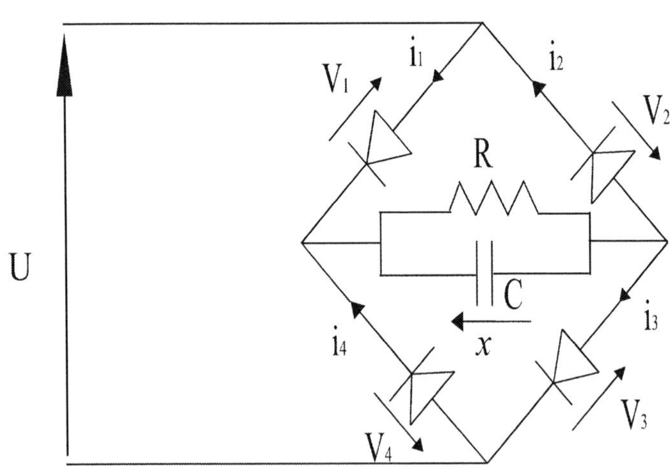

Figure 27. Four diodes bridge.

Let us denote by V_i the voltage of diode D_i $(1 \leq i \leq 4)$, x the voltage of the capacitor and *use the other notation indicated on Figure 27. Kirchhoff's laws yield the system:*

$$\begin{cases} i_1 + i_4 = \frac{x}{R} + C\frac{dx}{dt} \\ -V_4 = x + V_3 \\ i_3 = i_4 + i_1 - i_2 \\ -V_1 = x + V_3 - u \\ -V_2 = -V_3 + u. \end{cases}$$

We have

$$-V_4 \in -\partial\varphi_{D_4}(i_4), -V_1 \in -\partial\varphi_{D_1}(i_1), -V_2 \in -\partial\varphi_{D_2}(i_2).$$

Moreover

$$V_3 \in \partial\varphi_{D_3}(i_3) \Leftrightarrow i_3 \in \partial\varphi^*_{D_3}(V_3).$$

We set

$$\theta_{D_3}(x) = \varphi^*_{D_3}(-x), \forall x \in \mathbb{R},$$

and we assume the existence of a point $x_0 \in \mathbb{R}$ at which θ_{D_3} is finite and continuous. Then

$$\partial\theta_{D_3}(x) = -\partial\varphi^*_{D_3}(-x), \forall x \in \mathbb{R}.$$

Therefore

$$V_3 \in \partial\varphi_{D_3}(i_3) \Leftrightarrow i_3 \in -\partial\theta_{D_3}(-V_3).$$

We set

$$\Phi(x) = \varphi_{D_4}(x_1) + \theta_{D_3}(x_2) + \varphi_{D_1}(x_3) + \varphi_{D_2}(x_4), \ \forall x \in \mathbb{R}^4.$$

It results that the dynamical behavior of the circuit in Figure 27 are descibed by the system:

$$\frac{dx}{dt} = \frac{-1}{RC}x + \overbrace{\begin{pmatrix} \frac{1}{C} & 0 & \frac{1}{C} & 0 \end{pmatrix}}^{B} \begin{pmatrix} i_4 \\ -V_3 \\ i_1 \\ i_2 \end{pmatrix}, \tag{132}$$

$$\overbrace{\begin{pmatrix} -V_4 \\ i_3 \\ -V_1 \\ -V_2 \end{pmatrix}}^{y} = \overbrace{\begin{pmatrix} 1 \\ 0 \\ 1 \\ 0 \end{pmatrix}}^{C} x + \overbrace{\begin{pmatrix} 0 & -1 & 0 & 0 \\ 1 & 0 & 1 & -1 \\ 0 & -1 & 0 & 0 \\ 0 & 1 & 0 & 0 \end{pmatrix}}^{N} \overbrace{\begin{pmatrix} i_4 \\ -V_3 \\ i_1 \\ i_2 \end{pmatrix}}^{y_L} + \overbrace{\begin{pmatrix} 0 \\ 0 \\ -1 \\ 1 \end{pmatrix}}^{F} u \tag{133}$$

and

$$y \in -\partial\Phi(y_L). \tag{134}$$

Assuming that u remains constant, i.e. $u(.) \equiv u$, the stationary solutions (or fixed points) of (132)-(134) satisfy the problem:

$$\begin{cases} -ax + By_L = 0 \\ \langle Ny_L + Cx + Fu, v - y_L\rangle + \Phi(v) - \Phi(y_L) \geq 0, \ \ \forall\, v \in \mathbb{R}^4, \end{cases} \tag{135}$$

with $a := \frac{1}{RC} > 0$.
From the first equation of (135) one deduces that $x = \frac{1}{a}By_L$, *so that* $y = (N + \frac{1}{a}CB)y_L + Fu$ *and our problem reduces to problem* $\mathbf{VI}((\mathbf{N} + \frac{1}{\mathbf{a}}\mathbf{CB}), \mathbf{\Phi}, \mathbf{Fu})$, *i.e:*

$$\langle (N + \frac{1}{a}CB)y_L + Fu, v - y_L\rangle + \Phi(v) - \Phi(y_L) \geq 0, \;\; \forall\, v \in \mathbb{R}^4. \tag{136}$$

The matrix $N \in \mathbb{R}^{4\times 4}$ *is skew-symmetric and the matrix*

$$M := N + \frac{1}{a}CB = \begin{pmatrix} R & -1 & R & 0 \\ 1 & 0 & 1 & -1 \\ R & -1 & R & 0 \\ 0 & 1 & 0 & 0 \end{pmatrix}$$

is positive semidefinite since

$$\langle x, Mx\rangle = \frac{1}{a}\langle x, CBx\rangle = R(x_1 + x_3)^2, \;\forall x \in \mathbb{R}^4.$$

We have:

$$\ker\{M + M^T\} = \{x \in \mathbb{R}^4 : x_1 = -x_3\}.$$

Note also that:

$$x \in \ker\{M + M^T\} \implies Mx = \begin{pmatrix} -x_2 \\ -x_4 \\ -x_2 \\ +x_2 \end{pmatrix}.$$

Case 1. *Suppose that all diodes are ideal, i.e.*

$$\varphi_{D_i}(x) = \Psi_{\mathbb{R}_+}(x), \forall x \in \mathbb{R}\; (1 \leq i \leq 4).$$

Then

$$\begin{aligned}\Phi(x) &= \varphi_{D_4}(x_1) + \theta_{D_3}(x_2) + \varphi_{D_1}(x_3) + \varphi_{D_2}(x_4) \\ &= \Psi_{\mathbb{R}_+}(x_1) + \Psi_{\mathbb{R}_-}(-x_2) + \Psi_{\mathbb{R}_+}(x_3) + \Psi_{\mathbb{R}_+}(x_4) = \Psi_{(\mathbb{R}_+)^4}.\end{aligned}$$

Then $D(\Phi)_\infty = (\mathbb{R}_+)^4$, $D(\Phi_\infty) = (\mathbb{R}_+)^4$, $D(\Phi_\infty)^* = (\mathbb{R}_+)^4$ *and:*

$$D(\Phi)_\infty \cap \ker\{M + M^T\} \cap K(M, \Phi) = \{0\}.$$

Indeed, we have $x_1 \geq 0, x_2 \geq 0, x_3 \geq 0, x_4 \geq 0$ *and* $x \in \ker\{M + M^T\}$ *yields* $x_1 = -x_3$ *and thus* $x_1 = x_3 = 0$. *Using then* $x \in \mathcal{K}(M, \Phi)$, *we get also* $-x_2 \geq 0$ *and* $-x_4 \geq 0$ *and thus* $x_2 = x_4 = 0$. *It results from Corollary 7 that for each* $u \in \mathbb{R}$, *problem* $\mathbf{VI}((\mathbf{N} + \frac{1}{\mathbf{a}}\mathbf{CB}), \mathbf{\Phi}, \mathbf{Fu})$ *has at least one solution.*

Case 2. *Suppose that diodes* D_1, D_3 *and* D_4 *are ideal and consider for diode* D_2 *the practical Zener diode model with electrical superpotential* φ_Z *given by:*

$$\varphi_Z(x) = \begin{cases} \frac{(V_1 - V_3)}{2I_1}x^2 + V_3 x \;\; \text{if } x < 0 \\ \\ \frac{(V_2 - V_4)}{2I_2}x^2 + V_4 x \;\; \text{if } x \geq 0 \end{cases}, \;\; (\forall\, x \in \mathbb{R})$$

with $I_1 < 0 < I_2$, $V_1 < V_3 < 0 < V_4 < V_2$. *Here*

$$\Phi(x) = \varphi_{D_4}(x_1) + \theta_{D_3}(x_2) + \varphi_{D_1}(x_3) + \varphi_{D_2}(x_4)$$

$$= \Psi_{\mathbb{R}_+}(x_1) + \Psi_{\mathbb{R}_+}(x_2) + \Psi_{\mathbb{R}_+}(x_3) + \varphi_Z(x_4),\ \forall x \in \mathbb{R}^4$$

We have $D(\Phi) = D(\Phi)_\infty = \mathbb{R}_+ \times \mathbb{R}_+ \times \mathbb{R}_+ \times \mathbb{R}$, $D(\Phi_\infty) = \mathbb{R}_+ \times \mathbb{R}_+ \times \mathbb{R}_+ \times \{0\}$ *and* $D(\Phi_\infty)^* = \mathbb{R}_+ \times \mathbb{R}_+ \times \mathbb{R}_+ \times \mathbb{R}$. *We check that*

$$\mathcal{B}(M,\Phi) = D(\Phi)_\infty \cap \ker\{M + M^T\} \cap \mathcal{K}(M,\Phi) = \{0\} \times \{0\} \times \{0\} \times \mathbb{R}_-.$$

Indeed, let $x \in \mathcal{B}(M,\Phi)$ *be given. Here* $x \in \ker\{M + M^T\}$ *yields* $x_1 = -x_3$ *and thus* $x_1 = x_3 = 0$ *since* $x_1 \geq 0$ *and* $x_3 \geq 0$. *Using then* $x \in \mathcal{K}(M,\Phi)$, *i.e.* $Mx \in D(\Phi_\infty)^*$ *we get also* $x_4 \leq 0$ *and* $x_2 \leq 0$ *and thus* $x_2 = 0$ *since* $x_2 \geq 0$.

Moreover, if $v \in \mathcal{B}(M,\Phi), v \neq 0$ *then* $v_1 = v_2 = v_3 = 0$, $v_4 < 0$ *and* $\Phi_\infty(v) + \langle Fu, v\rangle = (\varphi_Z)_\infty(v_4) + v_4 u = \Psi_{\{0\}}(v_4) + v_4 u = +\infty$. *It results from Corollary 10 that for each* $u \in \mathbb{R}$, *problem* $\mathbf{VI}((\mathbf{N} + \frac{1}{\mathbf{a}}\mathbf{CB}), \mathbf{\Phi}, \mathbf{Fu})$ *has at least one solution.*

Various cases can be so studied and this shows that studying the fixed points of an interesting class of electrical circuits can be cast into a variational inequality problem such as (17), written as: Find $y_L \in \mathbb{R}^n$ such that

$$\langle My_L + q, v - y_L\rangle + \Phi(v) - \Phi(y_L) \geq 0,\ \ \forall\, v \in \mathbb{R}^n. \tag{137}$$

References

[1] Anonymous, *Electro-Principles I, ETEC* **1120** (http://www.technology.niagarac.on.ca").

[2] Anonymous, Electronic Devices and Circuits, *Engineering Sciences* **154**, Diode i-v Characteristics (http://people.deas.harvard.edu).

[3] Anonymous, 10ETS.. *SAFEIR Series*, Bulletin 12120 rev. A 07/97, (http://www.digchip.com).

[4] Anonymous, *Voltage regulator diodes*, (http://www.datasheetcatalog.com/).

[5] Anonymous, *Diode Basics*, (http://www.powerdesigners.com/).

[6] Anonymous, *Littelfuse varistors - Basic properties, Terminology and theory,* Application Note AN9767.1, 1999.

[7] *Anonymous*, www.pspice.com.

[8] Acary V., Brogliato B., Goeleven D., *Higher order Moreau's sweeping process: Mathematical formulation and numerical simulation*, to appear in *Mathematical Programming.*

[9] Addi K., Adly S., Brogliato B. and Goeleven D., A method using the approach of Moreau and Panagiotopoulos for the mathematical formulation of non-regular circuits in electronics, to appear in *Nonlinear Analysis, Series C : Hybrid Systems and Applications*.

[10] Addi K., Brogliato B. and Goeleven D., A qualitative mathematical analysis of a class of linear variational inequalities via semi-complementarity problems. Applications in electronics. INRIA technical report XXXX, (2006) (http://hal.inria.fr/inria/docs/00/10/33/37/PDF/RR-5590.pdf).

[11] Adly S. and Goeleven D., A Discretization Theory for a Class of Semi-coercive Unilateral Problems, *Numerische Mathematik*, **87**, (2000), 1-34.

[12] Adly S. and Goeleven D., A Stability Theory for Second-Order Nonsmooth Dynamical Systems with Applications to Friction Problems, *Journal de Mathématiques Pures et Appliquées*, **83**, (2004), 17-51.

[13] Brogliato B., Some Perspectives on the Analysis and Control of Complementarity Systems, *IEEE Transactions on Automatic Control*, **48**, (2003), 918-935.

[14] Brogliato B., Absolute Stability and the Lagrange-Dirichlet Theorem with Monotone Multivalued Mappings, *Systems and Control Letters*, **51**, (2004), 343-353.

[15] Brogliato B. and Goeleven D., The Krakovskii-LaSalle Invariance Principle for a Class of Unilateral Dynamical Systems, *Mathematics of Control, Signals and Systems*, **17**, (2005), 57-76.

[16] Brogliato B., Lozano R., Maschke B.M., Egeland O., *Dissipative Systems Analysis and Control*, Springer CCE series, London (2006).

[17] Camlibel M.K., Heemels W.P.M.H. and Schumacher H., On Linear Passive Complementarity Systems, *Eur. J. Control*, **8**, (2002), 220-237.

[18] Camlibel M.K. and Schumacher J.M., Existence and uniqueness of solutions for a class of piecewise linear dynamical systems, *Linear Algebra and its Applications*, **351-352**, (2002), 147-184.

[19] Denoyelle P. and Acary V., The Non-Smooth Approach applied to simulating Integrated Circuits and Power Electronics. Evolution of electronic circuit simulators towards fast-SPICE performance, INRIA technical report RT-0321, (2006) (http://hal.inria.fr/inria-00080839/en/).

[20] Duvaut G. and Lions J.L., *Les Inéquations en Mécanique et en Physique*, Dunod, Paris, (1972)

[21] Facchinei F. and Pang J.-S., *Finite Dimensional Variational Inequalities and Complementarity Problems*, Springer Verlag, Berlin, (2003).

[22] Fichera G., Boundary Value Problems in Elasticity with Unilateral Constraints, In: *Encyclopedia of Physics* (ed. by S. Flügge), Springer Verlag, Berlin, Vol. VI a/2, (1972).

[23] Goeleven D., Brogliato B., Stability and Instability matrices for Linear Evolution Variational Inequalities, *IEEE Transactions on Automatic Control*, **49**, (2004), 521-534.

[24] Goeleven D., On the Solvability of Linear Noncoercive Variational Inequalities in Separable Hilbert Spaces, *Journal of Optimization Theory and Applications*, 79, (1993), 493-511.

[25] Goeleven D., Stavroulakis G.E. and Panagiotopoulos P.D., Solvability Theory for a Class of Hemivariational Inequalities Involving Copositive Plus Matrices, Applications in Robotics, *Mathematical Programming*, 75, (1996), 441-465.

[26] Goeleven D., *Noncoercive Variational Problems and Related Results*, Pitman Research Notes in Mathematics Series, N357, Longman, 1996.

[27] Goeleven D. and Gwinner J., On Semicoerciveness of a Class of Variational Inequalities, and an Application to a Friction Problem for von Karman Plates, *Mathematische Nachrichten*, 144, (2002) 89-109.

[28] Goeleven D., Motreanu D., Dumont Y. and Rochdi M., *Variational and Hemivariational Inequalities – Theory, Methods and Applications. Vol.1: Unilateral Analysis and Unilateral Mechanics*, Kluwer Academic Publishers, Boston, (2003).

[29] Goeleven D. and Motreanu D., *Variational and Hemivariational Inequalities – Theory, Methods and Applications. Vol.2: Unilateral Problems and Unilateral Mechanics*, Kluwer Academic Publishers, Boston, (2003).

[30] Goeleven D. and Motreanu D., On the Solvability of Variational Inequalities via Relaxed Complementarity Problems, *Communications in Applied Analysis*, 4, 533-546, (2000).

[31] Gowda M.S., Complementarity Problems over Locally Compact Cones, *SIAM Journal on Control and Optimization*, **27**, (1989), 836-841.

[32] Gowda M.S. and Seidman I., Generalized Linear Complementarity Problems, *Mathematical Programming*, **46**, (1990) 329-340.

[33] Heemels W.P.M.H. and Schumacher J.M. and Weiland S., Linear complementarity systems, *SIAM J. Applied Math.*, **60**, (2000), 1234-1269.

[34] Hiriart-Urruty J.B. and Lemaréchal C., *Fundamentals of Convex Analysis*, Springer Grundlehren Text Editions, Heidelberg, (2001).

[35] Hlavacek I., Haslinger J., Necas J. and Lovisek J., *Solutions of Variational Inequalities in Mechanics*, Applied Mathematical Sciences 66, Springer Verlag, New York 1982.

[36] Horn A.H. and Johnson C.R., *Topics in Matrix Analysis*, Cambridge University Press 1991.

[37] Hu S.C. and Papageorgiou N.S., *Handbook of Set-Valued Analysis*, Kluwer Academic Publishers, Boston, (1999).

[38] Isac G., *Complementarity Problems*, Lecture Notes in Mathematics 1528, Springer Verlag, Heidelberg, 1993.

[39] Jofré A., Rockafellar R.T. and Wets R. J.-B., Variational Inequalities and Economic Equilibrium, To appear in Mathematics of Operations Research.

[40] Kikuchi N. and Oden J.T., *Contact Problems in Elasticity: A study of Variational Inequalities and Finite Element Methods*, SIAM, Philadelphia, 1988.

[41] Konnov I.V., and Volotskaya E.O., *Mixed Variational Inequalities and Economic Equilibrium Problems*, Journal of Applied Mathematics, Vol. 6, pp. 289-314, 2002.

[42] Leenaerts D. and van Bokhoven W.M.G., *Piecewise Linear Modeling and Analysis*, Kluwer Academic Publishers, Norwell, MA, USA (1998).

[43] Lloyd N.G., *Degree Theory*, Cambridge Tracts in Mathematics 73, Cambridge University Press, Cambridge (1978).

[44] Millman J. and Halkias C.C., *Integrated Electronics*, McGraw-Hill Kogakusha, LTD, Sydney, (1985).

[45] Murakami Y., A Method for the Formulation and Solution of Circuits Composed of Switches and Linear RLC networks, *IEEE Transactions on Circuits and Systems-I*, **49**, (2002), 315-327.

[46] Murty K.G., *Linear Complementarity, Linear and Nonlinear Programming*, (1997), available at http://www-personal.engin.umich.edu/$\sim$ murty/book/LCPbook/

[47] Naniewicz Z. and Panagiotopoulos P.D., The mathematical Theory of Hemivariational Inequalities and Applications, Marcel Dekker, N. York 1994.

[48] Panagiotopoulos P.D., *Inequality Problems in Mechanics and Applications, Convex and Nonconvex Energy Functions*, Birkhaüser, Basel, (1985).

[49] Panagiotopoulos P.D., *Hemivariational Inequalities, Applications in Mechanics and Engineering*, Springer Verlag Berlin Heidelberg (1993).

[50] Pang J.S., *Complementarity Problems* in R. Horst and P. Pardalos, editors, Handbook in Global Optimization, Kluwer Academic Publishers, Boston, (1994).

[51] Pascali D. and Sburlan S., *Nonlinear Mappings of Monotone Type*, Sijthoff and Noordhoff International Publishers, Amsterdam, The Netherlands, (1978).

[52] Rockafellar R.T., *Convex Analysis*, Princeton Univ. Press, Princeton, (1970).

Peer reviewers:

- **Prof. Aleksandar Rodic**
  Mihajlo Pupin Institute, Robotics department
  University of Belgrade
  11060 Belgrade, Zvezdara, Volgina 15
  SERBIA
  roda@robot.imp.bg.ac.yu
- **Prof. Samir Adly**
  DMI-XLIM, University of Limoges
  87060 Limoges
  FRANCE
  adly@unilim.fr

In: Nonlinear Analysis Research Trends
Editor: Inès N. Roux, pp. 99-155
ISBN 978-1-60456-358-0

Chapter 4

NONLINEAR MATHEMATICAL PROBLEMS OF ELECTROMAGNETOELASTIC INTERACTIONS

Viatcheslav Priimenko [1] [*] ***and Mikhail Vishnevskii*** [2] [†]
[1] Laboratory of Oil Engineering and Exploration,
North Fluminense State University
27.925-310, Macae, RJ, Brazil
[2] Laboratory of Mathematical Sciences,
North Fluminense State University
28.013-600, Campos dos Goytacazes, RJ, Brazil

Abstract

There are studied nonlinear mathematical problems of the interaction of electromagnetic fields with deformable media. The models are based on combination of the Lamé and Maxwell systems coupled through so-called seismomagnetic effect. Several direct and associated with them inverse problems are studied. Then speaking about the inverse problems, electromagnetic and elastic characteristics of a medium are the subject of reconstruction. The values of physical fields are connected through electromagnetoelastic interactions. We consider the processes which are observed when elastic waves propagate in an elastic electroconducting medium. Variations of the seismic and electromagnetic fields in this case are called electromagnetoelastic waves. There are described different statements of mathematical model of the electromagnetoelastic interactions. Then the theoretical results of the analytical solution are discussed for various nonlinear direct and inverse problems for the equations of electromagnetoelasticity.

Introduction

The interaction of electromagnetic fields with deformable media is a subject of many theoretical and experimental investigations in the field of continuum mechanics and geophysics

[*] E-mail address: slava@lenep.uenf.br
[†] E-mail address: mikhail@uenf.br

in the recent decades. For description of simple enough interactions, the theories of magnetohydrodynamics, [1], electroelasticity, [2], [3], and magnetoelasticity, [4], [5], [6], were developed. These theories are, basically, a combination (without introducing new conceptions) of objects and phenomena considered in continuum mechanics and electrodynamics.

Investigation of more complex electromagnetoelastic interactions in a continuous medium requires to consider complex models. For a more profound acquaintance with the modern state of the theory of electromagnetoelastic interactions the reader is referred to, e.g., [7], [8].

The aim of this paper is to study some nonlinear direct and inverse problems connected with electromagnetoelastic interactions. The model considered here is based on a simple variant of combination of the Lamé and Maxwell equations.

Let us give a brief characteristic of basic types of electromagnetoelastic interactions. It is well known that when an electrical-conducting elastic body oscillates in an electromagnetic field, variations of the electrical and magnetic fields are observed as a result of this motion. Similar processes are also observed when seismic waves propagate in the Earth's crust. Variations of seismic and electromagnetic fields arising in this case are called *electromagnetoelastic waves*. Such waves contain a certain information about electromagnetic and elastic parameters of the medium. In this case, as a rule, the following types of electromagnetoelastic interactions are distinguished:

a) Interaction based on the *electrokinetic effect*. It is supposed that generation of electrical signals with elastic waves propagation is connected precisely with manifestation of electrokinetic properties of a medium.

b) Interaction based on the *piezoelectric effect*. This interaction is connected with propagation of elastic waves in crystal rocks when the elastic deformation of a lattice substance produces displacement of electrons and, as consequence, there arises an electrical field induced by such deformations.

c) Interaction based on *slow movement* of a body in an external electromagnetic field. Whereas, for example, the electrokinetic effect is connected with local interactions of elastic waves with a flow in the pore liquid, this effect is based on slow moving of particles in an external electromagnetic field.

In seismics and seismology the third type of interaction leads to so-called *seismomagnetic effect* describing interaction of seismic waves with the Earth's magnetic field. This interaction results in induced electromagnetic waves propagating with speeds commensurable with the speeds of seismic waves.

1. Mathematical Model of Electromagnetoelastic Interactions

The interaction of electromagnetic fields with deformable media is considered with point of view of linear elasticity connected with electrodynamic of elastic moving media by means of motion of particles in the electromagnetic field. We do not consider any effects of interactions, which could arise as a result of some kind of relations in constitutive equations besides velocity. We, basically, follow Dunkin and Eringen, [4], when defining a mathematical model for electromagnetoelastic effect.

1.1. Electromagnetic Theory

Let $\mathbb{R}^3$ be a three-dimensional Euclidean space of points $x = (x_1, x_2, x_3)$. The process of propagation of electromagnetic waves in $\mathbb{R}^3$ will be described by the following Maxwell system:

$$\frac{\partial \boldsymbol{D}}{\partial t} + \boldsymbol{J} = \operatorname{rot} \boldsymbol{H}, \qquad \frac{\partial \boldsymbol{B}}{\partial t} + \operatorname{rot} \boldsymbol{E} = 0, \tag{1.1.1}$$

$$\operatorname{div} \boldsymbol{D} = \rho_e, \qquad \operatorname{div} \boldsymbol{B} = 0. \tag{1.1.2}$$

Here $\boldsymbol{E}, \boldsymbol{H}, \boldsymbol{D}$ and $\boldsymbol{B}$ are the electromagnetic vectors, $\boldsymbol{J}$ is the current density, ρ_e is the charge density, and all quantities are expressed in the MKS units. When a medium is at rest, the electromagnetic constitutive equations of an isotropic medium are

$$\boldsymbol{D}^0 = \epsilon \boldsymbol{E}^0, \qquad \boldsymbol{B}^0 = \mu \boldsymbol{H}^0, \qquad \boldsymbol{J}^0 = \sigma \boldsymbol{E}^0, \tag{1.1.3}$$

where ϵ, μ are called the electric and magnetic permeabilities and σ is the electrical conductivity. The same equations are assumed to be valid at each point in the reference frame moving with the velocity of a material point, i.e. the *proper* frame, but they are expressed in terms of the field measured in the *laboratory* frame in which motion is observed. For small velocities the proper quantities are related to the laboratory ones by the equations, see [6]

$$\begin{aligned}
\boldsymbol{E}^0 &= \boldsymbol{E} + \frac{\partial \boldsymbol{u}}{\partial t} \times \boldsymbol{B}, & \boldsymbol{D}^0 &= \boldsymbol{D} + c^{-2}\frac{\partial \boldsymbol{u}}{\partial t} \times \boldsymbol{H}, \\
\boldsymbol{H}^0 &= \boldsymbol{H} - \frac{\partial \boldsymbol{u}}{\partial t} \times \boldsymbol{D}, & \boldsymbol{B}^0 &= \boldsymbol{B} - c^{-2}\frac{\partial \boldsymbol{u}}{\partial t} \times \boldsymbol{E}, \\
\boldsymbol{J}^0 &= \boldsymbol{J} - \rho_e \frac{\partial \boldsymbol{u}}{\partial t}, & \rho_e^0 &= \rho_e, \qquad c \equiv (\epsilon_0 \mu_0)^{-1/2},
\end{aligned}$$

where ϵ_0, μ_0 are the dielectric and magnetic permeabilities of the vacuum and $\boldsymbol{u}$ is the displacement vector. Let us substitute these relations into constitutive equations (1.1.3). If the terms of order $|\frac{\partial \boldsymbol{u}}{\partial t}|^2/c^2$ and higher are dropped, the results are as follows, see [6]

$$\boldsymbol{D} = \epsilon \boldsymbol{E} + \alpha \frac{\partial \boldsymbol{u}}{\partial t} \times \boldsymbol{H}, \qquad \alpha \equiv \epsilon\mu - \epsilon_0\mu_0, \qquad \boldsymbol{B} = \mu \boldsymbol{H} - \alpha \frac{\partial \boldsymbol{u}}{\partial t} \times \boldsymbol{E}, \tag{1.1.4}$$

$$\boldsymbol{J} = \rho_e \frac{\partial \boldsymbol{u}}{\partial t} + \sigma\left(\boldsymbol{E} + \frac{\partial \boldsymbol{u}}{\partial t} \times \boldsymbol{B}\right). \tag{1.1.5}$$

For more details of electromagnetic theory, the reader is referred to many textbooks that treat this research field, e.g., [7], [9], [10].

Thus, we have obtained a complete system for freely moving media. They are Maxwell's equations (1.1.1)–(1.1.2) and the constitutive relations (1.1.4)–(1.1.5). Equation (1.1.5) is a modification of Ohm's law, where appears a term reflecting the influence of particles moving in the magnetic field with a current density.

The electromagnetic matching conditions are obtained in the following manner. First rewrite equations (1.1.1)–(1.1.2) in the equivalent form

$$\begin{aligned} \operatorname{rot}(\boldsymbol{E} + \frac{\partial \boldsymbol{u}}{\partial t} \times \boldsymbol{B}) &= -\frac{\partial \boldsymbol{B}}{\partial t} - \frac{\partial \boldsymbol{u}}{\partial t} \operatorname{div} \boldsymbol{B} + \operatorname{rot}(\frac{\partial \boldsymbol{u}}{\partial t} \times \boldsymbol{B}), \\ \operatorname{rot}(\boldsymbol{H} - \frac{\partial \boldsymbol{u}}{\partial t} \times \boldsymbol{D}) &= \frac{\partial \boldsymbol{D}}{\partial t} + \frac{\partial \boldsymbol{u}}{\partial t} \operatorname{div} \boldsymbol{D} - \operatorname{rot}(\frac{\partial \boldsymbol{u}}{\partial t} \times \boldsymbol{D}) + \boldsymbol{J} - \rho_e \frac{\partial \boldsymbol{u}}{\partial t}, \end{aligned} \tag{1.1.6}$$

$$\operatorname{div} \boldsymbol{B} = 0, \qquad \operatorname{div} \boldsymbol{D} = \rho_e. \tag{1.1.7}$$

Then integral analogues of these equations can be obtained by integrating (1.1.6) over the surface S' composed of material particles and bounded by a curve C and (1.1.7) over a volume V of material particles bounded by the surface S, see Fig. 1. Note that C, S', S and V move with the material. After applying Stokes' theorem on the left-hand sides of (1.1.6), we obtain

$$\begin{aligned} \int_C (\boldsymbol{E} + \frac{\partial \boldsymbol{u}}{\partial t} \times \boldsymbol{B}) \cdot d\boldsymbol{c} &= -\frac{d}{dt} \int_{S'} \boldsymbol{B} \cdot d\boldsymbol{s}', \\ \int_C (\boldsymbol{H} - \frac{\partial \boldsymbol{u}}{\partial t} \times \boldsymbol{D}) \cdot d\boldsymbol{c} &= \frac{d}{dt} \int_{S'} \boldsymbol{D} \cdot d\boldsymbol{s}' + \int_{S'} (\boldsymbol{J} - \rho_e \frac{\partial \boldsymbol{u}}{\partial t}) \cdot d\boldsymbol{s}', \end{aligned} \tag{1.1.8}$$

Applying the Gauss-Ostrogradskii theorem to (1.1.7) yields:

$$\begin{aligned} \int_S \boldsymbol{B} \cdot d\boldsymbol{s} &= 0, \\ \int_S \boldsymbol{D} \cdot d\boldsymbol{s} &= \int_V \rho_e dx, \end{aligned} \tag{1.1.9}$$

where we have also used the well-known relation

$$\frac{d}{dt} \int_{S'} \boldsymbol{F} \cdot d\boldsymbol{s}' = \int_{S'} [\frac{\partial \boldsymbol{F}}{\partial t} + \frac{\partial \boldsymbol{u}}{\partial t} \operatorname{div} \boldsymbol{F} - \operatorname{rot}(\frac{\partial \boldsymbol{u}}{\partial t} \times \boldsymbol{F})] d\boldsymbol{s}'.$$

Select S' to be a small rectangular area oriented perpendicular to the discontinuity surface such that one side lies in the part of material with one material properties and other one lies in the part with another material properties.

As the dimension of S', perpendicular to the boundary, tends to zero, equations (1.1.8) now look like:

$$[\boldsymbol{E} + \frac{\partial \boldsymbol{u}}{\partial t} \times \boldsymbol{B}]_t = 0, \qquad [\boldsymbol{H} - \frac{\partial \boldsymbol{u}}{\partial t} \times \boldsymbol{D}]_t = J_m^S - \rho_e^S \frac{\partial u_m}{\partial t}, \tag{1.1.10}$$

where the symbol $[\boldsymbol{F}]_t$ means a jump of the tangential components of the vector $\boldsymbol{F}$ across the surface, where the coefficients of equations have breaks, and J_m^S, ρ_e^S represent the surface current and the charge, respectively. Here and in the sequel the subscripts t, m, n denote the vector components in the directions $\boldsymbol{t}, \boldsymbol{m}, \boldsymbol{n}$ which form the right-hand orthogonal triad (Fig. 1). Now, let us choose V to be a small cylindrical volume whose axis is perpendicular to the discontinuity surface such that one of the circular ends lies in the part of the material with one material properties and another one lies in the part with another

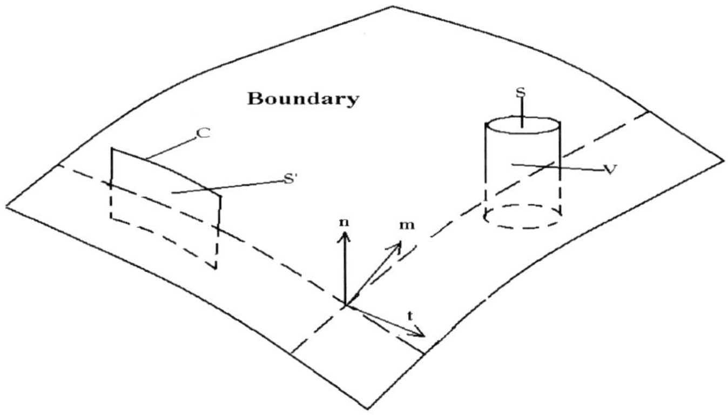

Figure 1. Areas selected for integration.

material properties. As the height of the volume tends to zero, equations (1.1.9) take the form:

$$[\boldsymbol{B}]_n = 0, \qquad [\boldsymbol{D}]_n = \rho_e^S, \tag{1.1.11}$$

where $[\boldsymbol{F}]_n$ means a jump in the normal component of $\boldsymbol{F}$.

Equations (1.1.10)–(1.1.11) constitute the complete electromagnetic matching conditions on a discontinuity surface.

1.2. Elastic Theory

Consider now the equations of motion of a deformable medium. The mechanical equations will be derived by applying the conservation of momentum to the volume of a material, V, with the bounding surface S in the absence of a mechanical force, using an assumption that only a mechanical effect of the electromagnetic fields is the introduction of the Lorentz force

$$\boldsymbol{f}^e = \rho_e \boldsymbol{E} + \boldsymbol{J} \times \boldsymbol{B}. \tag{1.2.1}$$

Thus the equation of global conservation of momentum in the rectangular coordinates is the following

$$\int_S T \cdot \boldsymbol{n} ds + \int_V \boldsymbol{f}^e dx = \frac{d}{dt} \int_V \boldsymbol{g}^m dx, \tag{1.2.2}$$

where T is a stress tensor and $\boldsymbol{g}^m$ is momentum per unit volume. Using the Gauss-Ostrogradskii theorem for the surface integral and differentiation of the volume integral according to [11, Eq. 20.9],

$$\frac{d}{dt} \int_V \boldsymbol{g}^m dv = \int_V (\frac{\partial \boldsymbol{g}^m}{\partial t} + \mathrm{Div}(\boldsymbol{g}^m \otimes \frac{\partial \boldsymbol{u}}{\partial t}))dx \tag{1.2.3}$$

we obtain

$$\int_V (\mathrm{Div}\, T + \boldsymbol{f}^e - \mathrm{Div}(\boldsymbol{g}^m \otimes \frac{\partial \boldsymbol{u}}{\partial t}) - \frac{\partial \boldsymbol{g}^m}{\partial t})dx = 0,$$

where

$$\operatorname{Div} T = \left(\sum_{j=1}^{3} \frac{\partial}{\partial x_j} T_{ij} \right)_{i=1}^{3}. \tag{1.2.4}$$

If the mechanical momentum is locally conserved, then

$$\operatorname{Div} T + \boldsymbol{f}^e = \operatorname{Div}(\boldsymbol{g}^m \otimes \frac{\partial \boldsymbol{u}}{\partial t}) + \frac{\partial \boldsymbol{g}^m}{\partial t}. \tag{1.2.5}$$

In an elastic solid $\boldsymbol{g}^m = \rho \frac{\partial \boldsymbol{u}}{\partial t}$, where ρ is the material density, and the assumption of infinitesimal strains and rotations (1.2.5) reduces to

$$\rho \frac{\partial^2 \boldsymbol{u}}{\partial t^2} = \operatorname{Div} T + \boldsymbol{f}^e. \tag{1.2.6}$$

The elastic matching conditions on stress are obtained by applying (1.2.6) to appropriate differential elements. Introduce Maxwell's stresses T^e and the electromagnetic momentum $\boldsymbol{g}^e$ according to Minkowski, see [12]

$$\begin{gathered} T^e = \boldsymbol{E} \otimes \boldsymbol{D} + \boldsymbol{H} \otimes \boldsymbol{B} - \frac{1}{2}(\boldsymbol{E} \cdot \boldsymbol{D} + \boldsymbol{H} \cdot \boldsymbol{B}) I, \\ \boldsymbol{g}^e = \boldsymbol{D} \times \boldsymbol{B}, \end{gathered} \tag{1.2.7}$$

where I is the unit matrix of order 3×3. Let us show that the Lorentz force $\boldsymbol{f}^e$ can be represented in the following form

$$\boldsymbol{f}^e = \operatorname{Div} T^e - \frac{\partial \boldsymbol{g}^e}{\partial t}. \tag{1.2.8}$$

It is easy to check the correctness of formula (1.2.8) taking into account equations (1.1.1)–(1.1.2) and constitutive relations

$$\boldsymbol{D} = \epsilon \boldsymbol{E}, \qquad \boldsymbol{B} = \mu \boldsymbol{H}.$$

After simple transformations we come to the equation

$$\operatorname{Div} T = \rho_e \boldsymbol{E} + \operatorname{rot} \boldsymbol{E} \times \boldsymbol{D} + \operatorname{rot} \boldsymbol{H} \times \boldsymbol{B}.$$

Using Maxwell's equations (1.1.1)–(1.1.2) we obtain

$$\operatorname{Div} T = \rho_e \boldsymbol{E} + \boldsymbol{J} \times \boldsymbol{B} + \frac{\partial \boldsymbol{D}}{\partial t} \times \boldsymbol{B} - \frac{\partial \boldsymbol{B}}{\partial t} \times \boldsymbol{D}$$

from which next formula is followed

$$\operatorname{Div} T = \rho_e \boldsymbol{E} + \boldsymbol{J} \times \boldsymbol{B} + \frac{\partial}{\partial t}(\boldsymbol{D} \times \boldsymbol{B}) = \boldsymbol{f}^e + \frac{\partial \boldsymbol{g}^e}{\partial t}.$$

This formula proves the representation (1.2.8).

The volume integral containing $\boldsymbol{f}^e$ can be written down as

$$\int_V \boldsymbol{f}^e dx = \int_S (T + \boldsymbol{g}^e \otimes \frac{\partial \boldsymbol{u}}{\partial t}) \cdot \boldsymbol{n} ds - \int_V (\frac{\partial \boldsymbol{g}^e}{\partial t} + \operatorname{Div}(\boldsymbol{g}^e \otimes \frac{\partial \boldsymbol{u}}{\partial t})) dx,$$

where $\mathrm{Div}(\boldsymbol{g}^e \otimes \frac{\partial \boldsymbol{u}}{\partial t})$ was added and subtracted, and the Gauss-Ostrogradskii theorem was used to convert the volume integral to the surface one. Using this expression and (1.2.3) in (1.2.2) gives us

$$\int_S (T + T^e + \boldsymbol{g}^e \otimes \frac{\partial \boldsymbol{u}}{\partial t}) \cdot \boldsymbol{n} ds = \frac{d}{dt} \int_V (\boldsymbol{g}^m + \boldsymbol{g}^e) dx, \tag{1.2.9}$$

which is the form appropriated for obtaining the matching conditions on surface fractions.

Now let S and V be the surface and the volume of a small cylindrical element, whose axis is perpendicular to the discontinuity surface such that one end of the cylinder lies in the part of a material with certain material properties and another one lies in the part with other material properties (Fig. 1). Applying (1.2.9) to this cylindrical region and allowing the axial dimension to approach zero, (1.2.9) becomes

$$[T + T^e + \boldsymbol{g}^e \otimes \frac{\partial \boldsymbol{u}}{\partial t}] \cdot \boldsymbol{n} = 0. \tag{1.2.10}$$

In the case of a body surrounded by vacuum $T = 0$ outside the body and (1.2.10) reduces to

$$T \cdot \boldsymbol{n} = -[T^e + \boldsymbol{g}^e \otimes \frac{\partial \boldsymbol{u}}{\partial t}] \cdot \boldsymbol{n}, \quad \text{on} \quad \Omega, \tag{1.2.11}$$

where Ω is the body surface.

The mechanical constitutive equations are taken to be the usual Hook's Law for an isotropic elastic medium, i.e.

$$T = \lambda \,\mathrm{tr}\, S \cdot I + 2\varkappa S, \tag{1.2.12}$$

where S is the strain tensor defined by the formula

$$S_{ij} = \frac{1}{2}\Big(\frac{\partial u_i}{\partial x_j} + \frac{\partial u_j}{\partial x_i}\Big), \quad i, j = 1, 2, 3.$$

In the above formulas, $\lambda, \varkappa$ are the Lamé coefficients. When using these relations it is assumed that the stresses and strains for this combined system in proper and laboratory frames are the same. Due to the fact that the system has been split to two parts, the mechanical part and the electromagnetic part, as expressed by the Minkowski energy-momentum tensor, this question needs further consideration. For the present purposes we simply assume that constitutive relations (1.2.12) for purely elastic medium are unaffected by the electromagnetic fields. For very large fields or finite deformations the interaction terms will enter in the constitutive relations thereby coupling together the elastic and electromagnetic constitutive equations, see [7].

1.3. Summary of Equations and Matching Conditions

Here we summarize the basic field equations and matching conditions for an electromagnetoelastic medium.

Field equations

$$\frac{\partial \boldsymbol{D}}{\partial t} + \boldsymbol{J} = \operatorname{rot} \boldsymbol{H}, \qquad \operatorname{div} \boldsymbol{D} = \rho_e,$$
$$\frac{\partial \boldsymbol{B}}{\partial t} + \operatorname{rot} \boldsymbol{E} = 0, \qquad \operatorname{div} \boldsymbol{B} = 0, \tag{1.3.1}$$

$$\rho \frac{\partial^2 \boldsymbol{u}}{\partial t^2} = \operatorname{Div} T + \rho_e \boldsymbol{E} + \boldsymbol{J} \times \boldsymbol{B}. \tag{1.3.2}$$

Constitutive equations

$$\boldsymbol{D} = \epsilon \boldsymbol{E} + \alpha \frac{\partial \boldsymbol{u}}{\partial t} \times \boldsymbol{H}, \qquad \alpha \equiv \epsilon\mu - \epsilon_0\mu_0,$$
$$\boldsymbol{B} = \mu \boldsymbol{H} - \alpha \frac{\partial \boldsymbol{u}}{\partial t} \times \boldsymbol{E}, \qquad \boldsymbol{J} = \rho_e \frac{\partial \boldsymbol{u}}{\partial t} + \sigma(\boldsymbol{E} + \frac{\partial \boldsymbol{u}}{\partial t} \times \boldsymbol{B}), \tag{1.3.3}$$

$$T = \lambda \operatorname{tr} S \cdot I + 2\varkappa S,$$
$$S_{ij} = \frac{1}{2}\Big(\frac{\partial u_i}{\partial x_j} + \frac{\partial u_j}{\partial x_i}\Big), \quad i, j = 1, 2, 3. \tag{1.3.4}$$

Matching conditions

$$[\boldsymbol{E} + \frac{\partial \boldsymbol{u}}{\partial t} \times \boldsymbol{B}]_t = 0, \qquad [\boldsymbol{H} - \frac{\partial \boldsymbol{u}}{\partial t} \times \boldsymbol{D}]_t = J_m^S - \rho_e^S \frac{\partial u_m}{\partial t},$$
$$[\boldsymbol{B}]_n = 0, \qquad [\boldsymbol{D}]_n = \rho_e^S, \tag{1.3.5}$$

$$[T + \boldsymbol{E} \otimes \boldsymbol{D} + \boldsymbol{H} \otimes \boldsymbol{B} - \frac{1}{2}(\boldsymbol{E} \cdot \boldsymbol{D} + \boldsymbol{H} \cdot \boldsymbol{B})I + (\boldsymbol{D} \times \boldsymbol{B}) \otimes \frac{\partial \boldsymbol{u}}{\partial t}] \cdot \boldsymbol{n} = 0. \tag{1.3.6}$$

2. Direct Problems

In this section we present some results of the solution of direct problems for the system of equations describing linear and nonlinear processes of the interaction of electromagnetic and elastic waves based on motion of particles.

2.1. The Cauchy Problem for the Equations of Electromagnetoelasticity for Weakly Conducting Media

Following the work [13], we present some results of solution of the Cauchy problem for a system of equations describing the linear interaction process of electromagnetic and elastic waves in a weakly conducting elastic medium. The linear interaction of electromagnetic field with an elastic isotropic medium based on motion of particles is described by the following equations:

$$\frac{\partial \boldsymbol{D}}{\partial t} + \boldsymbol{J} + \boldsymbol{j} = \operatorname{rot} \boldsymbol{H},$$
$$\frac{\partial \boldsymbol{B}}{\partial t} + \operatorname{rot} \boldsymbol{E} = 0, \tag{2.1.1}$$

$$\rho \frac{\partial^2 \boldsymbol{u}}{\partial t^2} = \operatorname{Div} T + \mu \boldsymbol{J} \times \boldsymbol{h}^0 + \boldsymbol{f}, \tag{2.1.2}$$

where the vectors $\boldsymbol{j}$ and $\boldsymbol{f}$ characterize the external source of currents and the external source of elastic oscillations. In this section $\boldsymbol{j}, \boldsymbol{f}$ are supposed to be distributions with finite supports and

$$(\boldsymbol{j}, \boldsymbol{f}) \equiv 0, \quad (x,t) \in \mathbb{R}^3 \times \mathbb{R}_-, \tag{2.1.3}$$

where $\mathbb{R}_- = \{t \in \mathbb{R} \mid t < 0\}$. The defining relations for the stress tensor T and components of the electric induction $\boldsymbol{D}$ and the magnetic induction $\boldsymbol{B}$ for an elastic isotropic inhomogeneous space are as follows:

$$\begin{gathered}\boldsymbol{D} = \varepsilon \boldsymbol{E} + \alpha \frac{\partial \boldsymbol{u}}{\partial t} \times \boldsymbol{h}^0, \qquad \boldsymbol{J} = \sigma \boldsymbol{E} + \sigma\mu \frac{\partial \boldsymbol{u}}{\partial t} \times \boldsymbol{h}^0, \qquad \boldsymbol{B} = \mu \boldsymbol{H}, \\ T = \lambda \cdot \operatorname{tr} S \cdot I + 2\varkappa S, \\ S: \quad S_{ij} = \frac{1}{2}\Big(\frac{\partial u_i}{\partial x_j} + \frac{\partial u_j}{\partial x_i}\Big), \quad i,j = 1,2,3.\end{gathered} \tag{2.1.4}$$

Here $\boldsymbol{h}^0$ is the magnetic intensity vector characterizing the external constant magnetic field. In this case we assume that μ is a positive constant and $\boldsymbol{h}^0$ is a constant nonzero vector. The material properties of a medium are described by the smooth bounded functions

$$\varepsilon, \rho, \lambda, \varkappa : \mathbb{R}^3 \to \mathbb{R}_+, \qquad \sigma : \mathbb{R}^3 \to \overline{\mathbb{R}}_+.$$

For system (2.1.1)–(2.1.4), we consider the Cauchy problem with the initial data

$$(\boldsymbol{H}, \boldsymbol{E}, \boldsymbol{u},) \equiv 0, \quad (x,t) \in \mathbb{R}^3 \times \mathbb{R}_-. \tag{2.1.5}$$

We shall treat the solution of the Cauchy problem (2.1.1)–(2.1.5) as a generalized function defined over the space of infinitely differentiable compactly supported functions. Let us now make a general note about terminology. Usually, system (2.1.1) is supplemented with the two equations

$$\operatorname{div} \boldsymbol{B} = 0, \qquad \operatorname{div} \boldsymbol{D} = \rho_e. \tag{2.1.6}$$

According to the classical theory, it is system (2.1.1), (2.1.6) which is used as a system of Maxwell's equations. However, it is often used to regard system (2.1.1) as an independent object, ignoring equations (2.1.6). This treatment is based on the following reasons. The first equation in (2.1.6) is a direct corollary of (2.1.1); so it is fulfilled for any solution to problem (2.1.1)–(2.1.5). The second equation in (2.1.6) can be naturally considered as an independent equation for determining the charge density ρ_e, which problem is beyond our interest here. At the same time, the electric strength vector $\boldsymbol{E}$ can be found from (2.1.1)–(2.1.5). Thus, equations (2.1.1) are the major and quite independent part of Maxwell's equations.

System (2.1.1)–(2.1.4) arises in result of linearization of a more complicated nonlinear system (1.3.1)–(1.3.4). These equations show that interaction of elastic medium with electromagnetic field is uniliteral interaction for nonconductive media ($\sigma = 0$): the displacement vector $\boldsymbol{u}$ is independent of electromagnetic field, but at the same time the vectors $\boldsymbol{E}, \boldsymbol{H}$ are dependent on the vector $\boldsymbol{u}$. The assumption about weak electrical conductivity of the medium ($\sigma \approx 0$) makes it possible to linearize the original equations (2.1.1)–(2.1.5) with respect to σ calculating the Frechèt derivative on $\sigma = 0$.

Let

$$\boldsymbol{H} = \boldsymbol{H}^0 + \boldsymbol{H}^1, \qquad \boldsymbol{E} = \boldsymbol{E}^0 + \boldsymbol{E}^1, \qquad \boldsymbol{u} = \boldsymbol{u}^0 + \boldsymbol{u}^1, \tag{2.1.7}$$

where $(\boldsymbol{H}^0, \boldsymbol{E}^0, \boldsymbol{u}^0)$ is the solution of the Cauchy problem (2.1.1)–(2.1.5) with $\sigma \equiv 0$:

$$\varepsilon \frac{\partial \boldsymbol{E}^0}{\partial t} + \alpha \frac{\partial^2 \boldsymbol{u}^0}{\partial t^2} \times \boldsymbol{h}^0 + \boldsymbol{j} = \operatorname{rot} \boldsymbol{H}^0, \quad (x,t) \in \mathbb{R}^3 \times \mathbb{R}, \tag{2.1.8}$$

$$\mu \frac{\partial \boldsymbol{H}^0}{\partial t} + \operatorname{rot} \boldsymbol{E}^0 = 0, \quad (x,t) \in \mathbb{R}^3 \times \mathbb{R}, \tag{2.1.9}$$

$$\rho \frac{\partial^2 \boldsymbol{u}^0}{\partial t^2} - \operatorname{Div} T(\boldsymbol{u}^0) = \boldsymbol{f}, \quad (x,t) \in \mathbb{R}^3 \times \mathbb{R}, \tag{2.1.10}$$

$$(\boldsymbol{H}^0, \boldsymbol{E}^0, \boldsymbol{u}^0) \equiv 0, \quad (x,t) \in \mathbb{R}^3 \times \mathbb{R}_-, \tag{2.1.11}$$

Thus, the problem corresponding to linearization with respect to σ in the neighborhood of $\sigma \equiv 0$ has the following form:

$$\varepsilon \frac{\partial \boldsymbol{E}^1}{\partial t} + \alpha \frac{\partial^2 \boldsymbol{u}^1}{\partial t^2} \times \boldsymbol{h}^0 + \sigma \boldsymbol{E}^0 + \mu\sigma \frac{\partial \boldsymbol{u}^0}{\partial t} \times \boldsymbol{h}^0 = \operatorname{rot} \boldsymbol{H}^1, \quad (x,t) \in \mathbb{R}^3 \times \mathbb{R}, \tag{2.1.12}$$

$$\mu \frac{\partial \boldsymbol{H}^1}{\partial t} + \operatorname{rot} \boldsymbol{E}^1 = 0, \quad (x,t) \in \mathbb{R}^3 \times \mathbb{R}, \tag{2.1.13}$$

$$\rho \frac{\partial^2 \boldsymbol{u}^1}{\partial t^2} - \operatorname{Div} T(\boldsymbol{u}^1) - \sigma\mu\Big(\boldsymbol{E}^0 + \mu \frac{\partial \boldsymbol{u}^0}{\partial t} \times \boldsymbol{h}^0\Big) \times \boldsymbol{h}^0 = 0, \quad (x,t) \in \mathbb{R}^3 \times \mathbb{R}, \tag{2.1.14}$$

$$(\boldsymbol{H}^1, \boldsymbol{E}^1, \boldsymbol{u}^1) \equiv 0, \quad (x,t) \in \mathbb{R}^3 \times \mathbb{R}_-, \tag{2.1.15}$$

where as $T(\boldsymbol{u}^k)$, $k = 0, 1$, we marked the value of the stress tensor on the functions $\boldsymbol{u}^k$, $k = 0, 1$, respectively.

Thus, the interaction of electromagnetic field with weakly conductive elastic medium is described by equations (2.1.7)–(2.1.15). Obtained in such way system is the basic subject of our investigation in this section.

Problem Statement and Basic Result

The basic question, which we are going to study here is the structure of the solution to the Cauchy problem (2.1.7)–(2.1.15). First of all we are interested in singular part of the Cauchy problem solution, and as well that regular part, which defined discontinuities on characteristic cones of the solution. The structure of the solution of the Cauchy problem (2.1.7)–(2.1.15) will be studied under the condition that the vector-functions $\boldsymbol{j}$ and $\boldsymbol{f}$ have the form

$$\boldsymbol{j} = \boldsymbol{j}^0 \, \delta(x - x^0, t), \qquad \boldsymbol{f} = \boldsymbol{f}^0 \, \delta(x - x^0, t), \tag{2.1.16}$$

where $\boldsymbol{j}^0$ and $\boldsymbol{f}^0$ are constant vectors and $x^0 \in \mathbb{R}^3$ is a fixed point.

Let c_1, c_2, and c_3 be velocities of electromagnetic, longitudinal and transverse elastic waves, respectively:

$$c_1 = \sqrt{\frac{1}{\varepsilon\mu}}, \qquad c_2 = \sqrt{\frac{\lambda + 2\varkappa}{\rho}}, \qquad c_3 = \sqrt{\frac{\varkappa}{\rho}}. \tag{2.1.17}$$

For each of them we introduce the Riemannian metric with the length element $d\tau_k$ defined by the formula

$$d\tau_k = c_k(x)ds, \quad k = 1, 2, 3, \tag{2.1.18}$$

where ds is the length element in the Euclidean metric. Let us denote by $\Gamma_k(x^0, x)$ the geodesic of metric (2.1.18) connecting the points x^0 and x and by $\tau_k(x^0, x)$ its length. It is well-known that $\tau_k(x^0, x)$ as a function of the variable x satisfies the relations

$$\begin{gathered} |\text{grad}_x\, \tau_k(x^0, x)| = c_k^{-2}(x), \\ \tau_k(x^0, x) = O(|x - x^0|) \quad \text{for} \quad x \to x^0. \end{gathered} \tag{2.1.19}$$

In what follows we will assume that each of metrics (2.1.18) is simple, i.e., each pair of points x^0, x is connected by one and only one geodesic $\Gamma_k(x^0, x)$. In addition, we assume that $c_k(x)$, $k = 1, 2, 3$, satisfy the conditions

$$0 < m_3 \le c_3(x) < c_2(x) < c_1(x) \le M_1 < \infty, \tag{2.1.20}$$

where m_3 and M_1 are constants defined by the formulas

$$m_3 = \inf_{x \in \mathbb{R}^3} c_3, \qquad M_1 = \sup_{x \in \mathbb{R}^3} c_1.$$

Let $\theta_0(t)$ be the Heaviside function:

$$\theta_0(t) = 1 \quad \text{for} \quad t \ge 0, \qquad \theta_0(t) = 0 \quad \text{for} \quad t < 0.$$

Introduce the functions

$$\theta_n(t) = \frac{t^n}{n!}\theta_0(t), \qquad \theta_{-n}(t) = \frac{d^n}{dt^n}\theta_0(t), \quad n = 1, 2, 3, \ldots, \tag{2.1.21}$$

$$S_k = S_k(x, t, x^0) \equiv t - \tau_k(x^0, x), \quad k = 1, 2, 3. \tag{2.1.22}$$

The differentiation in (2.1.21) is understood in the sense of the theory of distributions. The equalities $S_k = 0$, $k = 1, 2, 3$, define the characteristic cones corresponding to the velocities c_k, $k = 1, 2, 3$.

We now introduce new vector functions $\boldsymbol{V}, \boldsymbol{V}^0, \boldsymbol{V}^1$, defined by the formulas

$$\boldsymbol{V} = \boldsymbol{V}^0 + \boldsymbol{V}^1, \ \boldsymbol{V}^k = (\boldsymbol{V}^{1k}, \boldsymbol{V}^{2k}, \boldsymbol{V}^{3k}) \equiv (\boldsymbol{H}^k, \boldsymbol{E}^k, \boldsymbol{u}^k), \ k = 0, 1. \tag{2.1.23}$$

The following theorem holds.

Theorem 2.1.1 *Let there be a certain number $\delta_0 > 0$ such that the coefficients $\varepsilon, \sigma, \rho, \lambda$, and $\varkappa$ are constants in the domain*

$$D_0 = \{x \in \mathbb{R}^3 \mid |x - x^0| < \delta_0\}$$

and belong to $C^m(\mathbb{R}^3)$, $m > 3(N+10)+10$, for a certain integer $N \geq 5$. Then the solution to the Cauchy problem (2.1.7)–(2.1.15) *can be represented in the form*

$$\boldsymbol{V}^{ik}(x,t) = \sum_{n=-2}^{N} \{\boldsymbol{\alpha}_n^{ik}(x)\,\theta_n(S_1) + \boldsymbol{\beta}_n^{ik}(x)\,\theta_n(S_2) + \boldsymbol{\gamma}_n^{ik}(x)\,\theta_n(S_3)\} + \boldsymbol{V}_N^{ik}(x,t), \quad i = 1,2,3; \quad k = 0,1, \quad (2.1.24)$$

where $\boldsymbol{V}_N^{ik}(x,t) \in C^N(\mathbb{R}^3 \times \mathbb{R})$ and the coefficients $\boldsymbol{\alpha}_n^{ik}(x)$, $\boldsymbol{\beta}_n^{ik}(x)$, and $\boldsymbol{\gamma}_n^{ik}(x)$ have the following properties:

a) $\boldsymbol{\alpha}_{-2}^{3k}(x) = \boldsymbol{\beta}_{-2}^{3k}(x) = \boldsymbol{\gamma}_{-2}^{3k}(x) \equiv 0, \qquad k = 0,1,$

b) *$\boldsymbol{\alpha}_n^{ik}(x)$, $\boldsymbol{\beta}_n^{ik}(x)$, and $\boldsymbol{\gamma}_n^{ik}(x)$ are analytic real-valued functions in the domain $D_0 \setminus \{x^0\}$ and smooth outside of D_0 (more precisely, functions of the class $C^{m-2n-8}(\mathbb{R}^3 \setminus D_0)$). Moreover, there exists a positive constant $C > 0$ depending only on the values of the coefficients $\varepsilon, \mu, \sigma, \rho, \lambda$, and $\varkappa$ in the domain D_0 and on the values of $|\boldsymbol{j}^0|$, $|\boldsymbol{f}^0|$, and $|\boldsymbol{h}^0|$ such that*

$$\left(|\boldsymbol{\alpha}_n^{i0}|, |\boldsymbol{\beta}_n^{i0}|, |\boldsymbol{\gamma}_n^{i0}|\right) \leq C \cdot \begin{cases} |x - x^0|^{-(3+n)}, & i = 1,2, \\ |x - x^0|^{-(2+n)}, & i = 3; \end{cases} \quad (2.1.25)$$

$$\left(|\boldsymbol{\alpha}_n^{i1}|, |\boldsymbol{\beta}_n^{i1}|, |\boldsymbol{\gamma}_n^{i1}|\right) \leq C \cdot \begin{cases} |x - x^0|^{-(2+n)}, & i = 1,2, \\ |x - x^0|^{-(1+n)}, & i = 3; \end{cases} \quad (2.1.26)$$

c) *in the domain $\{(x,t) \mid S_3 > 0,\ x \in D_0\}$, the functions*

$$\widetilde{\boldsymbol{V}}_N^{ik}(x,t) = \sum_{n=-2}^{N} \{\boldsymbol{\alpha}_n^{ik}(x)\,\theta_n(S_1) + \boldsymbol{\beta}_n^{ik}(x)\,\theta_n(S_2) + \boldsymbol{\gamma}_n^{ik}(x)\,\theta_n(S_3)\}$$

satisfy the estimates

$$\begin{gathered} \widetilde{\boldsymbol{V}}_N^{10} = \widetilde{\boldsymbol{V}}_N^{30} \equiv 0, \quad |\widetilde{\boldsymbol{V}}_N^{2k}| \leq C\,|x - x^0|^{-3}, \quad k = 0,1, \\ |\widetilde{\boldsymbol{V}}_N^{11}| \leq C\,|x - x^0|^{-2}, \quad |\widetilde{\boldsymbol{V}}_N^{31}| \leq C\,|x - x^0|^{-1} \end{gathered} \quad (2.1.27)$$

with the same constant C as in (2.1.25) *and* (2.1.26).

Remark 2.1.1 *For the correctness of representation* (2.1.24) *at some fixed point x^1 for values $t \leq T$, for any $T > \tau_1(x^0, x^1)$, it is sufficient, by virtue of hyperbolic system* (2.1.7)–(2.1.15), *to suppose the condition of simplicity of the metrics* (2.1.18), *and the conditions* *of smoothness of the coefficients be valid in a finite domain of the space* $\mathbb{R}^3$ *bounded by the Riemannian ellipsoid*

$$\tau_1(x^0, x) + \tau_1(x, x^1) = T.$$

Representation (2.1.24) *is also correct for all the points satisfying the inequalities*

$$\tau_1(x^0, x) \leq t \leq T - \tau_1(x, x^1).$$

Remark 2.1.2 *Apparently, the above theorem is valid without hypothesis about homogeneity of medium in the domain* D_0.

The above theorem is a basis for formulating inverse problems for weakly conducting media.

Sketch of the Proof

The basic idea of the proof of the theorem is the following. We construct for a homogeneous medium the solution to problem (2.1.7)–(2.1.15) having the form of (2.1.24) with $N = 5, \boldsymbol{V}_N^{ik} \equiv 0$. The thus obtained solution is equal to the required one for $t < t_0$, where $t_0 = \delta_0/M_1$, by virtue of the hyperbolic system and an assumption about constancy of the coefficients in the theorem. By this reason we can consider the problem for $t > t_0$ only. Representing for this case the functions $\boldsymbol{V}^{ik}(x, t)$ in the form of (2.1.24) with $N \geq 5$, for $\alpha_n^{ik}, \beta_n^{ik}, \gamma_n^{ik}$ a system of algebraic and ordinary differential equations along geodesic metrics (2.1.18) is obtained. The solution to such equations is selected from the condition of their coincidence with the solution constructed in the domain D_0. As a result of such action we obtain for the functions $\boldsymbol{V}_N^{ik}(x, t)$ the Cauchy problem with zero initial data for $t < t_0$ and a smooth right-hand side. Using the method of energetic estimates enables us attain necessary smoothness of the functions $\boldsymbol{V}_N^{i}(x, t)$. The proof utilizes the methods developed in [17] for investigation of the structure of fundamental solutions to the Cauchy problem for hyperbolic equations.

For the complete proof the reader is referred to the original work [13].

2.2. Initial Boundary-Value Problem for the Equations of Electromagnetoelasticity with Partially Nonlinear Interaction

In this section, following the original work [20], we present some results of solution of the first initial boundary-value problem for the electromagnetoelasticity system in the case when the nonlinear term describing interaction of the electromagnetic and elastic fields is presented in Maxwell's system only.

We will consider one possible statement of the problem which arises in the theory of electromagnetoelasticity under the following assumptions:

1. Ω_1, Ω_2, and Ω are three bounded connected open sets in $\mathbb{R}^3$ such that Ω_2 and Ω belong to the classes C^3 and C^2, respectively, and the following conditions are fulfilled:

$$\overline{\Omega}_2 \subset\subset \Omega, \qquad \Omega_1 = \Omega \setminus \overline{\Omega}_2 . \tag{2.2.1}$$

2. An oscillating inhomogeneous isotropic electrical-conducting elastic body B, which occupies the domain $\Omega_2 \subset \mathbb{R}^3$, is placed into the domain Ω where the process of propagation of electromagnetic waves occurs.

3. The electromagnetic field arises as a result of propagation of elastic oscillations. Moreover, we neglect the reverse effect of the electromagnetic field on the process of elastic waves propagation .

4. We neglect the transport currents in the domain Ω.

5. The motion of the medium occurs with velocities which are lower than those of electromagnetic waves in the elastic medium.

By virtue of the previous assumptions the constitutive relations (1.3.3) take the form

$$\boldsymbol{D} = \varepsilon \boldsymbol{E}, \qquad \boldsymbol{B} = \mu \boldsymbol{H}, \qquad \boldsymbol{J} = \sigma \left[\boldsymbol{E} + \mu \frac{\partial \widetilde{\boldsymbol{u}}}{\partial t} \times \boldsymbol{H} \right], \tag{2.2.2}$$

where $\widetilde{\boldsymbol{u}}$ is an continuation of the function $\boldsymbol{u}$ by zero over the whole of the domain $(0,T) \times \Omega, T > 0$. Using the relations obtained, we can write down the Maxwell system (1.3.1) in the domain $(0,T) \times [\Omega_1 \cup \Omega_2]$ in the form

$$\begin{aligned} &\varepsilon \frac{\partial \boldsymbol{E}}{\partial t} + \sigma \boldsymbol{E} + \sigma \mu \frac{\partial \widetilde{\boldsymbol{u}}}{\partial t} \times \boldsymbol{H} = \operatorname{rot} \boldsymbol{H}, \\ &\mu \frac{\partial \boldsymbol{H}}{\partial t} + \operatorname{rot} \boldsymbol{E} = 0, \qquad \operatorname{div} \mu \boldsymbol{H} = 0. \end{aligned} \tag{2.2.3}$$

According to assumptions 2 and 3, the propagation of elastic waves in the body B is governed by the ordinary system of the Lamé equations

$$\rho \frac{\partial^2 \boldsymbol{u}}{\partial t^2} = \operatorname{Div} T + \boldsymbol{f}, \quad (t,x) \in (0,T) \times \Omega_2, \tag{2.2.4}$$

where $\rho : \Omega_2 \to \mathbb{R}_+$ and $\boldsymbol{f}, \boldsymbol{u} : (0,T) \times \Omega_2 \to \mathbb{R}^3$, and the stress tensor T is defined by the formula (1.3.4) with $\lambda, \varkappa : \Omega_2 \to \mathbb{R}_+$. In this section, we assume the function $\boldsymbol{f}$ to have the following representation:

$$\boldsymbol{f}(t,x) = f(t)\, \boldsymbol{g}(t,x), \tag{2.2.5}$$

where $\boldsymbol{g} : [0,T] \times \Omega_2 \to \mathbb{R}^3$ and $f : [0,T] \to \mathbb{R}$ are known functions.

Our main problem consists in determining the functions $\boldsymbol{E}, \boldsymbol{H}, \boldsymbol{u}$. To this end, we need to supplement differential equations (2.2.3) and (2.2.4) with appropriate initial and boundary conditions and with the gluing conditions for the solution of the problem on the surfaces, where the coefficients of differential equations have breaks.

Now we can formulate the direct problem.

Direct Problem 2.2.1 *Determine a set of the functions*

$$\boldsymbol{u} : [0,T] \times \Omega_2 \to \mathbb{R}^3, \quad \boldsymbol{E}, \boldsymbol{H} : [0,T] \times \Omega \to \mathbb{R}^3,$$

such that

$$\rho\frac{\partial^2 \boldsymbol{u}}{\partial t^2} = \operatorname{Div} T + f(t)\,\boldsymbol{g}(t,x), \quad (t,x) \in (0,T)\times\Omega_2, \tag{2.2.6}$$

$$\boldsymbol{u}(0,x) = \boldsymbol{u}_0(x), \quad x \in \Omega_2, \tag{2.2.7}$$

$$\frac{\partial \boldsymbol{u}}{\partial t}(0,x) = \boldsymbol{u}_1(x), \quad x \in \Omega_2, \tag{2.2.8}$$

$$\boldsymbol{u}(t,x) = 0, \quad (t,x) \in (0,T)\times\partial\Omega_2, \tag{2.2.9}$$

$$\varepsilon\frac{\partial \boldsymbol{E}}{\partial t} + \sigma\boldsymbol{E} + \sigma\mu\frac{\partial \widetilde{\boldsymbol{u}}}{\partial t}\times\boldsymbol{H} = \operatorname{rot}\boldsymbol{H}, \quad (t,x) \in (0,T)\times[\Omega_1\cup\Omega_2], \tag{2.2.10}$$

$$\mu\frac{\partial \boldsymbol{H}}{\partial t} + \operatorname{rot}\boldsymbol{E} = 0, \qquad \operatorname{div}\mu\boldsymbol{H} = 0, \quad (t,x) \in (0,T)\times[\Omega_1\cup\Omega_2], \tag{2.2.11}$$

$$\boldsymbol{E}(0,x) = \boldsymbol{E}_0(x), \quad x \in \Omega, \tag{2.2.12}$$

$$\boldsymbol{H}(0,x) = \boldsymbol{H}_0(x), \quad x \in \Omega, \tag{2.2.13}$$

$$\boldsymbol{n}\times\boldsymbol{E} = 0, \quad (t,x) \in (0,T)\times\partial\Omega, \tag{2.2.14}$$

$$[\boldsymbol{E}\times\boldsymbol{n}]_{\partial\Omega_2} = [\boldsymbol{H}\times\boldsymbol{n}]_{\partial\Omega_2} = 0, \quad (t,x) \in (0,T)\times\partial\Omega_2, \tag{2.2.15}$$

It is assumed that the functions $\varepsilon, \mu : \overline{\Omega} \to \mathbb{R}_+$, $\sigma : \overline{\Omega} \to \overline{\mathbb{R}}_+$, and $\boldsymbol{E}_0, \boldsymbol{H}_0 : \Omega \to \mathbb{R}^3$ are continuous in the domain $\overline{\Omega} \setminus \partial\Omega_2$ with possible jumps on the surface $\partial\Omega_2$. We also assume that the functions $f : [0,T] \to \mathbb{R}$, $\boldsymbol{g} : [0,T]\times\Omega_2 \to \mathbb{R}^3$, $\boldsymbol{u}_0, \boldsymbol{u}_1 : \Omega_2 \to \mathbb{R}^3$ are given and have sufficient smoothness.

As is easy to see, the solution of Direct Problem 2.2.1 can be divided into two steps because we can separate the solution of problem (2.2.6)–(2.2.9) from the solution of problem (2.2.10)–(2.2.15) (the coupling term is presented in Maxwell's system only). For this reason we divide this process into two ones: the first - to solve the direct problem (2.2.6)–(2.2.9), and the second - to solve the direct problem (2.2.10)–(2.2.15).

Solution of Direct Problem (2.2.6)–(2.2.9)

It is worth noting that the term $\partial\widetilde{\boldsymbol{u}}/\partial t \times \boldsymbol{H}$ occurring in equation (2.2.10) creates certain difficulties in the course of solution of direct problem (2.2.6)–(2.2.9). In fact, we cannot confine ourselves to consideration of the weak solution $(\boldsymbol{E}, \boldsymbol{H}, \boldsymbol{u})$ of the problem since in this case the product $\partial\widetilde{\boldsymbol{u}}/\partial t \times \boldsymbol{H}$ may fail to be an element of the space $L^2(\Omega_2;\mathbb{R}^3)$. In order that imposing too severe constraints on the function be avoided we should require at least, that both the multipliers $\partial\widetilde{\boldsymbol{u}}/\partial t$ and $\boldsymbol{H}$ be elements of the space $L^4(\Omega_2;\mathbb{R}^3)$. For this, we assume the density ρ, the Lamé coefficients λ and $\varkappa$, the free term $f\boldsymbol{g}$, and the initial data $\boldsymbol{u}_0$ and $\boldsymbol{u}_1$ of problem (2.2.6)–(2.2.9) to satisfy the following conditions:

$$\rho \in H^2(\Omega_2;\mathbb{R}), \qquad \varkappa, \lambda \in W^{2,\infty}(\Omega_2;\mathbb{R}), \tag{2.2.16}$$

$$\min\big(\rho(x), \lambda(x), \varkappa(x)\big) \geq \rho_0 > 0, \quad \forall x \in \overline{\Omega_2}, \tag{2.2.17}$$

$$f \in L^p((0,T);\mathbb{R}), \quad \boldsymbol{g} \in L^{p'}\big((0,T); H^2(\Omega_2;\mathbb{R}^3)\big) \cap H_0^1(\Omega_2;\mathbb{R}^3) \\ \Big(\frac{1}{p} + \frac{1}{p'} = 1\Big), \tag{2.2.18}$$

$$\boldsymbol{U}_0 \in H^3(\Omega_2;\mathbb{R}^3) \cap H_0^1(\Omega_2;\mathbb{R}^3), \qquad \operatorname{Div} T \in H_0^1(\Omega_2;\mathbb{R}^3), \tag{2.2.19}$$

$$\boldsymbol{U}_1 \in H^2(\Omega_2;\mathbb{R}^3) \cap H_0^1(\Omega_2;\mathbb{R}^3). \tag{2.2.20}$$

Theorem 2.2.1 *Let ρ, λ, $\varkappa$, f, $\boldsymbol{g}$, $\boldsymbol{u}_0$, and $\boldsymbol{u}_1$ satisfy conditions* (2.2.16)–(2.2.20). *Then there exists a unique solution $\mathbf{u}(f)$ to problem* (2.2.6)–(2.2.9)*. This solution satisfies the conditions*

$$\boldsymbol{u}(f) \in C\big([0,T]; H^3(\Omega_2;\mathbb{R}^3)\big) \cap H_0^1(\Omega_2;\mathbb{R}^3) \cap C^1\big([0,T]; H^2(\Omega_2;\mathbb{R}^3)\big) \\ \cap W^{2,1}\big((0,T); H^1(\Omega_2;\mathbb{R}^3)\big), \tag{2.2.21}$$

$$\Big(\|\boldsymbol{u}(f)(t)\|_{3,2}^2 + \Big\|\frac{\partial}{\partial t}\boldsymbol{u}(f)(t)\Big\|_{2,2}^2 + \Big\|\frac{\partial^2}{\partial t^2}\boldsymbol{u}(f)(t)\Big\|_{1,2}^2\Big)^{1/2} \\ \leq C_1\big(\rho_0^{-1}, \|\rho\|_{2,2}, \|\varkappa\|_{2,\infty}, \|\lambda\|_{2,\infty}\big) \\ \times \Big[\big(\|\boldsymbol{u}_0\|_{3,2}^2 + \|\boldsymbol{u}_1\|_{2,2}^2\big)^{1/2} + \|\boldsymbol{g}\|_{t,0,p',2,2} \cdot \|f\|_{t,0,p}\Big], \\ \forall t \in [0,T], \quad \forall f \in L^p((0,T);\mathbb{R}), \tag{2.2.22}$$

$$\Big(\|\boldsymbol{u}(f_2)(t) - \boldsymbol{u}(f_1)(t)\|_{3,2}^2 + \Big\|\frac{\partial}{\partial t}\boldsymbol{u}(f_2)(t) - \frac{\partial}{\partial t}\boldsymbol{u}(f_1)(t)\Big\|_{2,2}^2 \\ + \Big\|\frac{\partial^2}{\partial t^2}\boldsymbol{u}(f_2)(t) - \frac{\partial^2}{\partial t^2}\boldsymbol{u}(f_1)(t)\Big\|_{1,2}^2\Big)^{1/2} \\ \leq C_1\big(\rho_0^{-1}, \|\rho\|_{2,2}, \|\varkappa\|_{2,\infty}, \|\lambda\|_{2,\infty}\big)\, \|\boldsymbol{g}\|_{t,0,p',2,2} \cdot \|f_2 - f_1\|_{t,0,p}, \\ \forall t \in [0,T], \quad \forall f_1, f_2 \in L^p((0,T);\mathbb{R}). \tag{2.2.23}$$

Here $\|\cdot\|_{j,2}$, $\|\cdot\|_{t,0,q,j,2}$, and $\|\cdot\|_{t,0,p}$ are the norms in the spaces $H^j(\Omega_2;\mathbb{R}^3)$, $L^q((0,t); H^j(\Omega_2;\mathbb{R}^3))$, and $L^p((0,t);\mathbb{R})$, respectively, and C_1 is a non-negative function continuous and nondecreasing in each of its arguments.

Solution of direct problem (2.2.10)–(2.2.15)

Now we are able to solve direct problem (2.2.10)–(2.2.15). We make the following assumptions about the coefficients and initial data of the problem:

$$\begin{gathered}\varepsilon, \mu, \sigma \in W^{1,\infty}(\Omega_1;\mathbb{R}) \cap W^{1,\infty}(\Omega_2;\mathbb{R}),\\ \min\left\{\varepsilon(x), \mu(x)\right\} \geq \gamma^{-1} > 0, \quad \forall x \in \Omega_1 \cup \Omega_2;\end{gathered} \tag{2.2.24}$$

$$\begin{gathered}\boldsymbol{E}_0 \in H(\mathrm{rot}, \Omega), \quad \boldsymbol{H}_0 \in H(\mathrm{rot}, \Omega) \cap H^1(\Omega_1;\mathbb{R}^3) \cap H^1(\Omega_2;\mathbb{R}^3),\\ \mu \boldsymbol{H}_0 \in H(\mathrm{div};\Omega);\end{gathered} \tag{2.2.25}$$

$$\begin{gathered}\boldsymbol{n} \times \boldsymbol{E}_0 = 0, \quad x \in \partial\Omega; \qquad \mathrm{div}\, \mu \boldsymbol{H}_0 = 0, \quad x \in \Omega;\\ \boldsymbol{n} \cdot \boldsymbol{H}_0 = 0, \quad x \in \partial\Omega.\end{gathered} \tag{2.2.26}$$

The following theorem is valid under these assumptions.

Theorem 2.2.2 *Let the vector-functions* $\boldsymbol{g}$, $\boldsymbol{u}_0$, $\boldsymbol{u}_1$, $\boldsymbol{E}_0$, *and* $\boldsymbol{H}_0$ *satisfy conditions* (2.2.18)–(2.2.20), (2.2.25), *and* (2.2.26). *Then for every function* $f \in L^p((0,T);\mathbb{R})$ *problem* (2.2.10)–(2.2.15) *has a unique solution* $(\boldsymbol{E}, \boldsymbol{H}) = (\boldsymbol{E}(f), \boldsymbol{H}(f))$ *satisfying the conditions*

$$\boldsymbol{E}(f) \in C\big([0,T]; H(\mathrm{rot};\Omega)\big) \cap C^1\big([0,T]; L^2(\Omega;\mathbb{R}^3)\big), \tag{2.2.27}$$

$$\boldsymbol{H}(f) \in C\big([0,T]; H(\mathrm{rot};\Omega) \cap H^1(\Omega_2;\mathbb{R}^3)\big) \cap C^1\big([0,T]; L^2(\Omega;\mathbb{R}^3)\big), \tag{2.2.28}$$

$$\begin{aligned}&\left\|\frac{\partial}{\partial t}\boldsymbol{E}(f)(t)\right\|_{0,2,\Omega} + \left\|\frac{\partial}{\partial t}\boldsymbol{H}(f)(t)\right\|_{0,2,\Omega} + \|\mathrm{rot}\, \boldsymbol{E}(f)(t)\|_{0,2,\Omega}\\ &\quad + \|\boldsymbol{E}(f)(t)\|_{0,2,\Omega} + \|\boldsymbol{H}(f)(t)\|_{1,2,\Omega_1} + \|\boldsymbol{H}(f)(t)\|_{1,2,\Omega_2}\\ &\qquad \leq C_2(T) + TC_3(T, \|f\|_{T,0,p}), \quad \forall t \in (0,T),\end{aligned} \tag{2.2.29}$$

where C_2 *and* C_3 *are positive nondecreasing continuous functions depending also on the norms of the data of the problem. Moreover, for every pair of the functions* $f_1, f_2 \in L^p((0,T);\mathbb{R})$ *the following estimate holds:*

$$\begin{aligned}&\left\|\frac{\partial}{\partial t}\boldsymbol{E}(f_2)(t) - \frac{\partial}{\partial t}\boldsymbol{E}(f_1)(t)\right\|_{0,2,\Omega} + \left\|\frac{\partial}{\partial t}\boldsymbol{H}(f_2)(t) - \frac{\partial}{\partial t}\boldsymbol{H}(f_1)(t)\right\|_{0,2,\Omega}\\ &\quad + \|\mathrm{rot}\, \boldsymbol{E}(f_2)(t) - \mathrm{rot}\, \boldsymbol{E}(f_1)(t)\|_{0,2,\Omega} + \|\boldsymbol{E}(f_2)(t) - \boldsymbol{E}(f_1)(t)\|_{0,2,\Omega}\\ &\quad + \|\boldsymbol{H}(f_2)(t) - \boldsymbol{H}(f_1)(t)\|_{1,2,\Omega_1} + \|\boldsymbol{H}(f_2)(t) - \boldsymbol{H}(f_1)(t)\|_{1,2,\Omega_2}\\ &\qquad \leq C_4(T, \|f_1\|_{T,0,p}) \int_0^t h(f_2)(t-s)\, \|f_2 - f_1\|_{s,0,p} ds,\end{aligned} \tag{2.2.30}$$

where

$$h(f)(t) = \exp\left[t\left(\gamma\, \|\sigma\|_{0,\infty,\Omega} + C_5(T)\, \|f\|_{T,0,p} \cdot \|g\|_{T,0,p',2,2,\Omega_2}\right)\right] \tag{2.2.31}$$

and C_4 *and* C_5 *are positive nondecreasing continuous functions depending also on the norms of the data of the problem.*

The proofs of Theorems 2.2.1 and 2.2.2 are rather bulky and therefore are omitted. For their complete proofs the reader is referred to the original paper [20].

2.3. Initial Boundary-Value Problem for the Electromagnetoelasticity Equations with Complete Nonlinear Interaction

Following basically to the original work [15], we present some results of the solution of the first initial boundary-value problem for a system of electromagnetoelasticity, when nonlinear terms describing the interaction of electromagnetic and elastic fields are presented in both Maxwell's and Lamè systems. The results represented in this section are quite new and by this reason there will be done their complete demonstration.

Basic Equations

Consider the case of diffusion approximation of Maxwell's system. This means that in field equations (1.3.1)–(1.3.2) we neglect by displacement current $\frac{\partial \boldsymbol{D}}{\partial t}$ formally assuming $\epsilon = 0$, and set $\rho_e = 0$. Simultaneously we put in constitutive equations (1.3.3)–(1.3.4) $\alpha = 0$ and $\rho_e = 0$, too. It is easy to show that in this case in the presence of external electromagnetic $\boldsymbol{j}$ and elastic $\boldsymbol{f}$ sources of oscillations we can form the following electromagnetoelasticity system

$$\begin{aligned} \sigma \boldsymbol{E} + \sigma\mu \frac{\partial \boldsymbol{u}}{\partial t} \times \boldsymbol{H} + \boldsymbol{j} &= \operatorname{rot} \boldsymbol{H}, \\ \mu \frac{\partial \boldsymbol{H}}{\partial t} + \operatorname{rot} \boldsymbol{E} = 0, &\qquad \operatorname{div} \mu \boldsymbol{H} = 0, \end{aligned} \tag{2.3.1}$$

$$\rho \frac{\partial^2 \boldsymbol{u}}{\partial t^2} = \operatorname{Div} T + \mu \operatorname{rot} \boldsymbol{H} \times \boldsymbol{H} + \boldsymbol{f}. \tag{2.3.2}$$

We make the following assumptions about functions $\boldsymbol{E}, \boldsymbol{H}, \boldsymbol{u}, \boldsymbol{j}, \boldsymbol{f}$:

$$\begin{aligned} \boldsymbol{E} = (0,1,0)E(z,t), \quad \boldsymbol{H} = (1,0,0)H(z,t), \quad \boldsymbol{u} = (0,0,1)u(z,t) \\ \boldsymbol{j} = (0,1,0)j(z,t), \qquad \boldsymbol{f} = (0,0,1)f(z,t), \end{aligned} \tag{2.3.3}$$

where the variable z stands to the variable x_3. Under such assumptions for the case $\rho = const$, $\mu = const$ we can form the following non-dimensional model system

$$h_t = (rh_z)_z - (hu_t)_z - (rj)_z, \tag{2.3.4}$$

$$u_{tt} = (\nu^2 u_z)_z - phh_z + f, \tag{2.3.5}$$

where h, u, j, f are dimensionless analogues of the functions introduced by formulas (2.3.3), $r^{-1} = \mu L V_0 \sigma$ is the magnetic Reynolds number, $p = \mu H_0^2 \rho^{-1} V_0^{-2}$, $\nu = \sqrt{(\lambda + 2\varkappa)/\rho V_0^2}$ is dimensionless velocity of the elastic waves propagation; and L, V_0, H_0 are characteristic values of length, seismic velocity and magnetic field, respectively.

The Problem Statement. Weak Solutions

Now we can formulate the direct problem. Let be $Q_T = (t,z) : t \in (0,T) \times \Omega$, where $\Omega = (-l, l)$.

Direct Problem 2.3.1 *Determine a set of the functions*

$$h : \overline{Q}_T \to \mathcal{R}, \quad u : \overline{Q}_T \to \mathcal{R}$$

such that

$$h_t = (rh_z)_z - (hu_t)_z - (rj)_z, \quad (z,t) \in Q_T, \tag{2.3.6}$$

$$u_{tt} = (\nu^2 u_z)_z - phh_z + f, \quad (z,t) \in Q_T, \tag{2.3.7}$$

$$h(z,0) = h_0(z), \quad z \in \Omega,, \tag{2.3.8}$$

$$u(z,0) = u_0(z), \quad u_t(z,0) = u_1(z), \quad z \in \Omega, \tag{2.3.9}$$

$$h(\pm l,t) = 0, \quad t \in (0,T), \tag{2.3.10}$$

$$u(\pm l,t) = 0, \quad t \in (0,T). \tag{2.3.11}$$

Here $r(z), \nu(z)$ are positive piecewise smooth functions and $j(z,t), f(z,t)$ are piecewise smooth functions, discontinuous at the points $z = z_k, k = 1, 2, \ldots, m, -l < z_1 < z_2 < \cdots < z_m < l$; p is a positive number.

Direct Problem 2.3.1 can be considered as a diffraction problem, i.e., as the problem with Q_T partitioned into several domains $Q_T^{(k)}, Q_T^{(k)} = \Omega^{(k)} \times (0,T), \Omega^{(k)} = (z_k, z_{k+1}), k = 0, 1, \ldots, m, z_0 = -l, z_{m+1} = l$, in each of which there is given parabolic-hyperbolic system (2.3.6)–(2.3.7) with smooth coefficients and free terms. We wish to find in $\overline{Q}_T$ a solution of this system satisfying:

- in $Q_T^{(k)}, k = 1, 2, \ldots, m$, the corresponding equations (2.3.6)–(2.3.7);
- on the lower base of Q_T the initial conditions (2.3.8)–(2.3.9);
- on the lateral surface of Q_T the boundary conditions (2.3.10)–(2.3.11);
- at the jump points $z_k, k = 1, 2, \ldots, m$, the following compatibility conditions

$$[h] = [u] = 0, \tag{2.3.12}$$

$$[r(h_z - j)] = \left[\nu^2 u_z\right] = 0\,. \tag{2.3.13}$$

The symbol $[v]$ denotes the jump of the function v as it passes through z_k.

Problems of this type can be reduced by means of a simple technique to problems for the determination of weak (generalized) solutions of ordinary initial boundary-value problems with discontinuous coefficients, see [21, pp. 224-232]. This fact will be used for the analysis of Direct Problem 2.3.1.

Direct Problem 2.3.1 can be formulated in weak terms defining its solutions as follows

Definition 2.3.1 *Functions* $h(z,t) \in \overset{o}{V}_2 (Q_T), u(z,t) \in \overset{o}{W}{}_2^{1,1} (Q_T)$ *are called a weak solution of Direct Problem 2.3.1 if they satisfy the identities*

$$-\int_{Q_T} h\eta_t dzdt + \int_{Q_T} rh_z\eta_z dzdt - \int_{Q_T} hu_t\eta_z dzdt \tag{2.3.14}$$
$$= \int_{Q_T} rj\eta_z dzdt + \int_\Omega h_0(z)\eta(z,0)dz,$$

$$-\int_{Q_T} u_t\zeta_t dzdt + \int_{Q_T} \nu^2 u_z\zeta_z dzdt + \int_{Q_T} phh_z\zeta dzdt \tag{2.3.15}$$
$$= \int_{Q_T} f\zeta dzdt + \int_\Omega u_1(z)\zeta(z,0)dz, \quad u(z,0) = u_0(z),\ z \in \Omega,$$

for all $\eta(z,t), \zeta(z,t)$ *from* $\overset{o}{W}{}_2^{1,1}(Q_T)$ *that are equal to zero for t=T.*

It is possible to define the weak solution of (2.3.6)–(2.3.13) somewhat differently.

Definition 2.3.2 *Functions* $h(z,t) \in \overset{o}{V}_2(Q_T), u(z,t) \in \overset{o}{W}{}_2^{1,1}(Q_T)$ *are called a weak solution of Direct Problem 2.3.1 if they satisfy for almost all* $t_1 \in [0,T]$ *the identities*

$$-\int_{Q_{t_1}} h\eta_t dzdt + \int_{Q_{t_1}} rh_z\eta_z dzdt - \int_{Q_{t_1}} hu_t\eta_z dzdt \tag{2.3.16}$$

$$= \int_{Q_{t_1}} rj\eta_z dzdt + \int_\Omega h_0(z)\eta(z,0)dz - \int_\Omega h(z,t_1)\eta(z,t_1)dz,$$

$$-\int_{Q_{t_1}} u_t\zeta_t dzdt + \int_{Q_{t_1}} \nu^2 u_z\zeta_z dzdt + \int_{Q_{t_1}} phh_z\zeta dzdt \tag{2.3.17}$$

$$= \int_{Q_{t_1}} f\zeta dzdt + \int_\Omega u_1(z)\zeta(z,0)dz - \int_\Omega u_t(z,t_1)\zeta(z,t_1)dz,$$

$u(z,0) = u_0(z), z \in \Omega$, *where* $Q_{t_1} = \Omega \times (0,t_1)$, *and* $\eta(z,t), \zeta(z,t) \in \overset{o}{W}{}_2^{1,1}(Q_T)$.

Both of the definitions are equivalent to each other, see [14]. The fulfillment of transmission conditions (2.3.13) is understand in the sense of the identities considered in Definition 2.3.1. We follow Ladyzhenskaia, Solonnikov and Uralceva, [21], when defining the functional spaces.

Main Results

Suppose that the functions r, ν, j, f, the constant p and the initial data h_0, u_0, u_1 in Direct Problem 2.3.1 enjoy the properties

(a) r, ν, j, f are supposed to be piecewise smooth functions with jumps at the points $z_m : -l < z_1 < z_2 < \cdots < z_m < l$; $0 < r_0 \le r(z) \le r_1 < \infty$, $0 < \nu_0 \le \nu(z) \le \nu_1 < \infty$ and p is a positive number;

(b) $h_0 \in C^\alpha(\overline{\Omega}), \alpha \in (0,1), h_0(\pm l) = 0$, and $u_0 \in \overset{o}{W}{}_2^{1}(\Omega), u_1 \in L_2(\Omega)$.

Let us show that Direct Problem 2.3.1 is solvable; namely, let us prove the following proposition.

Theorem 2.3.1 *If conditions* $(a) - (b)$ *are fulfilled, then Direct Problem 2.3.1 has a weak solution*

$$h(z,t) \in \overset{o}{V}_2(Q_T), \qquad u(z,t) \in \overset{o}{W}{}_2^{1,1}(Q_T).$$

Proof For a proof we make use of Faedo-Galerkin's method. Let us check that the functions h, u satisfy identities (2.3.16)–(2.3.17). Suppose $\{\psi_k(z)\}_{k=1}^\infty$ is a set of generalized eigenfunctions of the problem

$$(r\psi_{kz})_z = \lambda_k\psi_k, \tag{2.3.18}$$

$$\psi_k(\pm l) = 0, \quad [\psi_k]_{z=z_i} = [r\psi_{kz}]_{z=z_i} = 0, \quad i = 1, \ldots, m, \tag{2.3.19}$$

such that $\|\psi_k\|_{L_2(\Omega)} = 1$. Repeating reasoning giving in Chapter VIII of [23] we can prove that the set $\{\psi_k(z)\}_{k=1}^{\infty}$ is complete and orthogonal in $L_2(\Omega)$. The set $\{\psi_k(z)\}_{k=1}^{\infty}$ is complete and orthogonal in $\overset{o}{W}{}_2^1(\Omega)$ with a scalar product defined by formula

$$(u, v)_1 = \int_{\Omega} (r u_z v_z + uv) dz.$$

The eigenfunctions $\psi_k, k = 1, 2, \dots,$ are simple and admit (2.3.18)–(2.3.19) in the classical sense.

Any function $g(z) \in L_2(\Omega)$ can be expanded in a converged in $L_2(\Omega)$ Fourier series

$$g = \sum_{k=1}^{\infty} g_k \psi_k, \quad g_k = (g_0, \psi_k).$$

The following inequality is valid

$$\|\sum_{k=1}^{N} g_k \psi_k\|_2 = \|g^N\|_2 \leq \|g\|_2.$$

Additionally suppose $g \in \overset{o}{W}{}_2^1(\Omega)$. The $\{g^N\}_{N=1}^{\infty}$ is converged in $\overset{o}{W}{}_2^1(\Omega)$ and the following inequality is valid, see [23]

$$\|g^N\|_{\overset{o}{W}{}_2^1(\Omega)} \leq \|g\|_{\overset{o}{W}{}_2^1(\Omega)}.$$

The same properties has the set $\{\phi_k(z)\}_{k=1}^{\infty}$, where ϕ_k are the generalized eigenfunctions of the problem

$$(\nu^2 \phi_{kz})_z = \mu_k \phi_k, \tag{2.3.20}$$

$$\phi_k(\pm l) = 0, \qquad [\phi_k]_{z=z_i} = [\nu^2 \phi_{kz}]_{z=z_i} = 0, \qquad i = 1, \dots, m. \tag{2.3.21}$$

We will seek an approximating solution in the form

$$h^N(z,t) = \sum_{k=1}^{N} a_k^N(t)\psi_k(z) + J(z,t) \equiv \widetilde{h}^N(z,t) + J(z,t), \tag{2.3.22}$$

$$u^N(z,t) = \sum_{k=1}^{N} b_k^N(t)\phi_k(z), \tag{2.3.23}$$

where $J_z' = j$, $J(0,t) = 0$. Functions $a_k^N = (\widetilde{h}^N, \psi_k)$, $b_k^N = (u^N, \phi_k)$, $k = 1, \dots, N$, are determined from the equations

$$(h^N, \psi_k)_t = (-r h_z^N + u_t^N h^N + rj, \psi_{kz}), \tag{2.3.24}$$

$$(h^N(\cdot, 0), \psi_k) = h_{0k},$$

$$(u^N, \phi_k)_{tt} = (-\nu^2 u_z^N + \frac{p}{2}(h^N)^2, \phi_{kz}) + (f, \phi_k), \tag{2.3.25}$$

$$(u^N(\cdot, 0), \phi_k) = u_{0k}, \quad (u^N(\cdot, 0), \phi_k)_t = u_{1k},$$

and h_{0k}, u_{0k}, u_{1k} are the Fourier coefficients in $L_2(\Omega)$ of the corresponding initial data with respect to the systems of functions $\{\psi_k(z)\}_{k=1}^{\infty}$ and $\{\phi_k(z)\}_{k=1}^{\infty}$. Equations (2.3.24)–(2.3.25) are a system of nonlinear ordinary differential equations. Its solution exists on a maximal interval $[0, \tau)$ if $\tau \le T$, and

$$\lim_{t\to\tau} \max_k(|a_k^N|, |b_k^N|) \to \infty\ .$$

The function

$$\widetilde{h}^N(z,0) = \sum_{k=1}^{N} a_k^N(0)\psi_k(z) = \widetilde{h}_0^N(z) = \sum_{k=1}^{N} h_{0k}\psi_k(z)$$

is the orthogonal projection of $h_0 - J(\cdot,0)$ on the span$\{\psi_k\}_{k=1}^{k=N}$. At that the following inequality takes place

$$\|\widetilde{h}^N(\cdot,0)\|_{2,\Omega} \le \|h_0 - J(\cdot,0)\|_{2,\Omega} \le \|h_0\|_{2,\Omega} + \|J(\cdot,0)\|_{2,\Omega} \le C_1,$$

where the positive constant C_1 does not depend on N.

Analogously we have

$$u^N(z,0) = \sum_{k=1}^{N} b_k^N(0)\phi_k(z) = u_0^N(z) = \sum_{k=1}^{N} u_{0k}\phi_k(z).$$

At that the $\{u_0^N\}_{N=1}^{\infty}$ is converged because of $u_0 \in \overset{o}{W}{}_2^1(\Omega)$, see [22]. It is easy to show that there is valid the following estimate

$$\|u_0^N\|_{\overset{o}{W}{}_2^1(\Omega)} \le \|u_0\|_{\overset{o}{W}{}_2^1(\Omega)}.$$

To prove the latter inequality we used that the $\{\phi_k\}_{k=1}^{\infty}$ is an orthogonal basis in $\overset{o}{W}{}_2^1(\Omega)$ and u_0^N is the orthogonal projection of u_0 on the span$\{\phi_k\}_{k=1}^{k=N}$.

Analogously we have

$$u_t^N(z,0) = \sum_{k=1}^{N} b_{kt}^N(0)\phi_k(z) = u_1^N(z) = \sum_{k=1}^{N} u_{1k}\phi_k(z).$$

There is valid the following estimate

$$\|u_1^N\|_{2,\Omega} \le \|u_1\|_{2,\Omega}.$$

Using (2.3.18) we can represent problem (2.3.24) in a equivalent form

$$\begin{aligned} (\widetilde{h}^N, \psi_k)_t &= (-r\widetilde{h}_z^N + u_t^N(\widetilde{h}^N + J), \psi_{kz}) - (J_t, \psi_k), \\ (\widetilde{h}^N(\cdot,0), \psi_k) &= h_{0k} - (J(\cdot,0), \psi_k), \end{aligned} \tag{2.3.26}$$

We will prove that $|a_k^N|, |b_k^N|, k = 1, \dots, N$, are bounded functions for $t \in [0,T]$, and therefore system (2.3.24)–(2.3.25) will have a unique solution on $[0,T]$ for any $T > 0$. For this purpose we multiply the differential equation in (2.3.26) by pa_k^N and the differential equation in (2.3.25) - by b_{kt}^N, then sum the obtained equalities over all k from 1 to N; and integrate the result with respect to t from 0 to $t_1, 0 < t_1 \leq T$. The sum of the results obtained gives

$$\frac{p}{2}\|\widetilde{h}^N(\cdot,t)\|^2_{2,\Omega}\Big|_{t=0}^{t=t_1} + \frac{1}{2}\|u_t^N(\cdot,t)\|^2_{2,\Omega}\Big|_{t=0}^{t=t_1} + \frac{1}{2}\|\nu u_z^N(\cdot,t)\|^2_{2,\Omega}\Big|_{t=0}^{t=t_1}$$
$$+ \|\sqrt{pr}\widetilde{h}_z^N\|^2_{2,Q_{t_1}} = -\int_{Q_{t_1}} pj(\widetilde{h}^N + J)u_t^N dzdt + \int_{Q_{t_1}} fu_t^N dzdt - \int_{Q_{t_1}} J_t\widetilde{h}^N dzdt$$
$$= -\int_{Q_{t_1}} pj\widetilde{h}^N u_t^N dzdt - \int_{Q_{t_1}} J_t\widetilde{h}^N dzdt + \int_{Q_{t_1}} f_1 u_t^N dzdt, \quad (2.3.27)$$

where $f_1 = f - pjJ$. Note that

$$\|\widetilde{h}^N(\cdot,0)\|^2_{2,\Omega} = \sum_{k=1}^N a_k^2(0) \leq \|\widetilde{h}_0\|^2_{2,\Omega} \leq 2(\|h_0\|^2_{2,\Omega} + \|J(\cdot,0)\|^2_{2,\Omega});$$
$$\|u_t^N(\cdot,0)\|^2_{2,\Omega} = \sum_{k=1}^N b_{kt}^2(0) \leq \|u_1\|^2_{2,\Omega};\ \|u^N(\cdot,0)\|^2_{\overset{o}{W}{}^1_2(\Omega)} \leq \|u_0\|^2_{\overset{o}{W}{}^1_2(\Omega)};$$
$$\|\nu u_z^N(\cdot,0)\|^2_{2,\Omega} \leq C_2\nu_1^2\|u_{0z}\|^2_{2,\Omega},$$

where the positive constant C_2 does not depend on N. Therefore the following inequality is valid

$$\frac{p}{2}\|h^N(\cdot,t_1)\|^2_{2,\Omega} + \frac{1}{2}\|u_t^N(\cdot,t_1)\|^2_{2,\Omega} + \frac{1}{2}\|\nu u_z^N(\cdot,t_1)\|^2_{2,\Omega} + \frac{1}{2}\|\sqrt{pr}h_z^N\|^2_{2,Q_{t_1}}$$
$$\leq C_3 + p\int_{Q_{t_1}} |j\widetilde{h}^N u_t^N|dzdt + \int_{Q_{t_1}} |fu_t^N|dzdt + p\int_{Q_{t_1}} |J_t\widetilde{h}^N|dzdt, \quad (2.3.28)$$

where the positive constant

$$C_3 = \frac{p}{2}\|h_0\|^2_{2,\Omega} + \frac{1}{2}\|u_1\|^2_{2,\Omega} + \frac{C_2\nu_1^2}{2}\|u_{0z}\|^2_{2,\Omega} + \frac{1}{2}\int_{Q_T} prj^2 dz\,dt$$

does not depend on N. The first integral in the right-hand side of (2.3.28) can be estimated in the following way

$$p\int_{Q_{t_1}} |j\widetilde{h}^N u_t^N|dz\,dt \leq p\int_0^{t_1} \|\widetilde{h}^N(\cdot,\tau)\|^2_{2,\Omega}dt + C_4\int_0^{t_1} \|u_t^N(\cdot,\tau)\|^2_{2,\Omega}\,dt \quad (2.3.29)$$

where the positive constant C_4 does not depend on N, t_1. So, from (2.3.28) we have

$$p\|\widetilde{h}^N(\cdot,t_1)\|^2_{2,\Omega} + \|u_t^N(\cdot,t_1)\|^2_{2,\Omega} \leq$$
$$\leq C_5 + C_6\int_0^{t_1} [p\|\widetilde{h}^N(\cdot,\tau)\|^2_{2,\Omega} + \|u_t^N(\cdot,\tau)\|^2_{2,\Omega}]d\tau\,,$$

where the positive constants C_5, C_6 do not depend on N, t_1. The latter and Gronwall's inequality give us the estimate

$$p\|\widetilde{h}^N(\cdot,t_1)\|^2_{2,\Omega} + \|u^N_t(\cdot,t_1)\|^2_{2,\Omega} \le C_7 \qquad \forall t_1 \in [0,T], \tag{2.3.30}$$

where the positive constant C_7 does not depend on N. It means that

$$p\int_{Q_{t_1}} |j\widetilde{h}^N u^N_t| dzdt \le C_7(1+C_4)T.$$

Finally from the latter inequality and (2.3.28)–(2.3.30) we obtain the following estimate

$$\begin{aligned} p\|\widetilde{h}^N(\cdot,t_1)\|^2_{2,\Omega} + \|u^N_t(\cdot,t_1)\|^2_{2,\Omega} + \\ + \|\nu u^N_z(\cdot,t_1)\|^2_{2,\Omega} + \|\sqrt{pr}\widetilde{h}^N_z\|^2_{2,Q_{t_1}} \le C_8 \qquad \forall t_1 \in [0,T] \end{aligned} \tag{2.3.31}$$

where the positive constant C_8 does not depend on N. It follows from (2.3.31) that a^N_k, b^N_k, b^N_{kt} are uniformly bounded functions for $t \in [0,T]$. Let us show that for fixed k and arbitrary $N \ge k$ they are equicontinuous on $[0,T]$. Indeed, from (2.3.26) we have

$$a^N_k(t+\triangle t) - a^N_k(t) = -\int_{Q_{t,t+\triangle t}} (r\widetilde{h}^N_z - u^N_t(\widetilde{h}^N + J))\psi_{kz} dzdt - \int_{Q_{t,t+\triangle t}} J_t\psi_k dzdt,$$

where $Q_{t,t+\triangle t} = \Omega\times(t,t+\triangle t)$. For an estimate of the right-hand side we use the following inequalities

$$\begin{aligned} \int_{Q_{t,t+\triangle t}} |v_1v_2v_3| dzdt \le \|v_1\|_{q_1,\gamma_1,Q_{t,t+\triangle t}} \cdot \|v_2\|_{q_2,\gamma_2,Q_{t,t+\triangle t}} \cdot \|v_3\|_{q_3,\gamma_3,Q_{t,t+\triangle t}}, \\ q_i,\gamma_i \in [1,\infty),\ i=1,2,3, \quad \frac{1}{q_1}+\frac{1}{q_2}+\frac{1}{q_3}=1, \quad \frac{1}{\gamma_1}+\frac{1}{\gamma_2}+\frac{1}{\gamma_3}=1. \end{aligned} \tag{2.3.32}$$

$$\int_{Q_{t,t+\triangle t}} |r\widetilde{h}^N_z\psi_{kz}| dzdt \le r_1\|\psi_{kz}\|_{2,Q_{t,t+\triangle t}} \cdot \|\widetilde{h}^N_z\|_{2,Q_{t,t+\triangle t}}, \tag{2.3.33}$$

$$\int_{Q_{t,t+\triangle t}} |u^N_t(\widetilde{h}^N + J)\psi_{kz}| dzdt \le \|\psi_{kz}\|_{2,Q_{t,t+\triangle t}} \cdot \|u^N_t\|_{2,4,Q_{t,t+\triangle t}} \cdot \|\widetilde{h}^N + J\|_{\infty,4,Q_{t,t+\triangle t}}. \tag{2.3.34}$$

Using Theorem 2.2. in [21, pp.62-63] gives us

$$\|v\|_{q,\Omega} \le \beta\|v_z\|^{\frac{2}{\gamma}}_{2,\Omega} \cdot \|v\|^{1-\frac{2}{\gamma}}_{2,\Omega},$$

where

$$\beta = 2^{\frac{2}{\gamma}}, \quad q \in [2,+\infty], \quad \gamma \in [4,+\infty], \quad \frac{1}{\gamma}+\frac{1}{2q}=\frac{1}{4}.$$

After an integration of the latter inequality with respect to the time variable from t to $t+\triangle t$ we obtain

$$\|v\|_{q,\gamma,Q_{t,t+\triangle t}} \le \beta\|v_z\|^{\frac{2}{\gamma}}_{2,Q_{t,t+\triangle t}} \cdot \text{vrai} \max_{\tau\in[t,t+\triangle t]} \|v\|^{1-\frac{2}{\gamma}}_{2,\Omega}.$$

Using Young's inequality allows us to rewrite the latter as

$$\|v\|_{q,\gamma,Q_{t,t+\triangle t}} \leq \beta\frac{2}{\gamma}\|v_z\|_{2,Q_{t,t+\triangle t}} + \beta(1-\frac{2}{\gamma})\text{vrai}\max_{\tau\in[t,t+\triangle t]}\|v\|_{2,\Omega}\,. \tag{2.3.35}$$

The fact that integrals (2.3.33)–(2.3.34) tend to zero as $\triangle t \to 0$ gives

$$|a_k^N(t+\triangle t) - a_k^N(t)| \leq \epsilon(\triangle t)\|\psi_k\|_{2,\Omega}^{(1)}$$

with $\epsilon(\triangle t)$ not depending on N and tending to zero as $\triangle t \to 0$, i.e., the equicontinuity of the a_k^N, $N = k, k+1, \ldots$, in t.

Equicontinuity of functions b_k^N follows from the boundedness of their derivatives. The fact that b_{kt}^N are equicontinuous functions for $N \geq k$ is proved in same way as it was done for the functions a_k^N, $N \geq k$. By the usual diagonal process we select a subsequence $\{N_m\}_{m=1}^\infty$ such that the function sequences $\{a_k^{N_m}\}, \{b_k^{N_m}\}$ be converging uniformly on $[0,T]$ to some continuous functions $a_k(t), b_k(t)$. The functions a_k, b_k define two functions

$$h = \sum_{k=1}^{\infty} a_k\psi_k + J, \qquad u = \sum_{k=1}^{\infty} b_k\phi_k\,.$$

To the function $h(\cdot,t)$ the sequence $\{h^{N_m}\}$ converges weakly in $L_2(\Omega)$ and uniformly with respect to t in $[0,T]$. Indeed, for any function $\psi(z)$ from $L_2(\Omega)$ we have

$$(h^{N_m} - h, \psi) = \sum_{k=1}^{s}(\psi,\psi_k)(h^{N_m} - h, \psi_k) + (\sum_{k=s+1}^{\infty}(h^{N_m} - h, (\psi,\psi_k)\psi_k) \tag{2.3.36}$$

with

$$|(h^{N_m} - h, \sum_{k=s+1}^{\infty}(\psi,\psi_k)\psi_k)| \leq C_9(\sum_{k=s+1}^{\infty}(\psi,\psi_k)^2)^{1/2} \equiv C_9R(s),$$

where the positive constant C_9 does not depend on N_m, s. We choose s so large that $C_9R(s)$ becomes less than a preassigned $\epsilon > 0$. On the other hand, for fixed s and large enough N_m, the first sum in (2.3.36) will be less than ϵ for all t in $[0,T]$. Thus $|(h^{N_m} - h, \psi)|$ can be made less than ϵ for all t in $[0,T]$. It is shown that the sequence $\{h^{N_m}\}$ converges to h weakly in $L_2(\Omega)$, uniformly with respect to $t \in [0,T]$.

The sequence $\{u^{N_m}\}$ is bounded in $L_\infty(0,T;\overset{o}{W}{}_2^1(\Omega))$ and the sequence $\{u_t^{N_m}\}$ is bounded in $L_\infty(0,T;L_2(\Omega))$. For this reason the sequence $\{u^{N_m}\}$ converges to u $*$-weakly in $L_\infty(0,T;\overset{o}{W}{}_2^1(\Omega))$ and the sequence $\{u_t^{N_m}\}$ converges to u_t $*$-weakly in $L_\infty(0,T;L_2(\Omega))$. The functions u^{N_m} belong to $\overset{o}{W}{}_2^{1,1}(Q_T)$ and in virtue of $\overset{o}{W}{}_2^{1,1}(Q_T) \hookrightarrow L_2(Q_T)$ the sequence $\{u^{N_m}\}$ converges to the function u strongly in $L_2(Q_T)$ and pointwise a.e. in Q_T. Let us show that the sequence $\{u_{tt}^N\}$ is bounded in $L_2(0,T;H^{-1}(\Omega))$, where $H^{-1}(\Omega)$ is the dual space to $H_0^1(\Omega) = \overset{o}{W}{}_2^1(\Omega)$. For this purpose consider a function $\Phi(z)$ from $\overset{o}{W}{}_2^1(\Omega)$, such that $\|\Phi\|_{\overset{o}{W}{}_2^1(\Omega)} = 1$, $\Phi = \Phi_1 + \Phi_2$, where $\Phi_1 \in \text{span}\{\phi_k\}_{k=1}^{k=N}$, and

$(\Phi_2, \phi_k) = 0, k = 1, \dots, N$. Let us denote by $\langle a, b\rangle$ the pairing between $a \in H^{-1}(\Omega)$ and $b \in \overset{o}{W}{}_2^1(\Omega)$. From (2.3.25) we have

$$\langle u_{tt}^N, \Phi\rangle = (u_{tt}^N, \Phi_1) = -(\nu^2 u_z^N, \Phi_{1z}) + \frac{p}{2}((h^N)^2, \Phi_{1z}) + (f, \Phi_1)\,.$$

It is easy to check that $\|\Phi_1\|_{\overset{o}{W}{}_2^1(\Omega)} \le 1$. In view of this and estimates obtained above, we get

$$|\langle u_{tt}^N, \Phi\rangle| \le C_{10},$$

where the positive constant C_{10} does not depend on N. It proves that $\{u_{tt}^N\}$ is bounded in $L_2(0, T; H^{-1}(\Omega))$. Thus we can take for granted

$$u_{tt}^N \rightharpoonup u_{tt} \text{ weakly in } L_2(0, T; H^{-1}(\Omega))\,.$$

It follows from (2.3.31) that one can extract from the sequence $\{h^{N_m}\}$ a subsequence converging to h weakly in $L_2(Q_T)$ together with $\{h_z^{N_m}\}$. Let us show the sequence $\{h_t^N\}$ is bounded in $L_2(0, T; H^{-1}(\Omega))$. For this purpose consider a function $\Psi(z) \in \overset{o}{W}{}_2^1(\Omega)$ such that $\|\Psi\|_{\overset{o}{W}{}_2^1(\Omega)} = 1$, $\Psi = \Psi_1 + \Psi_2$, where $\Psi_1 \in \text{span}\{\psi_k\}_{k=1}^{k=N}$, and $(\Psi_2, \psi_k) = 0, k = 1, \dots, N$. Because of

$$h_t^N - J_t \equiv \widetilde{h}_t^N = \sum_{k=1}^{N} a_{kt}^N \psi_k \in \text{span}\{\psi_k\}_{k=1}^{k=N}$$

the following equality is valid

$$\langle \widetilde{h}_t^N, \Psi\rangle = \langle \widetilde{h}_t^N, \Psi_1\rangle = -(rh_z^N - u_t^N h^N - rj, \Psi_{1z}) - \langle J_t, \Psi_1\rangle\,.$$

Using $\|\Psi_1\|_{\overset{o}{W}{}_2^1(\Omega)} \le 1$ and (2.3.31)–(2.3.35) we obtain the inequality

$$|\langle h_t^N, \Psi\rangle| \le C_{11},$$

where the positive constant C_{11} does not depend on N. The latter proves that $\{h_t^N\}$ is bounded in $L_2(0, T; H^{-1}(\Omega))$. From (2.3.31) follows that $\{h^N\}$ is bounded in $L_2(0, T; \overset{o}{W}{}_2^1(\Omega))$ too. A subsequence of $\{h^N\}$ converges strongly in $L_2(Q_T)$, which means the convergence of a subsequence of the previous one a.e. in Q_T, see Theorem 5.1 in [24, p.58]. Without loss of generality we can assume that the sequences $\{h^{N_m}\}, \{u^{N_m}\}$ converge to the limit functions h, u in the sense mentioned above. By this reason the sequence $\{h^N\}$ converges a.e. in Q_T.

Let us prove now that the limit functions h, u satisfy equalities (2.3.14)–(2.3.15). First, we will show that the function h satisfies equality (2.3.14). For this purpose we multiply each equation of (2.3.24) by a smooth function $\alpha_k(t)$ that is equal to zero for $t = T$, then sum over all k from 1 to $N' \le N$, and integrate the result with respect to t from 0 to T. After an integration by parts we obtain

$$\int_0^T (h^N, \Psi_t^{N'})dt = \int_0^T [(rh_z^N, \Psi_z^{N'}) - (u_t^N h^N, \Psi_z^{N'}) - (rj, \Psi_z^{N'})]dt + (h_0^N, \Psi_z^{N'}(\cdot, 0)), \quad (2.3.37)$$

where $\Psi^{N'}(z,t) = \sum_{k=1}^{N'} \alpha_k(t)\psi_k(z)$. We claim that we can pass to the limit along the subsequence $\{N_m\}$ selected above, assuming $\Psi^{N'}$ fixed, and thereby arrive at (2.3.37) with h^{N_m}, u^{N_m} being replaced by h, u. Indeed, let us prove that $\int_0^T (u_t^{N_m} h^{N_m}, \Psi_z^{N'})dt$ tends to $\int_0^T (u_t h, \Psi_z^{N'})dt$, i.e.,

$$\int_0^T (u_t^{N_m} h^{N_m} - u_t h, \Psi_z^{N'})dt \to 0, \quad \text{as} \quad m \to \infty .$$

Observe that the integral can be represented in the equivalent form

$$\int_0^T (u_t^{N_m} h^{N_m} - u_t h, \Psi_z^{N'})dt = \int_0^T (u_t^{N_m}(h^{N_m} - h), \Psi_z^{N'})dt + \int_0^T (u_t^{N_m} - u_t, h\Psi_z^{N'})dt .$$

The first term in the right hand side of the latter equality tends to zero as $m \to \infty$. Indeed, from (2.3.31)–(2.3.32), (2.3.35) we can conclude that $u_t h \in L_\infty(0,T; L_2(\Omega))$ and $\{u_t^{N_m} h^{N_m}\}$ is bounded in the same space, by this reason, the sequence $\{u_t^{N_m} h^{N_m} - u_t h\}$ converges $*$-weakly to a function ψ in $L_\infty(0,T; L_2(\Omega))$. Applying Lemma 1.3 of [24, p.12], we can show that $\psi \equiv 0$. Indeed, since $\{h^{N_m}\}$ converges to h in $L_2(Q_T)$ and $\{u_t^{N_m}\}$ satisfies (2.3.31), we get that $\{(h^{N_m} - h)u_t^{N_m}\}$ converges to 0 in $L_1(Q_T)$ and tends a.e. to 0 in Q_T. From (2.3.35) we have

$$\begin{aligned}\|u_t^{N_m} h^{N_m}\|_{2,Q_T} &\le \|u_t^{N_m}\|_{2,4} \cdot \|h^{N_m}\|_{\infty,4} \\ &\le \|u_t^{N_m}\|_{2,4} \cdot (\|h_z^{N_m}\|_{2,Q_T} + \text{vrai} \max_{t\in[0,T]} \|h^{N_m}\|_{2,\Omega}) \le C_{12},\end{aligned}$$

where a positive constant C_{12} does not depend on N. Therefore $\{u_t^{N_m} h^{N_m}\}$ is bounded in $L_2(Q_T)$. Analogously we can prove that the function $u_t h$ and the sequences $\{u_t^{N_m} h\}, \{u_t h^{N_m}\}$ are bounded in $L_2(Q_T)$ too. Therefore $\{u_t^{N_m}(h^{N_m} - h)\}$ tends to 0 $*$-weakly. Obviously that the second integral tends to 0 too. Thus, $\{u_t^{N_m} h^{N_m}\}$ tends to the function $u_t h$ $*$-weakly and we can pass in formula (2.3.37) to the limit as $m \to \infty$. But the $\Psi^{N'}$ are dense in the space of all functions required in the first definition of a weak solution, see Lemma 4.12 in [21, p.89]. In view of this h satisfies equality (2.3.14) and belongs to be the space $\overset{o}{V}_2(Q_T)$. Since

$$u_t^2 \in L_{1,\frac{2}{1-2\kappa}}(Q_T), \quad \kappa \in (0, \frac{1}{2}),$$

we deduce from Theorems 7.1 and 10.1 of [21, p.181, p.204] that

$$\max_{Q_T} |h| \le C_{13}, \qquad h \in C^{\alpha,\alpha/2}(\overline{Q}_T)$$

with some positive constant C_{13}. Let us show now that the function u satisfies equality (2.3.15). For this purpose we multiply each equation of (2.3.25) by the $\alpha_k(t)$, then sum over all k from 1 to $N' \le N$, and integrate the result with respect to t from 0 to T. After an integration by parts we arrive at

$$\begin{aligned}\int_0^T (u_t^N, \Phi_t^{N'})dt &= \int_0^T [(\nu^2 u_z^N, \Phi_z^{N'}) - \frac{p}{2}((h^N)^2, \Phi_z^{N'}) + (f, \Phi^{N'})]dt \\ &\quad - (u_1^N, \Phi^{N'}(\cdot, 0)),\ u^N(z,0) = u_0^N(z), \quad (2.3.38)\end{aligned}$$

where the $\Phi^{N'}(z,t) = \sum_{k=1}^{N'} \alpha_k(t)\phi_k(z)$. In this equality one can pass to the limit with respect to the subsequence $\{N_m\}$ selected above, assuming $\Phi^{N'}$ fixed, and thereby arrive at (2.3.38) with h^{N_m}, u^{N_m} being replaced by h, u. Lemma 1.3 of [24, p.12] was used to show that the sequence $\{(h^{N_m})^2\}$ converges to h^2 weakly; it allowed to pass to the limit in the nonlinear term in (2.3.38). Since $\max_{Q_T} |h| \leq C_{13}$ then $\{\int_0^T (h^2, \Phi_z^{N'})dt\}$ is bounded for any $\Phi^{N'} \in \overset{0}{W}{}_2^{1,1}(Q_T)$ and, as the $\Phi^{N'}$ are dense in the space considered in the definition of weak solution, we deduce that the function u satisfies equality (2.3.15) and is the weak solution from $\overset{o}{W}{}_2^{1,1}(Q_T)$, see Lemma 4.12 in [21, p.89]. Theorem 2.3.1 is proved.

From Lemma 4.1 of [21, p.158] and Theorem 2.3.1 we get

Corollary 2.3.1 *Any weak solution $h(z,t)$ of Direct Problem 2.3.1 from $\overset{o}{V}_2(Q_T)$ belongs to $\overset{o}{V}{}_2^{1,1/2}(Q_T)$.*

Corollary 2.3.2 *For any function $\phi(z,t) \in L_2(0,T; \overset{o}{W}{}_2^{1}(\Omega))$ is valid the following equality*

$$\int_0^T \langle u_{tt}, \phi\rangle dt = \int_0^T \{(f - phh_z, \phi) - (\nu^2 u_z, \phi_z)\}dt\,.$$

The latter allows us to conclude that the following equation holds for any function $\xi(z) \in \overset{o}{W}{}_2^{1}(\Omega)$ and for almost all $t \in [0,T]$

$$\langle u_{tt}, \xi\rangle = (f - phh_z, \xi) - (\nu^2 u_z, \xi_z)\,.$$

Moreover notice that

$$u \in C([0,T]; L_2(\Omega)),\ u_t \in C([0,T]; H^{-1}(\Omega))\,.$$

We need the following lemma to prove uniqueness theorem.

Lemma 2.3.1 *Suppose $h(z,t) \in \overset{o}{V}{}_2^{1,1/2}(Q_T), u(z,t) \in \overset{o}{W}{}_2^{1,1}(Q_T)$ are a weak solution of Direct Problem 2.3.1. Then the following inequality is valid for almost all $t_1 \in [0,T]$*

$$\begin{aligned}\frac{1}{2}\int_\Omega (ph^2(z,t_1) + u_t^2(z,t_1) + \nu^2 u_z^2(z,t_1))dz + \frac{1}{2}\int_{Q_{t_1}} prh_z^2 dzdt \\ \leq \int_\Omega (ph_0^2 + u_1^2 + \nu^2 u_{0z}^2)dz + \int_{Q_{t_1}} prj^2 dzdt + 2t_1\int_{Q_{t_1}} f^2 dzdt, \quad (2.3.39)\end{aligned}$$

where $Q_{t_1} = \Omega \times (0,t_1)$.

Proof Suppose h, u are a weak solution of Direct Problem 2.3.1. We take as η and ζ in (2.3.14)–(2.3.15) the functions

$$\hat{\eta}_{\bar{k}}(z,t) = \frac{1}{k}\int_{t-k}^t \hat{\eta}(z,\tau)\,d\tau, \qquad \hat{\zeta}_{\bar{k}}(z,t) = \frac{1}{k}\int_{t-k}^t \hat{\zeta}(z,\tau)\,d\tau,$$

where $\hat{\eta}(z,\tau), \hat{\zeta}(z,\tau) \in W_2^{1,1}(Q_{-k,T})$, and $\hat{\eta}(z,\tau) = 0, \hat{\zeta}(z,\tau) = 0$ for $\tau \in [-k,0] \cup [T-k,T], Q_{-k,T} = \Omega \times (-k,T), k \in (0,T)$. In view of $D_t(\hat{\eta}_{\bar{k}}) = (D_t\hat{\eta})_{\bar{k}}$ we transform the first term in (2.3.14) in the following manner

$$-\int_{Q_{T-k}} h\hat{\eta}_{\bar{k}t}dzdt = -\int_{Q_{T-k}} h_k\hat{\eta}_t dzdt = \int_{Q_{T-k}} h_{kt}\hat{\eta}dzdt.$$

Here we have used the notation

$$h_k(z,t) = \frac{1}{k}\int_t^{t+k} h(z,\tau)\,\mathrm{d}\tau$$

and the relation

$$\int_0^T f(t)\hat{g}_{\bar{k}}(t)dt = \int_0^{T-k} f_k(t)\hat{g}(t)dt,$$

valid for any piecewise summable functions $f(t), \hat{g}(t)$ on $[-k,T]$, one of each is equal to zero on the intervals $[-k,0]$ and $[T-k,T]$. The latter equality is the result of interchanging the order of integration with respect to t and τ. In a similar way we obtain

$$-\int_{Q_{T-k}} u_t\hat{\zeta}_{\bar{k}t}dzdt = \int_{Q_{T-k}} u_{ktt}\hat{\zeta}dzdt.$$

In all other terms of (2.3.14)–(2.3.15) we also transfer the averaging $()_{\bar{k}}$ from $\hat{\eta}, \hat{\zeta}$ to theirs coefficients. Taking into account the permutability of this averaging with differentiation with respect to z we obtain identities

$$\int_{Q_{T-k}} \{h_{kt}\hat{\eta} + (rh_z - hu_t - rj)_k\hat{\eta}_z\}dzdt = 0, \tag{2.3.40}$$

$$\int_{Q_{T-k}} \{\nu^2 u_{kz}\hat{\zeta}_z + (u_{tt} + phh_z - f)_k\hat{\zeta}\}dzdt = 0. \tag{2.3.41}$$

These identities are actually valid for a class of functions $\hat{\eta}, \hat{\zeta}$ that is more extended than the class just considered; namely, they are valid for any functions $\hat{\eta}, \hat{\zeta}$ that are equal to zero for $t \geq t_1$ $(0 \leq t_1 \leq T-k)$ and are equal to some functions $\eta \in \overset{0}{V}{}_2^{1,0}(Q_{t_1}), \zeta \in \overset{0}{W}{}_2^{1,1}(Q_{t_1})$ for $t \in [0,t_1]$. This property was proved in [14]. Thereby we have

$$\int_{Q_{t_1}} \{h_{kt}\eta + (rh_z - hu_t - rj)_k\eta_z\}dzdt = 0,$$

$$\int_{Q_{t_1}} \{\nu^2 u_{kz}\zeta_z + (u_{tt} + phh_z - f)_k\zeta\}dzdt = 0.$$

In the latter formulas we take $\eta = ph_k, \zeta = u_{kt}$ and represent the corresponding terms in the form

$$\int_{Q_{t_1}} h_{kt}h_k dzdt = \frac{1}{2}\int_\Omega h_k^2(z,t)dz\Big|_{t=0}^{t=t_1}, \quad \int_{Q_{t_1}} u_{ktt}u_{kt}dzdt = \frac{1}{2}\int_\Omega u_{kt}^2(z,t)dz\Big|_{t=0}^{t=t_1},$$

$$\int_{Q_{t_1}} \nu^2 u_{kz}u_{kzt}dzdt = \frac{1}{2}\int_\Omega \nu^2(z)u_{kz}^2(z,t)dz\Big|_{t=0}^{t=t_1},$$

after which let k tend to zero. By analogy with (2.3.27) we obtain

$$\frac{1}{2}\int_\Omega (ph^2(z,t_1)+u_t^2(z,t_1)+\nu^2u_z^2(z,t_1))dz - \frac{1}{2}\int_\Omega (ph_0^2+u_1^2+\nu^2u_{0,z}^2)dz + \int_{Q_{t_1}} prh_z^2dzdt = \int_{Q_{t_1}} (prjh_z+fu_t)dzdt.$$

Note that

$$\int_{Q_{t_1}} pr(\frac{1}{2}h_z^2 - jh_z + \frac{1}{2}j^2)dzdt = \frac{1}{2}\int_{Q_{t_1}} pr(h_z-j)^2dzdt \geq 0.$$

In view of this we have

$$\frac{1}{2}\int_\Omega (ph^2(z,t_1)+u_t^2(z,t_1)+\nu^2u_z^2(z,t_1))dz - \frac{1}{2}\int_\Omega (ph_0^2+u_1^2+\nu^2u_{0,z}^2)dz + \frac{1}{2}\int_{Q_{t_1}} prh_z^2dzdt \leq \frac{1}{2}\int_{Q_{t_1}} prj^2dzdt + \int_{Q_{t_1}} |fu_t|dzdt. \quad (2.3.42)$$

The latter gives us

$$\int_\Omega u_t^2(z,t_1)dz \leq \int_\Omega (ph_0^2+u_1^2+\nu^2u_{0z}^2)dz + \int_{Q_{t_1}} prj^2dzdt + 2|\int_{Q_{t_1}} fu_tdzdt|.$$

Integrating this inequality with respect to the time variable from 0 to t_1 and applying Cauchy's inequality with $\epsilon = 1/2t_1$, we obtain

$$\int_{Q_{t_1}} u_t^2dzdt \leq 4t_1\int_\Omega (ph_0^2+u_1^2+\nu^2u_{0z}^2)dz + 2t_1\int_{Q_{t_1}} prj^2dzdt + 4t_1^2\int_{Q_{t_1}} f^2dzdt.$$

Using the latter and Cauchy's inequality with $\epsilon = 1/2t_1$ in (2.3.42) gives us

$$\frac{1}{2}\int_\Omega (ph^2(z,t_1)+u_t^2(z,t_1)+\nu^2u_z^2(z,t_1))dz + \frac{1}{2}\int_{Q_{t_1}} prh_z^2dzdt \leq \int_\Omega (ph_0^2+u_1^2+\nu^2u_{0z}^2)dz + \int_{Q_{t_1}} prj^2dzdt + 2t_1\int_{Q_{t_1}} f^2dzdt.$$

Thus (2.3.39) is proved.

From (2.3.39) we obtain (see [14] for details)

$$\max_{Q_T}|h| + \|h_z\|_{2,Q_T} + \|u\|_{\overset{o}{W}{}_2^{1,1}(Q_T)} \leq C_{14}, \quad h \in C^{\alpha,\alpha/2}(\overline{Q}_T), \quad (2.3.43)$$

where C_{14} is a positive constant. Fix now a positive constant T_0. Let us prove that the constant C_{14} depends only on T_0 whenever $T \in [0, T_0]$: we will write $C_{14} = C_{14}(T_0)$. From inequality (2.3.39) it follows that $\|h_z\|_{2,Q_T}, \|u\|_{\overset{o}{W}{}_2^{1,1}(Q_T)}$ are well defined from above with a constant independent of T. To estimate $\max_{Q_T}|h|$ we consider (2.3.6) as the linear parabolic equation with coefficients satisfying additionally to conditions (7.1)-(7.2) in [21, p.181] with constants independent of T. It follows from Theorem 7.1 of [21, p.181] that vrai $\max_{Q_T} |h|$ is bounded from above with a constant independent of T. Let us use now Theorem 10.1 of [21, p.204] taking into account that on the parabolic boundary h is a function from the Hölder space with exponent α in z. Function h is equal to zero in $z = \pm l$ and by this reason has any smoothness on the boundary with respect to t. So, $h \in C^{\alpha, \alpha/2}(\overline{Q}_T)$ and the exponent α is independent of $T \in [0, T_0]$. By this reason vrai $\max_{Q_T} |h| = \max_{Q_T} |h|$.

Let us now prove the uniqueness result about solvability of Direct Problem 2.3.1.

Theorem 2.3.2 *Direct Problem 2.3.1 cannot have more then one weak solution.*

Proof Let $h_k, u_k, k = 1, 2$, be two solutions of Direct Problem 2.3.1. Introduce functions v, w by the formulas

$$v(z,t) = h_2(z,t) - h_1(z,t), \; w(z,t) = u_2(z,t) - u_1(z,t). \tag{2.3.44}$$

The functions v, w are a weak solution of the homogeneous problem

$$v_t = (rv_z)_z - (h_2 w_t + u_{1t} v)_z, \tag{2.3.45}$$

$$w_{tt} = (\nu^2 w_z)_z - p(h_2 v_z + h_{1z} v), \tag{2.3.46}$$

$$v(\pm l, t) = w(\pm l, t) = 0, \tag{2.3.47}$$

$$v(z,0) = w(z,0) = w_t(z,0) = 0, \tag{2.3.48}$$

satisfying at the jump points $z_k, k = 1, 2, \ldots, m$, the compatibility conditions (2.3.12)–(2.3.13). Therefore the functions v, w satisfy the integral equalities

$$\int_{Q_T} \{-v\eta_t + (rv_z - h_2 w_t - u_{1t} v)\eta_z\} dzdt = 0,$$

$$\int_{Q_T} \{-w_t \zeta_t + \nu^2 w_z \zeta_z + p(h_2 v_z + h_{1z} v)\zeta\} dzdt = 0.$$

By analogy with (2.3.40)–(2.3.41) we obtain

$$\begin{aligned} &\int_{Q_{T-k}} \{v_{kt}\hat{\eta} + (rv_z - h_2 w_t - u_{1t} v)_k \hat{\eta}_z\} dzdt = 0, \\ &\int_{Q_{T-k}} \{\nu^2 w_{kz} \hat{\zeta}_z + (w_{tt} + ph_2 v_z + ph_{1z} v)_k \hat{\zeta}\} dzdt = 0. \end{aligned} \tag{2.3.49}$$

Multiply the first equality in (2.3.49) by p and sum the result with the second one. In the result obtained we take

$$\hat{\eta}(z,t) = \begin{cases} \eta(z,t), & \text{if} \quad t \in (0, t_1] \\ 0, & \text{if} \quad t \notin (0, t_1] \end{cases}$$

and

$$\hat{\zeta}(z,t) = \begin{cases} \zeta(z,t), & \text{if} \quad t \in (0, t_1] \\ 0, & \text{if} \quad t \notin (0, t_1] \end{cases}$$

where $\eta = v_k, \zeta = w_{kt}, t_1 \in (0, T-k]$, and represent the corresponding terms in the form

$$\int_{Q_{T-k}} v_{kt} v_k dz dt = \frac{1}{2} \int_\Omega v_k^2 dz|_{t=0}^{t=t_1},$$

$$\int_{Q_{T-k}} w_{ktt} w_{kt} dz dt = \frac{1}{2} \int_\Omega w_{kt}^2 dz|_{t=0}^{t=t_1}, \quad \int_{Q_{T-k}} w_{kt} w_k dz dt = \frac{1}{2} \int_\Omega w_k^2 dz|_{t=0}^{t=t_1}.$$

Passing to the limit $k \to 0$ and using of the initial data (2.3.48) gives us

$$\frac{p}{2}\|v(\cdot, t_1)\|_{2,\Omega}^2 + \frac{1}{2}\|w_t(\cdot, t_1)\|_{2,\Omega}^2 + \frac{1}{2}\|\nu w_z(\cdot, t_1)\|_{2,\Omega}^2 + \int_{Q_{t_1}} prv_z^2 dz dt = \int_{Q_{t_1}} p(u_{1t} v_z v - h_{1z} v w_t) dz dt. \quad (2.3.50)$$

The integral in the right hand side is estimated using (2.3.43)

$$|\int_{Q_{t_1}} p(u_{1t} v_z v + h_{1z} v w_t) dz dt| \leq pC_{14}(\|v_z\|_{2,Q_{t_1}} + \|w_t\|_{2,Q_{t_1}}) \cdot \max_{Q_{t_1}} |v|.$$

Thus, for almost all $t_1 \in [0, T]$ the following inequality holds

$$\frac{p}{2}\|v(\cdot, t_1)\|_{2,\Omega}^2 + \frac{1}{2}\|w_t(\cdot, t_1)\|_{2,\Omega}^2 + \frac{1}{2}\|\nu w_z(\cdot, t_1)\|_{2,\Omega}^2 + \int_{Q_{t_1}} prv_z^2 dz dt \leq pC_{14}(\|v_z\|_{2,Q_{t_1}} + \|w_t\|_{2,Q_T}) \cdot \max_{Q_T} |v|. \quad (2.3.51)$$

Consider the first boundary-value problem for inhomogeneous parabolic equation (2.3.45) with the zero boundary and initial conditions and the free term $(h_2 w_t)_z$. If $\|h_2 w_t\|_{2,Q_T} = 0$, then $v \equiv 0$ and, consequently, $w \equiv 0$, see (2.3.46)-(2.3.48). Consider case $\|h_2 w_t\|_{2,Q_T} = q > 0$. From Theorem 7.1 of [21, p.181] we deduce that

$$\max_{Q_T} |\frac{v}{q}| \leq C_{15}(T_0), \quad \forall T \in (0, T_0],$$

where $C_{15}(T_0)$ depends on T_0 only, whenever $T \in (0, T_0]$. This is proved in the same way as the constant C_{14} was estimated previously. Then

$$\max_{Q_T} |v| \leq C_{14}(T_0) C_{15}(T_0) \|w_t\|_{2,Q_T}, \quad \forall (0, T_0].$$

This gives

$$\frac{p}{2}\|v(\cdot, t_1)\|_{2,\Omega}^2 + \frac{1}{2}\|w_t(\cdot, t_1)\|_{2,\Omega}^2 + \frac{1}{2}\|\nu w_z(\cdot, t_1)\|_{2,\Omega}^2 + \int_{Q_{t_1}} prv_z^2 dz dt \leq pC_{14}^2(T_0) C_{15}(T_0)(\|v_z\|_{2,Q_{t_1}} \cdot \|w_t\|_{2,Q_T} + \|w_t\|_{2,Q_T}^2).$$

Applying Cauchy's inequality with $\epsilon = 2r_0 C_{14}^{-2}(T_0) C_{15}^{-1}(T_0)$ to $\|v_z\|_{2,Q_{t_1}} \cdot \|w_t\|_{2,Q_T}$ with r_0 being the positive constant in assumption (a), we obtain the following inequality, valid for almost all $t_1 \in [0, T], T \in (0, T_0]$,

$$\frac{p}{2}\|v(\cdot, t_1)\|_{2,\Omega}^2 + \frac{1}{2}\|w_t(\cdot, t_1)\|_{2,\Omega}^2 + \frac{1}{2}\|\nu w_z(\cdot, t_1)\|_{2,\Omega}^2 \leq pC_{16}(T_0)\|w_t\|_{2,Q_T}^2,$$

where

$$C_{16}(T_0) = C_{14}^2(T_0)C_{15}(T_0)\big(1 + \frac{C_{14}^2(T_0)C_{15}(T_0)}{4r_0}\big).$$

Integrating the latter with respect to t_1 from 0 to T with $T \in (0, T_0]$, we have

$$\frac{p}{2}\|v\|_{2,Q_T}^2 + \frac{1}{2}\|w_t\|_{2,Q_T}^2 + \frac{\nu_0^2}{2}\|w_z\|_{2,Q_T}^2 \leq TpC_{16}(T_0)\|w_t\|_{2,Q_T}^2.$$

But it is impossible for $TpC_{16}(T_0) < 1/2$ in the case when w_t, w_z, v are non-zero functions. Therefore $v = w \equiv 0$ in Q_{T_1} with, e.g., $T_1 = \min\{(4pC_{16}(T_0))^{-1}, T_0\}$. Applying this result recursively, after a finite number of steps, we conclude that $v = w \equiv 0$ in Q_T, which proves Theorem 2.3.2.

Let us show that a weak solution of Direct Problem 2.3.1 is stable with respect to variations of the coefficients (except ν) and free terms of the equations, and also the initial conditions. This result will be established in the case when ν is supposed to be a smooth enough function. We have shown that Direct Problem 2.3.1 has the unique weak solution $h \in \overset{o}{V}_2 (Q_T)$, $u \in \overset{o}{W}{}_2^{1,1} (Q_T)$ having additional properties

$$\max_{Q_T} |h| + \|h_z\|_{2,Q_T} + \|u_t(\cdot,t)\|_{2,\Omega} + \|u_z(\cdot,t)\|_{2,\Omega} \leq C_{17}, \tag{2.3.52}$$

where the positive constant C_{17} does not depend on h, u. Along with Direct Problem 2.3.1 consider the family of the problems

$$h_t^m = (r^m h_z^m)_z - (h^m u_t^m)_z - (r^m j^m)_z, \ (z,t) \in Q_T, \tag{2.3.53}$$
$$u_{tt}^m = (\nu^2 u_z^m)_z - ph^m h_z^m - f^m, \ (z,t) \in Q_T, \tag{2.3.54}$$
$$h^m(z,0) = h_0^m(z), \ z \in \Omega, \tag{2.3.55}$$
$$u^m(z,0) = u_0^m(z), \ u_t^m(z,0) = u_1^m(z), \ z \in \Omega, \tag{2.3.56}$$
$$u^m(\pm l,t) = h^m(\pm l,t) = 0, \ t \in (0,T), \tag{2.3.57}$$

where $m \in \mathbb{N}$. Suppose that $r^m(z), j^m(z,t), h_0^m(z), u_0^m(z), u_1^m(z), f^m(z,t)$ are smooth functions satisfying the conditions of the uniqueness and existence theorems. In this case transmission conditions (2.3.12)–(2.3.13) can be dropped owing to the smoothness of the solution. Problems (2.3.53)–(2.3.57) have the unique weak solutions $h^m, u^m, m \in \mathbb{N}$.

Theorem 2.3.3 *Suppose the sequence $\{r^m\}$ is uniformly bounded and converges a.e. to r, while the sequences $\{j^m\}, \{f^m\}, \{h_0^m\}, \{u_0^m\}, \{u_1^m\}$ converge to j, f, h_0, u_0, u_1 in the norms of the spaces to which they belong according to the conditions of Theorem 2.3.1. Then the weak solutions $h^m \in \overset{o}{V}{}_2^{1,0} (Q_T)$, $u^m \in \overset{o}{W}{}_2^{1,1} (Q_T)$ converge in such spaces to the weak solution h, u of the limit Direct Problem 2.3.1.*

For a proof of this proposition we need the following lemma.

Lemma 2.3.2 *Let h, u and h^m, u^m be weak solutions of Direct Problem 2.3.1 and problems (2.3.53)–(2.3.57), and $v^m(z,t) = h^m(z,t) - h(z,t), w^m(z,t) = u^m(z,t) - u(z,t)$.*

Then there exists a positive number δ, independent of h, u, h^m, u^m, t_1, t_2, such that for any $t_1, t_2 \in [0, T], 0 \le t_2 - t_1 < \delta$, the following inequality holds true

$$\begin{aligned}\operatorname{vrai}\max_{t\in[t_1,t_2]}\{p\|v^m\|^2_{2,\Omega} + \|w^m_t\|^2_{2,\Omega} + 2\|\nu w^m_z\|^2_{2,\Omega}\} + 2pr_0\|v^m_z\|^2_{2,Q_{t_1,t_2}} \\ \le pC_{18}\{\|(r^m - r)h_z\|^2_{2,Q_{t_1,t_2}} + \|r^m j^m - rj\|^2_{2,Q_{t_1,t_2}}\} + \|f^m - f\|^2_{2,Q_{t_1,t_2}} \\ + 2p\|v^m(\cdot,t_1)\|^2_{2,\Omega} + 2\|w^m_t(\cdot,t_1)\|^2_{2,\Omega} + 2\|\nu w^m_z(\cdot,t_1)\|^2_{2,\Omega}, \quad (2.3.58)\end{aligned}$$

where $Q_{t_1,t_2} = \Omega \times (t_1, t_2)$, and C_{18} is a positive constant.

Proof For a proof of this proposition we subtract from the integral identities (2.3.14)–(2.3.15) for h^m, u^m the corresponding ones for h, u and write the result of this subtraction in the form of integral identities for the functions v^m, w^m in the following manner

$$\begin{aligned}-\int_{Q_T} v^m\eta_t dzdt + \int_{Q_T} r^m v^m_z \eta_z dzdt - \int_{Q_T} v^m u^m_t \eta_z dzdt - \int_{Q_T} hw^m_t\eta_z dzdt \\ = \int_{Q_T}(r^m j^m - rj)\eta_z dzdt + \int_{Q_T}(r^m - r)h_z\eta_z dzdt + \int_\Omega (h^m_0 - h_0)\eta(z,0)dz,\end{aligned} \quad (2.3.59)$$

$$\begin{aligned}-\int_{Q_T} w^m_t\zeta_t dzdt + \int_{Q_T}\nu^2 w^m_z\zeta_z dzdt + \int_{Q_T} ph^m_z v^m\zeta dzdt + \int_{Q_T} phv^m_z\zeta dzdt \\ = \int_{Q_T}(f^m - f)\zeta dzdt + \int_\Omega (u^m_1 - u_1)\zeta(z,0)dz,\ w^m(z,0) = u^m_0(z) - u_0(z).\end{aligned} \quad (2.3.60)$$

We take as η in (2.3.59) the function

$$\hat\eta_{\bar k}(z,t) = \frac{1}{k}\int_{t-k}^{t}\hat\eta(z,\tau)\,d\tau,$$

where $\hat\eta(z,t)$ is an arbitrary element of $\overset{o}{W}{}^{1,1}_2(Q_{-k,T})$ that is equal to zero for $t \ge T - k$ and $t \le 0$. In a manner analogous to that of Lemma 2.3.1 we obtain (the latter integral in formula (2.3.59) is equal to zero by the choice of the test function)

$$\begin{aligned}\int_{Q_{T-k}} v^m_{kt}\hat\eta dzdt + \int_{Q_{T-k}}(r^m v^m_z - v^m u^m_t - hw^m_t)_k\hat\eta_z dzdt \\ = \int_{Q_{T-k}}(r^m j^m - rj + (r^m - r)h_z)_k\hat\eta_z dzdt.\end{aligned} \quad (2.3.61)$$

This equality is actually valid for a class of functions $\hat\eta(z,t)$ that is more extensive than the class just considered; namely, it is valid for any function $\hat\eta$ that is equal to zero for $t \ge \tau$, $\tau \le T - k$ and is equal to some function $\eta(z,t)$ from $\overset{o}{V}{}^{1,0}_2(Q_\tau)$ for $t \in [0,\tau]$. Indeed, the set $\overset{o}{W}{}^{1,1}_2(Q_{-k,T})$ is dense in $\overset{o}{V}{}^{1,0}_2(Q_{-k,T})$. Thus for any η from $\overset{o}{V}{}^{1,0}_2(Q_{-k,T})$ there is a sequence of functions η_n from $\overset{o}{W}{}^{1,1}_2(Q_{-k,T})$ that is strongly convergent to it for

$n \to \infty$ in $\overset{o}{V}{}_2^{1,0}(Q_{-k,T})$. We denote by $\chi_l(t)$ the continuous piecewise-linear functions

$$\chi_l(t) = \begin{cases} 0, \text{ if } & t \leq 0 \\ lt, \text{ if } & t \in [0, \frac{1}{l}] \\ 1, \text{ if } & t \in [\frac{1}{l}, \tau - \frac{1}{l}] \\ l(\tau - t), \text{ if } & t \in [\tau - \frac{1}{l}, \tau] \\ 0, \text{ if } & t \geq \tau \end{cases}$$

with $\tau > 1/l$. Identity (2.3.61) is established for $\hat{\eta}_{nl}(z,t) = \eta_n(z,t)\chi_l(t)$ when $\tau \leq T-k$. It is easy to see that one can pass to the limit in it as n and $l \to \infty$ and thereby prove the identity

$$\int_{Q_\tau} v_{kt}^m \eta dz dt + \int_{Q_\tau} (r^m v_z^m - v^m u_t^m - h w_t^m)_k \eta_z dz dt$$
$$= \int_{Q_\tau} (r^m j^m - rj + (r^m - r)h_z)_k \eta_z dz dt \quad (2.3.62)$$

for any function $\eta \in \overset{o}{V}{}_2^{1,0}(Q_\tau)$ when $\tau \leq T-k$. In (2.3.62) we take $\eta = v_k^m$ and represent the first term in the form

$$\int_{Q_\tau} v_k^m (v_k^m)_t dz dt = \frac{1}{2} \int_\Omega (v_k^m)^2 dz |_{t=0}^{t=\tau},$$

after which we let k tend to zero. This gives

$$\frac{1}{2} \|v^m(\cdot,t)\|_{2,\Omega}^2 |_{t=0}^{t=\tau} + \int_{Q_\tau} (r^m v_z^m - v^m u_t^m - h w_t^m) v_z^m dz dt$$
$$= \int_{Q_\tau} (r^m j^m - rj + (r^m - r)h_z) v_z^m dz dt. \quad (2.3.63)$$

Analogously the following equality can be obtained from (2.3.60)

$$\frac{1}{2}(\|w_t^m(\cdot,t)\|_{2,\Omega}^2 + \|\nu w_z^m(\cdot,t)\|_{2,\Omega}^2)|_{t=0}^{t=\tau} + \int_{Q_\tau} p v^m h_z^m w_t^m dz dt$$
$$+ \int_{Q_\tau} p v_z^m h w_t^m dz dt = \int_{Q_\tau} (f^m - f) w_t^m dz dt. \quad (2.3.64)$$

Multiply (2.3.63) by p and sum the result with (2.3.64). This gives

$$\frac{1}{2}(p\|v^m(\cdot,t)\|_{2,\Omega}^2 + \|w_t^m(\cdot,t)\|_{2,\Omega}^2 + |\nu w_z^m(\cdot,t)\|_{2,\Omega}^2)|_{t=0}^{t=\tau}$$
$$+ \int_{Q_\tau} p(r^m (v_z^m)^2 + v^m h_z^m w_t^m - v^m v_z^m u_t^m) dz dt$$
$$= \int_{Q_\tau} p(v_z^m (r^m j^m - rj) + (r^m - r)h_z w_z^m) dz dt + \int_{Q_\tau} (f^m - f) w_t^m dz dt. \quad (2.3.65)$$

Subtract from (2.3.65) for $\tau = t$ the same equality for $\tau = t_1, t_1 < t < t_2$, and consider vrai $\max_{t\in[t_1,t_2]}$ of the result. We obtain

$$\frac{1}{2}\text{ vrai }\max_{t\in[t_1,t_2]}\{p\|v^m\|^2_{2,\Omega} + \|w^m_t\|^2_{2,\Omega} + \|\nu w^m_z\|^2_{2,\Omega}\} + p\int_{Q_{t_1,t_2}} r^m(v^m_z)^2 dzdt$$
$$\le p\int_{Q_{t_1,t_2}}\{|(r^m j^m - rj)v^m_z| + |(r^m - r)h_z v^m_z| + |v^m h^m_z w^m_t| + |v^m u^m_t v^m_z|\}dzdt$$
$$+\int_{Q_{t_1,t_2}}|(f^m - f)w^m_t|dzdt + \frac{1}{2}p\|v^m(\cdot,t_1)\|^2_{2,\Omega} + \frac{1}{2}\|w^m_t(\cdot,t_1)\|^2_{2,\Omega} + \frac{1}{2}\|\nu w^m_z(\cdot,t_1)\|^2_{2,\Omega}. \tag{2.3.66}$$

Using (2.3.32), (2.3.35) (with $q = \infty, \gamma = 4$) and (2.3.52) gives us

$$\int_{Q_{t_1,t_2}}|v^m u^m_t v^m_z|dzdt \le \|v^m_z\|_{2,Q_{t_1,t_2}}\cdot\|u^m_t\|_{2,4,Q_{t_1,t_2}}\cdot\|v^m\|_{\infty,4,Q_{t_1,t_2}}$$
$$\le \frac{C_{17}\delta^{1/4}}{\sqrt{2}}\|v^m_z\|_{2,Q_{t_1,t_2}}\cdot(\|v^m_z\|_{2,Q_{t_1,t_2}} + \text{vrai }\max_{t\in[t_1,t_2]}\|v^m\|_{2,\Omega}),$$

with $t_2 - t_1 \le \delta$. To prove the latter inequalities we used (2.3.52) and

$$\|u^m_t\|_{2,4,Q_{t_1,t_2}} = \{\int_{t_1}^{t_2}(\int_\Omega |u^m_t|^2 dz)^2 dt\}^{\frac{1}{4}} \le C_{17}\{\int_{t_1}^{t_2} dt\}^{\frac{1}{4}} \le C_{17}\delta^{\frac{1}{4}}.$$

The other terms are estimated in the similar way

$$\int_{Q_{t_1,t_2}}|v^m h^m_z w^m_t|dzdt \le \|h^m_z\|_{2,Q_{t_1,t_2}}\cdot\|w^m_t\|_{2,4,Q_{t_1,t_2}}\cdot\|v^m\|_{\infty,4,Q_{t_1,t_2}}$$
$$\le \frac{C_{17}\delta^{1/4}}{\sqrt{2}}(\|v^m_z\|_{2,Q_{t_1,t_2}} + \text{vrai }\max_{t\in[t_1,t_2]}\|v^m\|_{2,\Omega})\cdot\text{ vrai }\max_{t\in[t_1,t_2]}\|w^m_t\|_{2,\Omega},$$
$$\int_{Q_{t_1,t_2}}|(r^m j^m - rj)v^m_z|dzdt \le \frac{\epsilon}{2}\|v^m_z\|^2_{2,Q_{t_1,t_2}} + \frac{1}{2\epsilon}\|r^m j^m - rj\|^2_{2,Q_{t_1,t_2}},$$
$$\int_{Q_{t_1,t_2}}|(r^m - r)h_z v^m_z|dzdt \le \frac{\epsilon}{2}\|v^m_z\|^2_{2,Q_{t_1,t_2}} + \frac{1}{2\epsilon}\|(r^m - r)h_z\|^2_{2,Q_{t_1,t_2}},$$
$$\int_{Q_{t_1,t_2}}|(f^m - f)w^m_t|dzdt \le \|w^m_t\|^2_{2,Q_{t_1,t_2}} + \frac{1}{4}\|f^m - f\|^2_{2,Q_{t_1,t_2}}$$
$$\le \delta\text{ vrai }\max_{t\in[t_1,t_2]}\|w^m_t\|^2_{2,\Omega} + \frac{1}{4}\|f^m - f\|^2_{2,Q_{t_1,t_2}}.$$

We can choose ϵ, δ such that the following inequalities are valid

$$5C_{17}\delta^{1/4} \le \sqrt{2},\ p\sqrt{2}C_{17}\delta^{1/4} + 4\delta \le 1,\ 2\epsilon\sqrt{2} + 6C_{17}\delta^{1/4} \le r_0\sqrt{2}.$$

From (2.3.66) this gives

$$\text{vrai }\max_{t\in[t_1,t_2]}\{p\|v^m\|^2_{2,\Omega} + \|w^m_t\|^2_{2,\Omega} + 2\|\nu w^m_z\|^2_{2,\Omega}\} + 2pr_0\|v^m_z\|^2_{2,Q_{t_1,t_2}}$$
$$\le C_{18}p\{\|(r^m - r)h_z\|^2_{2,Q_{t_1,t_2}} + \|r^m j^m - rj\|^2_{2,Q_{t_1,t_2}}\} + \|f^m - f\|^2_{2,Q_{t_1,t_2}}$$
$$+ 2p\|v^m(\cdot,t_1)\|^2_{2,\Omega} + 2\|w^m_t(\cdot,t_1)\|^2_{2,\Omega} + 2\|\nu w^m_z(\cdot,t_1)\|^2_{2,\Omega}$$

with some positive constant C_{18}. Lemma 2.3.2 is proved.

Corollary 2.3.3 *Under the fulfilment of the conditions of Lemma 2.3.2 there is valid the following estimate*

$$p\|v^m(\cdot,t_2)\|^2_{2,\Omega} + \|w^m_t(\cdot,t_2)\|^2_{2,\Omega} + 2\|\nu w^m_z(\cdot,t_2)\|^2_{2,\Omega} + 2pr_0\|v^m_z\|^2_{2,Q_{t_1,t_2}}$$
$$\leq C_{18}p\{\|(r^m - r)h_z\|^2_{2,Q_{t_1,t_2}} + \|r^m j^m - rj\|^2_{2,Q_{t_1,t_2}}\} + \|f^m - f\|^2_{2,Q_{t_1,t_2}}$$
$$+ 2p\|v^m(\cdot,t_1)\|^2_{2,\Omega} + 2\|w^m_t(\cdot,t_1)\|^2_{2,\Omega} + 2\|\nu w^m_z(\cdot,t_1)\|^2_{2,\Omega}.$$

Proof of Theorem 2.3.3 Let us partition the interval $[0,T]$ into a finite number of intervals of length less than δ by the points $0 = t_0 < t_1 < t_2 < \cdots < t_s = T$ with t_i, δ satisfied to the conditions of Lemma 2.3.2. The s is a finite number and does not depend on h^m, u^m, h, u. An inequality of type (2.3.58) is valid for each cylinder $Q_{t_{i-1},t_i} = \Omega \times (t_{i-1},t_i), i = 1,2,\ldots,s$. For this reason we obtain in s steps

$$\operatorname{vrai}\max_{t\in[0,T]}\{p\|v^m\|^2_{2,\Omega} + \|w^m_t\|^2_{2,\Omega} + 2\nu_0^2\|w^m_z\|^2_{2,\Omega}\} + 2pr_0\|v^m_z\|^2_{2,Q_T}$$
$$\leq C_{19}\{p\|(r^m - r)h_z\|^2_{2,Q_T} + p\|r^m j^m - rj\|^2_{2,Q_T} + \|f^m - f\|^2_{2,Q_T}$$
$$+ \|u^m_0 - u_0\|^2_{2,\Omega} + \|u^m_1 - u_1\|^2_{W^1_2(\Omega)} + p\|h^m_0 - h_0\|^2_{2,\Omega}\},$$

where C_{19} is a positive constant. The sequence $\{r^m\}$ is bounded and tends to r a.e. in Ω. Then in view of $h_z \in L_2(Q_T)$ and the Dominated Convergence Theorem, see Theorem 5 in [22, p.648],

$$\|(r^m - r)h_z\|^2_{2,Q_T} \to 0 \text{ as } m \to \infty.$$

Analogously we can show that $\|r^m j^m - rj\|^2_{2,Q_T} \to 0$ as $m \to \infty$ too. This proves Theorem 2.3.3.

3. Inverse Problems

In this section we present some results of the solution to inverse problems for the system of equations describing nonlinear processes of interaction of electromagnetic and elastic waves based on motion of particles. As for the theory of inverse problems and their applications the reader is referred, for example, to [16–18].

3.1. Inverse Problems for the Equations of Electromagnetoelasticity for Weakly Conducting Media

Following the work [19], we present some results of solution of inverse problems for a system of equations describing the linear process of interaction of electromagnetic and elastic waves in a weakly conducting elastic medium. The results of the Cauchy problem solution represented in section 2.1. will be sufficiently used.

Formulation of an Inverse Problem and Main Results

We now consider the case, when the coefficients of equations (2.1.12)–(2.1.15) are known constants outside a finite domain D and are unknown functions of x in D. We also assume that the magnetic permeability μ is constant everywhere in $\mathbb{R}^3$. For simplicity we assume that all the coefficients are functions of the class $C^m(\mathbb{R}^3)$ with a sufficiently large m and are continuous together with all their derivatives of order up to m on the boundary of the domain D. Let S be some closed sufficiently smooth surface enclosing the domain D. Assume that the function $\boldsymbol{V}^1 = (\boldsymbol{H}^1, \boldsymbol{E}^1, \boldsymbol{u}^1)$ (the solution to the Cauchy problem (2.1.12)–(2.1.15)) is known on the set $S \times [0, T]$, where T is a sufficiently large positive number, and x^0 is an arbitrary point of the surface S, i.e.,

$$\boldsymbol{V}^1 = \boldsymbol{F}(x, t, x^0), \quad (x, t, x^0) \in S \times [0, T] \times S \equiv G. \tag{3.1.1}$$

Remark 3.1.1 *In what follows it is convenient for us to mark the dependence of functions on the variable x_0 because in the inverse problems under consideration the source position is a variable parameter which will be used.*

Inverse Problem 3.1.1 *Find the coefficients ε, σ, λ, ρ, and $\varkappa$ of equations (2.1.12)–(2.1.15) in the domain D if the function $\boldsymbol{F}(x, t, x^0)$ is given.*

In the sequel, we will assume that $T \geq d/m_3$, where d is diameter of the domain D and m_3 is the constant from formula (2.1.20). Using representation (2.1.24) from the function $\boldsymbol{F}(x, t, x^0)$ we can uniquely determine the following quantities:

a) $\tau_k(x^0, x), \ (x^0, x) \in S \times S, \ k = 1, 2, 3;$

b) $\boldsymbol{\alpha}_n^{i1}(x^0, x), \boldsymbol{\beta}_n^{i1}(x^0, x), \boldsymbol{\gamma}_n^{i1}(x^0, x), \ (x^0, x) \in S \times S, \ i = 1, 2, 3, \ n \geq -2.$

Inverse Problem 3.1.1 is reduced to the successive solution of the inverse kinematic problem of reconstructing the velocities $c_k(x)$, $k = 1, 2, 3$, inside the domain D from the functions $\tau_k(x^0, x)$, $(x^0, x) \in S \times S$, $k = 1, 2, 3$ and subsequent determination of the still unknown combinations of the sought functions from the functions $\boldsymbol{\alpha}_n^{i1}(x^0, x)$, $\boldsymbol{\beta}_n^{i1}(x^0, x)$, and $\boldsymbol{\gamma}_n^{i1}(x^0, x)$ given on $S \times S$. The theory of solution of the inverse kinematic problem in the case of isotropic media is developed sufficiently well (see, e.g., [17]). In particular, it was established that in the case of simple metrics, the values of the functions $\tau_k(x^0, x)$, $k = 1, 2, 3$, on the set $S \times S$ uniquely define the functions $c_k(x)$, $k = 1, 2, 3$, inside the domain D. Thus, we can think that the velocities $c_k(x)$, $k = 1, 2, 3$, were determined inside the domain D as a result of solution to the corresponding inverse kinematic problems. If we know the functions $c_k(x)$, $k = 1, 2, 3$, then we know three nonlinear combinations of the parameters of the medium (see formulas (2.1.17)). Since the coefficient μ is assumed to be known and constant everywhere, then the coefficient ε is uniquely defined from these relations, and the elastic moduli can be expressed in terms of the density of the medium ρ and the known velocities c_2 and c_3 only. Thus, among the sought parameters only the electrical conductivity σ and the density of the medium ρ remain unknown. To find them, we use the functions $\boldsymbol{\alpha}_n^{i1}(x^0, x)$, $\boldsymbol{\beta}_n^{i1}(x^0, x)$, and $\boldsymbol{\gamma}_n^{i1}(x^0, x)$.

It should be noted that this method of reducing an inverse problem to solution of the inverse kinematic problem and the subsequent use of amplitudes for determining the rest of the relations between the desired coefficients was earlier applied by Romanov ([17], [25]) for investigation of simpler inverse problems for hyperbolic equations of second order.

Consider one possible statement of the inverse problem of reconstructing the parameters σ and ρ; we will use the function $\boldsymbol{\alpha}_{-2}^{11}(x^0,x)$ for its solution. Assume that $\boldsymbol{f}^0 = 0$ and $\sigma(x) \equiv \sigma_0$ for $x \in \mathbb{R}^3 \setminus D$ and $\alpha \neq 0$ for $x \in D$. We will show that the problem under consideration can be reduced to the following integral geometry problem: Consider the integrals

$$\int_{\Gamma_1(x^0,x)} (a\eta_1 + b)ds = p_1(x^0,x), \quad (x^0,x) \in S \times S, \tag{3.1.2}$$

$$\int_{\Gamma_1(x^0,x)} a\eta_2 ds = p_2(x^0,x), \quad (x^0,x) \in S \times S. \tag{3.1.3}$$

Assume that we know the functions $p_k(x^0,x)$, $k=1,2$, with the weights η_k, $k=1,2$, also known. It is required to determine the coefficients a and b inside the domain D if they are known outside of D. The coefficients a and b are related to the parameters of the medium σ and ρ by the formulas

$$a = \frac{\sigma\alpha}{2\rho c_1 \epsilon^2 \left(c_1^2 - c_2^2\right)}, \qquad b = -\frac{\sigma}{2c_1\epsilon}.$$

The weight functions η_k, $k=1,2$, are defined by the relations

$$\begin{aligned} \eta_1 &= (\boldsymbol{h}^0 \cdot \boldsymbol{w}^0)^2 \, |\boldsymbol{w}^0|^{-2} + \xi\, (\boldsymbol{h}^0 \cdot \boldsymbol{\nu})^2, \\ \eta_2 &= (\boldsymbol{h}^0 \cdot \boldsymbol{w}^0) \left(\boldsymbol{h}^0 \cdot (\boldsymbol{\nu} \times \boldsymbol{w}^0)\right) |\boldsymbol{w}^0|^{-2}, \end{aligned} \tag{3.1.4}$$

where $\xi = (c_1^2 - c_2^2)/(c_1^2 - c_3^2)$, $\xi \in (0,1)$; $\boldsymbol{\nu}$ is an unit vector tangential to $\Gamma_1(x^0,x)$ at the point x; and $\boldsymbol{w}^0$ is the solution to problem (3.1.26)–(3.1.28); this solution is completely defined by specifying the quantities $\boldsymbol{j}^0$ and $c_1(x)$.

Remark 3.1.2 *The functions a and b uniquely define only σ and σ/ρ. Thus, the density of medium can be uniquely defined only at those points of the domain D, where $\sigma \neq 0$. This characteristic feature of the problem concerned is a consequence of the fact that we use the information about the coefficient $\boldsymbol{\alpha}_{-2}^{11}$. One can show that in the case, when $\boldsymbol{\alpha}_0^{31}(x^0,x)$, $(x^0,x) \in S \times S$ is used as additional information (provided $\boldsymbol{f}^0 \neq 0$), the density of medium is uniquely defined in the domain D.*

The fact that equalities (3.1.2) and (3.1.3) contain the weight functions η_k, $k=1,2$, complicates the solution of the corresponding integral geometry problem in the general case. So, we give only three special formulations of the original problem.

1. Let $c_1 = \text{const} = c_{10}$, $\boldsymbol{j}^0 = (1,0,0) \cdot 4\pi c_{10}^2$, $\boldsymbol{h}^0 = (1,0,1)$. Let $S(z)$ and $D(z)$ denote cross-sections of the surface S and the domain D by the plane $x_3 = z$. In this case, the integral geometry problem splits to a one-parameter family of planar problems for each cross-section of the domain D by the plane $x_3 = z$. The vector $\boldsymbol{\nu}$ of the corresponding line $\Gamma_1(x^0,x)$, $(x^0,x) \in S(z) \times S(z)$, can be characterized in this case by the angular coordinate φ: $\boldsymbol{\nu} = (\cos\varphi, \sin\varphi, 0)$. This coordinate is constant along $\Gamma_1(x^0,x)$ and, consequently, depends only on the boundary points $(x^0,x) \in S(z) \times S(z)$. Therefore, the weight functions η_k, $k = 1, 2$, can be written down as $\eta_1 = 1 + \xi \cos^2 \varphi$, $\eta_2 = \sin \varphi$. Since the weight function η_2 is constant along $\Gamma_1(x^0, x)$, equation (3.1.3) can be written down as

$$\int_{\Gamma_1(x^0,x)} a ds = p_2'(x^0, x), \quad (x^0, x) \in S \times S, \tag{3.1.5}$$

where $p_2' = p_2 / \sin \varphi$. The problem of determining the function a from equation (3.1.5) is a standard integral geometry problem. Having found the function a, we can calculate the integral of the product $a\eta_1$ along $\Gamma_1(x^0, x)$ in formula (3.1.2). This yields a similar integral geometry problem for determining the coefficient b. The uniqueness of its solution is evident. A conditional stability estimate is less evident because in the general case the weight function η_1 is not constant on $\Gamma_1(x^0, x)$.

The following theorem holds.

Theorem 3.1.1 *Let* (a, b) *and* $(\bar{a}, \bar{b})$ *be two solutions to equations* (3.1.2), (3.1.3) *corresponding to the data* (p_1, p_2) *and* $(\bar{p}_1, \bar{p}_2)$, *respectively, and coinciding outside the domain D. Then the following estimates hold for the differences* $\tilde{a} = a - \bar{a}$, $\tilde{b} = b - \bar{b}$, $\tilde{p}_1 = p_1 - \bar{p}_1$, *and* $\tilde{p}_2 = p_2 - \bar{p}_2$*:*

$$\int_{D(z)} \tilde{a}^2 d\bar{x} \le \frac{1}{2\pi} \int_{S(z)} dl' \int_{S(z)} \left| \frac{\partial \tilde{p}_2}{\partial l} \cdot \frac{\partial \tilde{p}_2}{\partial l'} \right| dl, \tag{3.1.6}$$

$$\int_{D(z)} \tilde{b}^2 d\bar{x} \le \frac{1}{\pi} \int_{S(z)} dl' \int_{S(z)} \left\{ \frac{23}{8} \left| \frac{\partial \tilde{p}_2}{\partial l} \cdot \frac{\partial \tilde{p}_2}{\partial l'} \right| + \left| \frac{\partial \tilde{p}_1}{\partial l} \cdot \frac{\partial \tilde{p}_1}{\partial l'} \right| \right\} dl, \tag{3.1.7}$$

where $d\bar{x} = dx_1 dx_2$*;* $\partial/\partial l$ *and* $\partial/\partial l'$ *denote derivatives in the directions tangential to* $S(z)$ *at the points* x^0 *and* x, *respectively; and* dl *and* dl' *are the length elements calculated at these points.*

It is more convenient to give the proof of Theorem 3.1.1 later.

2. Consider a slightly modified 2D version of the problem. Namely, let D be an infinite cylindrical domain with the ruling parallel to the axis x_3: $D = D_0 \times \mathbb{R}$, where D_0 is a cross-section of the domain D by the plane $x_3 = 0$. Assume that coefficients of the problem are independent of the variable x_3. Let Γ be a simple smooth closed curve enclosing the domain D_0; and let the vectors $\boldsymbol{j}^0$ and $\boldsymbol{h}^0$ be the same as in the previous case. In this case, the geodesics $\Gamma_1(x^0, x)$, $(x^0, x) \in S \times S$, are planar curves. Denoting the unit vector tangential to $\Gamma_1(x^0, x)$ at the point x by $\boldsymbol{\nu} = (\cos \varphi, \sin \varphi, 0)$, we conclude that in this case we also have $\eta_2 = \sin \varphi$ and $\eta_1 = 1 + \xi \cos^2 \varphi$. There is an essential distinction between this and the previous cases; namely, the weight function is not constant along a geodesic line. In this case, equation (3.1.3) can be written down as

$$\int_{\Gamma_1(x^0,x)} a dx_2 = p_2(x^0, x), \quad (x^0, x) \in S \times S. \tag{3.1.8}$$

Thus, p_2 is an integral along $\Gamma_1(x^0, x)$ of the simplest form of the first degree. The problems of finding a form of the first degree from its integrals along a family of geodesics of a Riemannian metric were considered in [17], [26], [27].

Assume that there are two solutions a and $\bar{a}$ to equation (3.1.8) corresponding to the function p_2 and coinciding outside the domain D_0. Then their difference $\tilde{a} = a - \bar{a}$ solves a homogeneous equation. It follows from [26] that $\partial\tilde{a}/\partial x_1 = 0$, $x \in D_0$. Since $\tilde{a} = 0$ outside the domain D_0, we conclude that $\tilde{a} = 0$, $x \in D_0$, i.e., $a = \bar{a}$. Thus, the following uniqueness theorem holds:

Theorem 3.1.2 *Specifying the functions p_1 and p_2 for $(x^0, x) \in S \times S$ uniquely defines the functions a and b in the domain D_0.*

It should be noted that certain estimates of the conditional stability of the solution to equations (3.1.2), (3.1.8) can be obtained from the estimates given in [27].

3. Let us turn to the original formulation of the problem. We consider the case, when the velocity c_1 essentially depends on all three variables x_1, x_2, and x_3. Let $\boldsymbol{j}^0$ be a constant nonzero vector. We also assume that the vector $\boldsymbol{h}^0$ may have two different values $\boldsymbol{h}^{0k}$, $k = 1, 2$, where $\boldsymbol{h}^{02} = r\boldsymbol{h}^{01}$, $r^2 \neq 1$, and that for each vector $\boldsymbol{h}^{0k}$ the function $\boldsymbol{\alpha}^1_{-2}(x^0, x)$ and, respectively, the functions $p_k(x^0, x)$, $(x^0, x) \in S \times S$, are known. We denote by p_{1k}, $k = 1, 2$, the function p_1 corresponding to the vectors $\boldsymbol{h}^{0k}$, $k = 1, 2$. From equality (3.1.2) for $\boldsymbol{h}^0 = \boldsymbol{h}^{0k}$, $k = 1, 2$, we readily obtain

$$\int_{\Gamma_1(x^0,x)} a\eta_0 ds = \bar{p}_1(x^0, x), \quad (x^0, x) \in S \times S, \tag{3.1.9}$$

$$\int_{\Gamma_1(x^0,x)} b ds = \bar{p}_2(x^0, x), \quad (x^0, x) \in S \times S, \tag{3.1.10}$$

where

$$\eta_0 = (\boldsymbol{h}^{01} \cdot \boldsymbol{w}^0)^2 \, |\boldsymbol{w}^0|^{-2} + \xi \, (\boldsymbol{h}^{01} \cdot \boldsymbol{\nu})^2,$$

$$\bar{p}_1 = \frac{p_{11} - p_{12}}{1 - r^2}, \qquad \bar{p}_2 = \frac{p_{12} - p_{11} r^2}{1 - r^2}.$$

Thus, using the data corresponding to two different values of the vector $\boldsymbol{h}^0$, we can split the original problem in two problems independent of each other. The problem of determining the function b from equation (3.1.10) is an integral geometry problem which has been studied in full (see [17].

The following conditional stability theorem (an analogue of estimate (3.54) from [17]) holds for this problem.

Theorem 3.1.3 *Let b and b' be two solutions to equation* (3.1.10) *corresponding to the right-hand sides $\bar{p}_2$ and $\bar{p}'_2$. Then the following inequality holds for the differences $\tilde{b} = b - b'$ and $\tilde{p}_2 = \bar{p}_2 - \bar{p}'_2$:*

$$\int_D c_1^{-1} \tilde{b}^2 dx \leq \frac{1}{8\pi} \int_S \int_S |\mathrm{grad}_x \tilde{p}_2 \times \boldsymbol{n}(x)| \cdot |\mathrm{grad}_{x^0} \tilde{p}_2 \times \boldsymbol{n}(x^0)|$$

$$\times \left(\sum_{k,j=1}^{3} \left(\frac{\partial^2 \tau_1}{\partial x_k \, \partial x_j^0} \right)^2 \right)^{1/2} dS_x dS_{x^0}, \tag{3.1.11}$$

where $\boldsymbol{n}(x)$ is a normal vector to the surface S at the point x.

As concerns the properties of equation (3.1.9), their investigation requires additional efforts.

Reduction to an Integral Geometry Problem

The differential and integral equations considered here correspond to inverse problem 3.1.1, which was formulated earlier. As the first step, let us derive a differential equation for finding the function $\boldsymbol{\alpha}^{11}_{-2}(x)$. Let $\boldsymbol{f}^0 = 0$. Then $\boldsymbol{u}^0 \equiv 0$ and, as consequence, $\boldsymbol{\alpha}^{31}_k = \boldsymbol{\beta}^{31}_k = \gamma^{31}_k \equiv 0$ for $k < 0$. This means that the function $\boldsymbol{u}$ has no singular part. Let us substitute the representation (2.1.24) into equations (2.1.8)–(2.1.15). Equating the coefficients at $\theta_n(S_1)$, $n = -3, -2$, we obtain the following relationships between $\boldsymbol{\alpha}^{10}_{-2}$, $\boldsymbol{\alpha}^{20}_{-2}$, $\boldsymbol{\alpha}^{10}_{-1}$, $\boldsymbol{\alpha}^{20}_{-1}$ and $\boldsymbol{\alpha}^{11}_{-2}$, $\boldsymbol{\alpha}^{21}_{-2}$, $\boldsymbol{\alpha}^{11}_{-1}$, $\boldsymbol{\alpha}^{21}_{-1}$, $\boldsymbol{\alpha}^{31}_0$:

$$\operatorname{grad}\tau_1 \times \boldsymbol{\alpha}^{10}_{-2} + \varepsilon\boldsymbol{\alpha}^{20}_{-2} = 0, \qquad \operatorname{grad}\tau_1 \times \boldsymbol{\alpha}^{20}_{-2} - \mu\boldsymbol{\alpha}^{10}_{-2} = 0, \tag{3.1.12}$$

$$\begin{aligned} &\operatorname{grad}\tau_1 \times \boldsymbol{\alpha}^{10}_{-1} + \varepsilon\boldsymbol{\alpha}^{20}_{-1} - \operatorname{rot}\boldsymbol{\alpha}^{10}_{-2} = 0,\\ &\operatorname{grad}\tau_1 \times \boldsymbol{\alpha}^{20}_{-1} - \mu\boldsymbol{\alpha}^{10}_{-1} - \operatorname{rot}\boldsymbol{\alpha}^{20}_{-2} = 0, \end{aligned} \tag{3.1.13}$$

$$\operatorname{grad}\tau_1 \times \boldsymbol{\alpha}^{11}_{-2} + \varepsilon\boldsymbol{\alpha}^{21}_{-2} = 0, \qquad \operatorname{grad}\tau_1 \times \boldsymbol{\alpha}^{21}_{-2} - \mu\boldsymbol{\alpha}^{11}_{-2} = 0, \tag{3.1.14}$$

$$\begin{aligned} &\operatorname{grad}\tau_1 \times \boldsymbol{\alpha}^{11}_{-1} + \varepsilon\boldsymbol{\alpha}^{21}_{-1} - \operatorname{rot}\boldsymbol{\alpha}^{11}_{-2} + \alpha\boldsymbol{\alpha}^{31}_0 \times \boldsymbol{h}^0 + \sigma\boldsymbol{\alpha}^{20}_{-2} = 0,\\ &\operatorname{grad}\tau_1 \times \boldsymbol{\alpha}^{21}_{-1} - \mu\boldsymbol{\alpha}^{11}_{-1} - \operatorname{rot}\boldsymbol{\alpha}^{21}_{-2} = 0, \end{aligned} \tag{3.1.15}$$

$$\begin{aligned} \rho\boldsymbol{\alpha}^{31}_0 - (\lambda + 2\varkappa)\,(\boldsymbol{\alpha}^{31}_0 \cdot \operatorname{grad}\tau_1)\operatorname{grad}\tau_1 + \varkappa\left(\operatorname{grad}\tau_1 \times (\operatorname{grad}\tau_1 \times \boldsymbol{\alpha}^{31}_0)\right)&\\ - \sigma\mu\boldsymbol{\alpha}^{20}_{-2} \times \boldsymbol{h}^0 = 0.& \end{aligned} \tag{3.1.16}$$

Using formulas (2.1.19), we can transform relations (3.1.12) and (3.1.14) to the equalities

$$\begin{aligned} &\operatorname{grad}\tau_1 \cdot \boldsymbol{\alpha}^{10}_{-2} = 0, \qquad \operatorname{grad}\tau_1 \cdot \boldsymbol{\alpha}^{20}_{-2} = 0,\\ &\operatorname{grad}\tau_1 \cdot \boldsymbol{\alpha}^{11}_{-2} = 0, \qquad \operatorname{grad}\tau_1 \cdot \boldsymbol{\alpha}^{21}_{-2} = 0, \end{aligned} \tag{3.1.17}$$

which express the principle of orthogonality of the electromagnetic field to the direction of propagation of its main singularity. To derive equations for the coefficient $\boldsymbol{\alpha}^{10}_{-2}$ we proceed as follows. We take the vector product of the first relation from (3.1.13) by $\operatorname{grad}\tau_1$ and then eliminate the term $\operatorname{grad}\tau_1 \times \boldsymbol{\alpha}^{20}_{-1}$ with the help of the second relation from (3.1.13). Using the first equality from (3.1.12), we eliminate $\boldsymbol{\alpha}^{20}_{-2}$ and, finally, obtain

$$\begin{aligned} \operatorname{grad}\tau_1 \times (\operatorname{grad}\tau_1 \times \boldsymbol{\alpha}^{10}_{-1}) + \varepsilon\mu\boldsymbol{\alpha}^{10}_{-1} - \varepsilon\operatorname{rot}\left(\varepsilon^{-1}(\operatorname{grad}\tau_1 \times \boldsymbol{\alpha}^{10}_{-2})\right)&\\ - \operatorname{grad}\tau_1 \times \operatorname{rot}\boldsymbol{\alpha}^{10}_{-2} = 0.& \end{aligned}$$

For further simplification, we use the following equalities, which are well-known from the vector analysis:

$$\begin{aligned} \boldsymbol{c} \times (\boldsymbol{a} \times \boldsymbol{b}) &= \boldsymbol{a}\,(\boldsymbol{c}\cdot\boldsymbol{b}) - \boldsymbol{b}\,(\boldsymbol{c}\cdot\boldsymbol{a}),\\ \operatorname{rot}(\boldsymbol{a} \times \boldsymbol{b}) &= \boldsymbol{a}\cdot\operatorname{div}\boldsymbol{b} - \boldsymbol{b}\cdot\operatorname{div}\boldsymbol{a} + (\boldsymbol{b}\cdot\operatorname{grad})\boldsymbol{a} - (\boldsymbol{a}\cdot\operatorname{grad})\boldsymbol{b},\\ \boldsymbol{a} \times \operatorname{rot}\boldsymbol{b} &= \operatorname{grad}(\boldsymbol{a}\cdot\boldsymbol{b}) - (\boldsymbol{b}\cdot\operatorname{grad})\boldsymbol{a} - (\boldsymbol{a}\cdot\operatorname{grad})\boldsymbol{b} - \boldsymbol{b} \times \operatorname{rot}\boldsymbol{a}. \end{aligned}$$

Taking into account formulas (2.1.19) for τ_1 and the equality $\operatorname{grad}\tau_1\cdot\boldsymbol{\alpha}_{-1}^{10}=\operatorname{div}\boldsymbol{\alpha}_{-2}^{10}$, which follows from the second relation from (3.1.13), we obtain

$$L\boldsymbol{\alpha}_{-2}^{10}\equiv 2\operatorname{grad}\boldsymbol{\alpha}_{-2}^{10}\cdot\operatorname{grad}\tau_1+\boldsymbol{\alpha}_{-2}^{10}(\Delta\tau_1-\operatorname{grad}\ln\varepsilon\cdot\operatorname{grad}\tau_1) + (\operatorname{grad}\ln\varepsilon\cdot\boldsymbol{\alpha}_{-2}^{10})\operatorname{grad}\tau_1=0. \quad (3.1.18)$$

The relations for $\boldsymbol{\alpha}_{-2}^{11}$ are derived in a similar way:

$$L\boldsymbol{\alpha}_{-2}^{11}=2c_1^{-1}\boldsymbol{F}, \quad (3.1.19)$$

where

$$\boldsymbol{F}\equiv-\frac{c_1}{2}\operatorname{grad}\tau_1\times(\alpha\boldsymbol{\alpha}_0^{31}\times\boldsymbol{h}^0+\sigma\boldsymbol{\alpha}_{-2}^{20}).$$

Using formula (3.1.16), we can find $\boldsymbol{\alpha}_0^{31}$:

$$\boldsymbol{\alpha}_0^{31}=\frac{\sigma}{\rho}\left[\frac{\boldsymbol{h}^0\cdot\boldsymbol{\alpha}_{-2}^{10}}{c_1^2-c_2^2}\operatorname{grad}\tau_1-\frac{\boldsymbol{h}^0\cdot\operatorname{grad}\tau_1}{c_1^2-c_3^2}\boldsymbol{\alpha}_{-2}^{10}\right]. \quad (3.1.20)$$

Thus, the final formula for the function $\boldsymbol{F}$ takes the form

$$\boldsymbol{F}=a\Big\{(\boldsymbol{h}^0\cdot\boldsymbol{\alpha}_{-2}^{10})\left[\boldsymbol{h}^0-c_1^2(\boldsymbol{h}^0\cdot\operatorname{grad}\tau_1)\operatorname{grad}\tau_1\right] + \xi c_1^2(\boldsymbol{h}^0\cdot\operatorname{grad}\tau_1)^2\boldsymbol{\alpha}_{-2}^{10}\Big\}+b\boldsymbol{\alpha}_{-2}^{10}, \quad (3.1.21)$$

where

$$a=\frac{\sigma\alpha}{2\rho\,(c_1^2-c_2^2)\,\varepsilon^2c_1},\qquad b=-\frac{\sigma}{2c_1\varepsilon},\qquad \xi=\frac{c_1^2-c_2^2}{c_1^2-c_3^2},\qquad \xi\in(0,1). \quad (3.1.22)$$

Equations (3.1.21) and (3.1.22) imply that the coefficient $\boldsymbol{\alpha}_{-2}^{11}$ linearly depends on σ and σ/ρ. Consequently, in this case the inverse problem can be treated as problem of determining σ and σ/ρ or the coefficients a and b, which is the same.

The equations for the geodesics $\Gamma_1(x^0,x)$ can be written down as

$$\frac{dx}{ds}=\boldsymbol{\nu},\qquad \frac{d\boldsymbol{\nu}}{ds}=-\operatorname{grad}\ln c_1+(\operatorname{grad}\ln c_1\cdot\boldsymbol{\nu})\,\boldsymbol{\nu}, \quad (3.1.23)$$

where s is the Euclidean length of the curve passing through the point x_0 and $\boldsymbol{\nu}=c_1\operatorname{grad}\tau_1$ is the unit vector tangential to $\Gamma_1(x^0,x)$ at the point x and oriented in the direction of increasing s.

Let us show that along the geodesic $\Gamma_1(x^0,x)$ equations (3.1.18) and (3.1.19) are ordinary differential equations. Indeed,

$$2\sum_{k=1}^{3}\frac{\partial\boldsymbol{\alpha}_{-2}^{10}}{\partial x_k}\frac{\partial\tau_1}{\partial x_k}=2c_1^{-1}\frac{d\boldsymbol{\alpha}_{-2}^{11}}{ds}.$$

Since the magnetic permeability μ is constant, we have $\operatorname{grad}\ln\varepsilon=-\operatorname{grad}\ln c_1^2$. Further in the treatment we use the equality

$$\Delta\tau_1+\operatorname{grad}\ln c_1^2\cdot\operatorname{grad}\tau_1=2c_1^{-1}\frac{d}{ds}\ln\left[\tau_1\Big(\det\frac{\partial\boldsymbol{g}(x^0,x)}{\partial x}\Big)^{-1/2}\right],$$

which follows immediately from [17, p. 116, formula (4.18)]. The function $\boldsymbol{g}(x^0, x)$ defines the Riemannian coordinates of the point x in the metric $d\tau_1 = c_1^{-1} ds$, which corresponds to the point x^0 and in our case can be found by the formula

$$\boldsymbol{g}(x^0, x) = -(2c_1^2(x^0))^{-1} \operatorname{grad}_{x^0} \tau_1^2(x^0, x).$$

Under the above assumptions, this function is smooth in both variables, namely, $\boldsymbol{g}(x^0, x) \in C^{m-2}(\mathbb{R}^6)$. In a homogeneous space we have $\boldsymbol{g}(x^0, x) \equiv x - x^0$.

Let us introduce the new functions

$$\boldsymbol{w}^0(x^0, x) = \boldsymbol{\alpha}_{-2}^{10}(x^0, x)\, \tau_1(x^0, x) \left(\det \frac{\partial \boldsymbol{g}(x^0, x)}{\partial x}\right)^{-1/2}, \tag{3.1.24}$$

$$\boldsymbol{w}(x^0, x) = \boldsymbol{\alpha}_{-2}^{11}(x^0, x)\, \tau_1(x^0, x) \left(\det \frac{\partial \boldsymbol{g}(x^0, x)}{\partial x}\right)^{-1/2}. \tag{3.1.25}$$

Then the differential equations for the functions $\boldsymbol{w}^0(x^0, x)$ and $\boldsymbol{w}(x^0, x)$ along the geodesic $\Gamma_1(x^0, x)$ take the form:

$$\frac{d\boldsymbol{w}^0}{ds} - (\boldsymbol{w}^0 \cdot \operatorname{grad} \ln c_1)\, \boldsymbol{\nu} = 0, \qquad \frac{d\boldsymbol{w}}{ds} - (\boldsymbol{w} \cdot \operatorname{grad} \ln c_1)\, \boldsymbol{\nu} = \boldsymbol{F}, \tag{3.1.26}$$

where

$$\boldsymbol{F} = a \left\{ (\boldsymbol{h}^0 \cdot \boldsymbol{w}^0) \left[\boldsymbol{h}^0 - \boldsymbol{\nu}\, (\boldsymbol{h}^0 \cdot \boldsymbol{\nu})\right] + \xi\, (\boldsymbol{h}^0 \cdot \boldsymbol{\nu})^2\, \boldsymbol{w}^0 \right\} + b\boldsymbol{w}^0. \tag{3.1.27}$$

The initial conditions at the point $s = 0$ can be found by comparing the singularities of the solutions $(\boldsymbol{H}^0, \boldsymbol{E}^0, \boldsymbol{u}^0)$ and $(\boldsymbol{H}, \boldsymbol{E}, \boldsymbol{u})$ in a homogeneous medium. Since the function $\boldsymbol{H}^0$ in a homogeneous medium is calculated by the formula

$$\boldsymbol{H}^0 = \operatorname{rot} \left[\frac{\boldsymbol{j}^0}{4\pi\, |x - x^0|}\, \theta_{-1}(S_1) \right],$$

then it follows that

$$\boldsymbol{w}^0 = (4\pi c_{10})^{-1} \boldsymbol{j}^0 \times \operatorname{grad} \tau_1 \big|_{x=x^0} = (4\pi c_{10}^2)^{-1} (\boldsymbol{j}^0 \times \boldsymbol{\nu}^0), \tag{3.1.28}$$

where $\boldsymbol{\nu}^0$ is the unit vector tangential to $\Gamma_1(x^0, x)$ at the point x^0 and c_{10} is the value of the velocity c_1 outside the domain D. Analyzing the solution to problem (2.1.12)–(2.1.15) corresponding to the homogeneous medium (see [17][Sec. 4]), we derive the following initial data for $\boldsymbol{w}$:

$$\boldsymbol{w}\big|_{s=0} = (4\pi c_{10}^2)^{-1} (\boldsymbol{j}^0 \times \boldsymbol{\nu}^0). \tag{3.1.29}$$

It should be noted that the vector-functions $\boldsymbol{w}^0$ and $\boldsymbol{w}$ are orthogonal to the vector $\boldsymbol{\nu}(x^0, x) = c_1 \operatorname{grad} \tau_1(x^0, x)$. The first equation from (3.1.26) implies that $d|\boldsymbol{w}^0|^2/ds = 0$, i.e., the function $|\boldsymbol{w}^0|$ is constant along the geodesic $\Gamma_1(x^0, x)$:

$$|\boldsymbol{w}^0| = (4\pi c_{10}^2)^{-1}\, |\boldsymbol{j}^0 \times \boldsymbol{\nu}^0|. \tag{3.1.30}$$

Let us represent the function $\boldsymbol{w}$ in the form

$$\boldsymbol{w} = (1 + p_1)\boldsymbol{w}^0 + p_2(\boldsymbol{\nu} \times \boldsymbol{w}^0). \tag{3.1.31}$$

Trivial calculations bring about the following equations for defining the functions p_1 and p_2:

$$\frac{dp_1}{ds} = a\eta_1 + b, \qquad p_1\big|_{s=0} = 0, \tag{3.1.32}$$

$$\frac{dp_2}{ds} = a\eta_2, \qquad p_2\big|_{s=0} = 0, \tag{3.1.33}$$

where

$$\begin{aligned} \eta_1 &= (\boldsymbol{h}^0 \cdot \boldsymbol{w}^0)^2 \, |\boldsymbol{w}^0|^{-2} + \xi \, (\boldsymbol{h}^0 \cdot \boldsymbol{w}^0)^2, \\ \eta_2 &= (\boldsymbol{h}^0 \cdot \boldsymbol{w}^0) \left(\boldsymbol{h}^0 \cdot (\boldsymbol{\nu} \times \boldsymbol{w}^0)\right) |\boldsymbol{w}^0|^{-2}. \end{aligned} \tag{3.1.34}$$

Integrating equations (3.1.32) and (3.1.33) along $\Gamma_1(x^0, x)$, $(x^0, x) \in S \times S$, we arrive at:

$$\int_{\Gamma_1(x^0,x)} (a\eta_1 + b)ds = p_1(x^0, x), \quad (x^0, x) \in S \times S, \tag{3.1.35}$$

$$\int_{\Gamma_1(x^0,x)} a\eta_2 ds = p_2(x^0, x), \quad (x^0, x) \in S \times S. \tag{3.1.36}$$

Since the function $\boldsymbol{\alpha}^{11}_{-2}$ is known for $(x^0, x) \in S \times S$, therefore, the functions $p_k(x^0, x)$, $k = 1, 2$, are also known for $(x^0, x) \in S \times S$. Consequently, the problem of determining the functions a and b from equations (3.1.35) and (3.1.36) is an integral geometry problem. In a special case, when $c_1 = \text{const} = c_{10}$ and the geodesics $\Gamma_1(x^0, x)$ are straight lines, this problem is a problem of tomography.

Proof of Theorem 3.1.1 Note that estimate (3.1.6) was obtained in [28], so we only need to prove estimate (3.1.7). Following the technique developed in [28], we can write down the inequality as

$$\int_{D(z)} \int_0^{2\pi} \left[(\tilde{a}\eta_1 + \tilde{b})^2 - \left(\tilde{a}\frac{\partial \eta_1}{\partial \varphi}\right)^2 \right] d\varphi d\bar{x} \le \int_{S(z)} \int_{S(z)} \left| \frac{\partial \tilde{p}_1}{\partial l} \cdot \frac{\partial \tilde{p}_1}{\partial l'} \right| dl dl', \tag{3.1.37}$$

which is an analogue of inequality (3.1.6). Since

$$\begin{aligned} (\tilde{a}\eta_1 + \tilde{b})^2 - \left(\tilde{a}\frac{\partial \eta_1}{\partial \varphi}\right)^2 &\ge \frac{1}{2}\tilde{b}^2 - \tilde{a}^2 \left(\eta_1^2 + \eta_{1\varphi}^2\right) \\ &\ge \frac{1}{2}\tilde{b}^2 - \tilde{a}^2 \left[(1 + \cos^2 \varphi)^2 + \sin^2 2\varphi\right], \end{aligned}$$

we have

$$\int_0^{2\pi} \left[(\tilde{a}\eta_1 + \tilde{b})^2 - \left(\tilde{a}\frac{\partial \eta_1}{\partial \varphi}\right)^2 \right] d\varphi \ge \pi \left(\tilde{b}^2 - \frac{23}{4}\tilde{a}^2 \right).$$

Consequently,

$$\int_{D(z)} \left(\tilde{b}^2 - \frac{23}{4}\tilde{a}^2 \right) d\bar{x} \le \frac{1}{\pi} \int_{S(z)} \int_{S(z)} \left| \frac{\partial \tilde{p}_1}{\partial l} \cdot \frac{\partial \tilde{p}_1}{\partial l'} \right| dl dl'. \tag{3.1.38}$$

Estimate (3.1.7) follows immediately from inequalities (3.1.6) and (3.1.38). This proves the theorem.

3.2. An Inverse Problem for Electromagnetoelasticity Equations with Partially Nonlinear Interaction

Following the original work [20], we present some results to solution of inverse problems for a system of equations of electromagnetoelasticity with partially nonlinear interaction between electromagnetic and elastic fields. The results of the first initial boundary-value problem solution presented in section 2.2. will be sufficiently used.

Formulation of an Inverse Problem

Our main inverse problem consists in determining the function f from equations (2.2.6)–(2.2.15) by appropriate additional information about the solution to Direct Problem 2.2.1.

Now we can formulate the inverse problem.

Inverse Problem 3.2.1 *Determine a set of the functions*

$$\boldsymbol{u} : [0,T] \times \Omega_2 \to \mathbb{R}^3, \quad \boldsymbol{E}, \boldsymbol{H} : [0,T] \times \Omega \to \mathbb{R}^3, \quad f : [0,T] \to \mathbb{R}$$

from equations (2.2.6)–(2.2.15) *and*

$$\Phi[\boldsymbol{E}] = \phi(t), \quad t \in [0,T]. \tag{3.2.1}$$

It is assumed that the functions $\varepsilon, \mu : \overline{\Omega} \to \mathbb{R}_+$, $\sigma : \overline{\Omega} \to \overline{\mathbb{R}}_+$, and $\boldsymbol{E}_0, \boldsymbol{H}_0 : \Omega \to \mathbb{R}^3$ are continuous in the domain $\overline{\Omega} \setminus \partial\Omega_2$ with possible jumps on the surface $\partial\Omega_2$. We also assume that the functions $\boldsymbol{g} : [0,T] \times \Omega_2 \to \mathbb{R}^3$, $\boldsymbol{u}_0, \boldsymbol{u}_1 : \Omega_2 \to \mathbb{R}^3$, and $\phi : [0,T] \to \mathbb{R}$ are given and have sufficient smoothness. As concerns the functional Φ, we assume that it is linear and depends only on the spatial variables. For example, we may assume that the functional Φ has the form

$$\Phi[\boldsymbol{E}] = \int_{\Omega} \boldsymbol{K}(x) \cdot \boldsymbol{E}(x) dx, \tag{3.2.2}$$

where $\boldsymbol{K} : \Omega \to \mathbb{R}^3$ is a given sufficiently smooth vector-function.

Remark 3.2.1 *It is the additional information* (3.2.1) *only that connects the solutions to the Lamé equations and the Maxwell equations. If instead of the additional information* (3.2.1) *we specify a certain information about the vector-function* $\boldsymbol{u}$*, for example, of the form*

$$\Phi[\boldsymbol{u}] = \phi(t), \quad t \in [0,T], \tag{3.2.3}$$

then Inverse Problem 3.2.1 splits to the inverse problem of determining the function $f(t)$ *from equations* (2.2.6)–(2.2.9) *and* (3.2.3) *and then the direct problem* (2.2.10)–(2.2.15).

Solution to Inverse Problem 3.2.1

We now can prove solvability of the inverse problem. We assume that the function f in the free term of equation (3.2.1) is unknown and the function $\boldsymbol{g}$ is known. By Theorems 2.2.1 and 2.2.2, our Inverse Problem 3.2.1 is equivalent to the following problem: find a function $f \in L^p((0,T);\mathbb{R})$ such that

$$\Phi[\boldsymbol{E}(f)(t,\cdot)] = \phi(t), \quad \forall t \in [0,T]. \tag{3.2.4}$$

To prove solvability of problem 3.2.4 we require the fulfilment of the following regularity conditions for the data ϕ, the function $\boldsymbol{g}$, and the kernel $\boldsymbol{K}$ of the functional Φ:

$$\phi \in W^{2,p}\big((0,T_0);\mathbb{R}\big), \quad \boldsymbol{g} \in L^{p'}\big((0,T_0);H^2(\Omega_2;\mathbb{R}^3)\cap H_0^1(\Omega_2;\mathbb{R}^3)\big), \quad T_0 > 0; \tag{3.2.5}$$

$$\|\boldsymbol{g}(t,\cdot)\|_{2,2,\Omega} \le C_6(T_1)\, t^{-\alpha} \tag{3.2.6}$$

for almost all $t \in (0,T_1)$ and certain $\alpha \in (0,1/p')$ and $T_1 \in (0,T_0)$; and

$$\varepsilon^{-1}\boldsymbol{K} \in H(\mathrm{rot};\Omega), \qquad \boldsymbol{n}\times\boldsymbol{K} = 0, \quad \forall x \in \partial\Omega. \tag{3.2.7}$$

Remark 3.2.2 *The first assumption in* (3.2.7) *is equivalent to the requirement that* $\boldsymbol{K} \in H(\mathrm{rot};\Omega_1)\cap H(\mathrm{rot};\Omega_2)$ *and* $\boldsymbol{n}\times[\varepsilon^{-1}\boldsymbol{K}]_{\partial\Omega_2} = 0$.

Using the results by [29] and formulas (3.2.2) and (3.2.7)) we easily obtain the relation

$$\Phi[\varepsilon^{-1}\,\mathrm{rot}\,\boldsymbol{H}(t,\cdot)] = \int_\Omega \boldsymbol{H}(t,x)\cdot \mathrm{rot}(\varepsilon^{-1}\boldsymbol{K})(x)dx := \Phi_1[\boldsymbol{H}(t,\cdot)]. \tag{3.2.8}$$

Replacing $\boldsymbol{u}$ in equations (2.2.6) and (2.2.10) by $\boldsymbol{u}(f)$, applying the linear functional Φ to both sides of equation 2.2.10, and using the information (3.2.4)), we obtain

$$\begin{aligned}\phi'(t) = \Phi_1[\boldsymbol{H}(f)(t,\cdot)] - \Phi\Big[\frac{\sigma}{\varepsilon}\boldsymbol{E}(f)(t,\cdot)\Big]\\ - \Phi_2\Big[\frac{\sigma\mu}{\varepsilon}\frac{\partial}{\partial t}\boldsymbol{U}(f)(t,\cdot)\times\boldsymbol{H}(f)(t,\cdot)\Big], \quad \forall t\in[0,T],\end{aligned} \tag{3.2.9}$$

where

$$\Phi_2[\boldsymbol{E}] = \int_{\Omega_2}\boldsymbol{K}(x)\cdot\boldsymbol{E}(x)dx. \tag{3.2.10}$$

Equations (3.2.4)) and (3.2.9)) imply that the data $(\boldsymbol{u}_1,\boldsymbol{E}_0,\boldsymbol{H}_0,\phi)$ should satisfy the following consistency conditions:

$$\phi(0) = \Phi[\boldsymbol{E}_0], \qquad \phi'(0) = \Phi_1[\boldsymbol{H}_0] - \Phi\Big[\frac{\sigma}{\varepsilon}\boldsymbol{E}_0\Big] - \Phi_2\Big[\frac{\sigma\mu}{\varepsilon}\boldsymbol{U}_1\times\boldsymbol{H}_0\Big]. \tag{3.2.11}$$

On the contrary, if the function $f\in L^p((0,T);\mathbb{R})$ solves equation (3.2.9)), then, using formulas (3.2.8) and (3.2.10)), we can conclude that the function f solves the equation

$$\frac{\partial}{\partial t}\big\{\Phi[\boldsymbol{E}(f)(t,\cdot)] - \phi(t)\big\} = 0, \quad \forall t\in[0,T]. \tag{3.2.12}$$

Using equation (3.2.12) and the first condition from (3.2.11), we can easily show that the function f is a solution to equation (3.2.4). Differentiating equation (3.2.9) once again with respect to the variable t and taking into account relation (2.2.6), we finally obtain

$$\begin{aligned}\phi''(t) = \Phi_1\Big[\frac{\partial}{\partial t}\boldsymbol{H}(f)(t,\cdot)\Big] - \Phi\Big[\frac{\sigma}{\varepsilon}\frac{\partial}{\partial t}\boldsymbol{E}(f)(t,\cdot)\Big]\\ - \Phi_2\Big[\frac{\sigma\mu}{\varepsilon}\frac{\partial}{\partial t}\boldsymbol{u}(f)(t,\cdot)\times\frac{\partial}{\partial t}\boldsymbol{H}(f)(t,\cdot)\Big]\\ - \Phi_2\Big[\frac{\sigma\mu}{\varepsilon\rho}\,\mathrm{Div}\,T(\boldsymbol{u}(f)(t,\cdot))\times\boldsymbol{H}(f)(t,\cdot)\Big]\\ - f(t)\,\Phi_2\Big[\frac{\sigma\mu}{\varepsilon\rho}\boldsymbol{g}(t,\cdot)\times\boldsymbol{H}(f)(t,\cdot)\Big], \quad \text{for a.a. } t\in(0,T).\end{aligned} \tag{3.2.13}$$

Note that according to the assumptions (3.2.5)–(3.2.7) and Theorems 2.2.1 and 2.2.2, each term in (3.2.13) has a sense.

On the contrary, if $f \in L^p((0,T);\mathbb{R})$ is a solution to equation (3.2.13), then f should be a solution to the equation

$$\frac{\partial}{\partial t}\Big\{\phi'(t) - \Phi_1[\boldsymbol{H}(f)(t,\cdot)] + \Phi\Big[\frac{\sigma}{\varepsilon}\boldsymbol{E}(f)(t,\cdot)\Big] + \Phi_2\Big[\frac{\sigma\mu}{\varepsilon}\frac{\partial}{\partial t}\boldsymbol{U}(f)(t,\cdot) \times \boldsymbol{H}(f)(t,\cdot)\Big]\Big\} = 0, \quad \text{for a.a. } t \in (0,T). \tag{3.2.14}$$

However, as follows from (3.2.14) and from the last condition in (3.2.11), the function f is a solution to equation (3.2.9). Thus, we have shown that the solutions to equations (3.2.4) and (3.2.13) are equivalent.

In order that it were possible to rewrite equation (3.2.13) in the form of an equation with a stationary point for f, we assume that $\boldsymbol{g}$ and $\boldsymbol{H}_0$ satisfy the condition

$$\Big|\int_{\Omega_2} \frac{\sigma(x)\,\mu(x)}{\varepsilon(x)\,\rho(x)}\boldsymbol{K}(x)\cdot\big[\boldsymbol{g}(t,x) \times \boldsymbol{H}_0(x)\big]\,dx\Big| = \Big|\Phi_2\Big[\frac{\sigma\mu}{\varepsilon\rho}\boldsymbol{g}(t,\cdot) \times \boldsymbol{H}_0\Big]\Big| \geq 2m, \quad \text{for a.a. } t \in (0,T), \tag{3.2.15}$$

where m is a certain positive constant.

Now we can formulate our main result.

Theorem 3.2.1 *Let $\rho, \lambda, \varkappa$ and ε, μ, σ satisfy conditions* (2.2.16), (2.2.17), *and* (2.2.24)*; and let the functions $\boldsymbol{u}_0, \boldsymbol{u}_1, \boldsymbol{E}_0, \boldsymbol{H}_0$ and $\boldsymbol{g}, \phi, \boldsymbol{K}$ satisfy conditions* (2.2.18)–(2.2.20), (2.2.25), (2.2.26), (3.2.5)–(3.2.7), *and* (3.2.11). *Then there exists a number $T^* \in (0, \min\{T_0, T_1\})$ for which Inverse Problem 3.2.1 has the unique solution $(\boldsymbol{u}, \boldsymbol{E}, \boldsymbol{H}, f)$ satisfying conditions* (2.2.18), (2.2.21), (2.2.27), *and* (2.2.28) *for every $T \in (0, T^*)$.*

Remark 3.2.3 *We can also show that the solution $(\boldsymbol{u}, \boldsymbol{E}, \boldsymbol{H}, f)$ of Inverse Problem 3.2.1 continuously depends on the data $(\boldsymbol{u}_0, \boldsymbol{u}_1, \boldsymbol{E}_0, \boldsymbol{H}_0, \boldsymbol{g}, \phi)$ in the norms of respective spaces. For this, we need to prove a continuous dependence of the solution on the data, just as was done in the case of solutions to direct problems* (2.2.6)–(2.2.9) *and* (2.2.10)–(2.2.15). *Since the proof is very bulky, we omit this part.*

Proof of Theorem 3.2.1 Now we proceed to the proof of Theorem 3.2.1. Introduce the family $X(M,T)$ of complete metric subspaces in $L^p((0,T);\mathbb{R})$, this family depending on the two positive constants M and T

$$X(M,T) = \big\{f \in L^p((0,T);\mathbb{R}) \;\big|\; \|f\|_{T,0,p} \leq M\big\}. \tag{3.2.16}$$

If we assume that $f \in X(M,T)$ is a solution to the operator equation (3.2.13), then we can write it in the following form convenient for application of the theorem about a stationary point

$$f(t) = \left(\Phi_2\Big[\frac{\sigma\mu}{\varepsilon\rho}\boldsymbol{g}(t,\cdot)\times\boldsymbol{H}(f)(t,\cdot)\Big]\right)^{-1}$$
$$\times\Big\{-\phi''(t)+\Phi_1\Big[\frac{\partial}{\partial t}\boldsymbol{H}(f)(t,\cdot)\Big]-\Phi\Big[\frac{\sigma}{\varepsilon}\frac{\partial}{\partial t}\boldsymbol{E}(f)(t,\cdot)\Big]$$
$$-\Phi_2\Big[\frac{\sigma\mu}{\varepsilon}\frac{\partial}{\partial t}\boldsymbol{u}(f)(t,\cdot)\times\frac{\partial}{\partial t}\boldsymbol{H}(f)(t,\cdot)\Big]$$
$$-\Phi_2\Big[\frac{\sigma\mu}{\varepsilon\rho}\operatorname{Div}T(\boldsymbol{u}(f)(t,\cdot))\times\boldsymbol{H}(f)(t,\cdot)\Big]\Big\} := N(f)(t) \quad (3.2.17)$$

for a.a. $t \in (0,T)$.

Our task is to show the local solvability of equation (3.2.17). For this, we set $T_3(M) = \min\{T_0, T_1, T_2(M)\}$, where $T_2(M)$ is the unique positive root of the equation

$$\|\boldsymbol{K}\|_{0,2,\Omega_2}\big[C_2(T_0)+TC_3(T_0,M)\big]\,C_6\,\gamma\rho_0^{-1}\,\|\sigma\|_{0,\infty,\Omega_2}\,\|\mu\|_{0,\infty,\Omega_2}\,T^{1-\alpha} = m. \quad (3.2.18)$$

Using estimate (2.2.19) and condition (3.2.15), we can show that each function $f \in X(M,T)$ satisfies the following basic inequality for a.a. $t \in (0,T) \subset (0,T_3(M))$

$$\left|\Phi_2\Big[\frac{\sigma\mu}{\varepsilon\rho}\boldsymbol{g}(t,\cdot)\times\boldsymbol{H}(f)(t,\cdot)\Big]\right|$$
$$\geq \left|\Phi_2\Big[\frac{\sigma\mu}{\varepsilon\rho}\boldsymbol{g}(t,\cdot)\times\boldsymbol{H}_0\Big]\right| - \left|\Phi_2\Big[\frac{\sigma\mu}{\varepsilon\rho}\boldsymbol{g}(t,\cdot)\times\int_0^t\frac{\partial}{\partial t}\boldsymbol{H}(f)(s,\cdot)ds\Big]\right|$$
$$\geq 2m - \|\boldsymbol{K}\|_{0,2,\Omega_2}\,\gamma\rho_0^{-1}\,\|\sigma\|_{0,\infty,\Omega_2}\,\|\mu\|_{0,\infty,\Omega_2}\,\|\boldsymbol{g}(t,\cdot)\|_{0,\infty,\Omega_2}$$
$$\times\left\|\frac{\partial}{\partial t}\boldsymbol{H}(f)(t,\cdot)\right\|_{T,\infty,0,2}\cdot t$$
$$\geq 2m - \|\boldsymbol{K}\|_{0,2,\Omega_2}\big[C_2(T_0)+TC_3(T_0,M)\big]\,C_6\,\gamma\rho_0^{-1}\,\|\sigma\|_{0,\infty,\Omega_2}$$
$$\times\|\mu\|_{0,\infty,\Omega_2}\,T^{1-\alpha} \geq m. \quad (3.2.19)$$

Taking into account formulas (2.2.21), (2.2.29), and (3.2.19) and the embedding $H^1(\Omega_2;\mathbb{R})\cdot H^1(\Omega_2;\mathbb{R}) \hookrightarrow L^2(\Omega_2;\mathbb{R})$ (where the dot denotes the functional product), we can estimate the nonlinear operator N

$$\|N(f)\|_{T,0,p} \leq \frac{1}{m}\Big\{\|\phi''\|_{T_0,0,p} + \|\operatorname{rot}(\varepsilon^{-1}\boldsymbol{K})\|_{0,2,\Omega}\left\|\frac{\partial}{\partial t}\boldsymbol{H}(f)(t,\cdot)\right\|_{0,2,\Omega}$$
$$+\gamma\,\|\sigma\|_{0,\infty,\Omega}\,\|\boldsymbol{K}\|_{0,2,\Omega}\left\|\frac{\partial}{\partial t}\boldsymbol{E}(f)(t,\cdot)\right\|_{0,2,\Omega}$$
$$+\gamma\,\|\sigma\|_{0,\infty,\Omega_2}\,\|\mu\|_{0,\infty,\Omega_2}\,\|\boldsymbol{K}\|_{0,2,\Omega}\bigg(\left\|\frac{\partial}{\partial t}\boldsymbol{u}(f)(t,\cdot)\right\|_{0,\infty,\Omega_2}\left\|\frac{\partial}{\partial t}\boldsymbol{H}(f)(t,\cdot)\right\|_{0,2,\Omega_2}$$
$$+\rho_0^{-1}C_7\big(\|\lambda\|_{0,\infty,\Omega_2},\|\varkappa\|_{0,\infty,\Omega_2},\Omega_2\big)\,\|\boldsymbol{u}(f)(t,\cdot)\|_{3,2,\Omega_2}\,\|\boldsymbol{H}(f)(t,\cdot)\|_{1,2,\Omega_2}\bigg)\Big\}$$
$$\leq \frac{1}{m}\Big\{\|\phi''\|_{T_0,0,p} + \big(1+\|\boldsymbol{g}\|_{T_0,p',2,2}+\|\boldsymbol{K}\|_{0,2,\Omega_2}+\|\operatorname{rot}(\varepsilon^{-1}\boldsymbol{K})\|_{0,2,\Omega}\big)$$
$$\times\big[C_8(T_0)+TC_9(T_0,M)\big]\Big\}, \quad \forall T\in(0,T_3(M)]. \quad (3.2.20)$$

Let us choose a pair of numbers (M, T^*) from the conditions

$$M = \frac{2}{m}\Big\{\|\phi''\|_{T_0,0,p} + C_8(T_0)\,\big(1 + \|\boldsymbol{g}\|_{T,0,p',2,2} + \|\boldsymbol{K}\|_{0,2,\Omega} + \|\mathrm{rot}(\varepsilon^{-1}\boldsymbol{K})\|_{0,2,\Omega}\big)\Big\}, \quad (3.2.21)$$

$$\frac{1}{m}\,\big(1 + \|\boldsymbol{g}\|_{T_0,p',2,2} + \|\boldsymbol{K}\|_{0,2,\Omega} + \|\mathrm{rot}(\varepsilon^{-1}\boldsymbol{K})\|_{0,2,\Omega}\big)\, C_9(T_0, M)\, T^* \le M, \quad T^* \in (0, T_3(M)]. \quad (3.2.22)$$

Formulas (3.2.20)–(3.2.22) imply that the operator N maps $X(M,T)$ into itself for all $T \in (0, T^*]$.

Now we estimate the difference $N(f_2) - N(f_1)$ for arbitrary functions $f_1, f_2 \in X(M,T)$, $T \in (0,T^*]$. To this end, we consider the equation

$$\begin{aligned}
&N(f_2)(t) - N(f_1)(t)\\
&= \Big(\Phi_2\Big[\frac{\sigma\mu}{\varepsilon\rho}\boldsymbol{g}(t,\cdot)\times \boldsymbol{H}(f_2)(t,\cdot)\Big]\Big)^{-1}\Big\{\Phi_1\Big[\frac{\partial}{\partial t}\boldsymbol{H}(f_2)(t,\cdot) - \frac{\partial}{\partial t}\boldsymbol{H}(f_1)(t,\cdot)\Big]\\
&\quad - \Phi\Big[\frac{\sigma}{\varepsilon}\Big(\frac{\partial}{\partial t}\boldsymbol{E}(f_2)(t,\cdot) - \frac{\partial}{\partial t}\boldsymbol{E}(f_1)(t,\cdot)\Big)\Big]\\
&\quad - \Phi_2\Big[\frac{\sigma\mu}{\varepsilon}\Big(\frac{\partial}{\partial t}\boldsymbol{U}(f_2)(t,\cdot) - \frac{\partial}{\partial t}\boldsymbol{u}(f_1)(t,\cdot)\Big)\times\frac{\partial}{\partial t}\boldsymbol{H}(f_2)(t,\cdot)\Big]\\
&\quad - \Phi_2\Big[\frac{\sigma\mu}{\varepsilon}\frac{\partial}{\partial t}\boldsymbol{u}(f_1)(t,\cdot)\times\Big(\frac{\partial}{\partial t}\boldsymbol{H}(f_2)(t,\cdot) - \frac{\partial}{\partial t}\boldsymbol{H}(f_1)(t,\cdot)\Big)\Big]\\
&\quad - \Phi_2\Big[\frac{\sigma\mu}{\varepsilon\rho}\,\mathrm{Div}\,T\big(\boldsymbol{u}(f_2)(t,\cdot) - \boldsymbol{u}(f_1)(t,\cdot)\big)\times\boldsymbol{H}(f_2)(t,\cdot)\Big]\\
&\quad ds - \Phi_2\Big[\tfrac{\sigma\mu}{\varepsilon\rho}\,\mathrm{Div}\,T(\boldsymbol{u}(f_1)(t,\cdot))\times\big[\boldsymbol{H}(f_2)(t,\cdot) - \boldsymbol{H}(f_1)(t,\cdot)\big]\Big]\Big\}\\
&- \Big(\Phi_2\Big[\frac{\sigma\mu}{\varepsilon\rho}\boldsymbol{g}(t,\cdot)\times\boldsymbol{H}(f_2)(t,\cdot)\Big]\Big)^{-1}\\
&\quad ds\times\Phi_2\Big[\tfrac{\sigma\mu}{\varepsilon\rho}\boldsymbol{g}(t,\cdot)\times\big[\boldsymbol{H}(f_2)(t,\cdot) - \boldsymbol{H}(f_1)(t,\cdot)\big]\Big]\,N(f_1)(t),\\
&\quad \text{for a.a. } t\in(0,T).
\end{aligned} \quad (3.2.23)$$

Using formulas (2.2.22), (2.2.23), (2.2.29), (2.2.30), (3.2.20), and (3.2.23), we can easily derive the estimate

$$\begin{aligned}
&\big|N(f_2)(t) - N(f_1)(t)\big|\\
&\le C_{10}(T_0, M, m^{-1})\,\big(1 + \|\boldsymbol{g}\|_{T_0,p',2,2} + \|\boldsymbol{K}\|_{0,2,\Omega_2} + \|\mathrm{rot}(\varepsilon^{-1}\boldsymbol{K})\|_{0,2,\Omega}\big)\\
&\quad\times\Big(\int_0^t h_1(t-s)\,\|f_2 - f_1\|_{s,0,p}ds + \|f_2 - f_1\|_{t,0,p}\Big), \quad \forall t\in[0,T],
\end{aligned} \quad (3.2.24)$$

where

$$h_1(t) = \exp\big\{t\,\|\sigma\|_{0,\infty,\Omega} + C_6(T_0)M\|\boldsymbol{g}\|_{T,0,p',2,2,\Omega_2}\big\}.$$

Estimate (3.2.24) implies the inequality

$$\begin{aligned}\|N(f_2) - N(f_1)\|_{t,0,p} &\\ \le C_{10}(T_0, M, m^{-1})\big(1 + \|\boldsymbol{g}\|_{T_0,p',2,2} &+ \|\boldsymbol{K}\|_{0,2,\Omega_2} + \|\mathrm{rot}(\varepsilon^{-1}\boldsymbol{K})\|_{0,2,\Omega}\big)\\ &\times [1 + T_0 h_1(T_0)] \int_0^t \|f_2 - f_1\|_{s,0,p} ds\\ &:= C_{11}(T_0, M, m^{-1}) \int_0^t \|f_2 - f_1\|_{s,0,p} ds, \quad \forall t \in [0, T]. \qquad (3.2.25)\end{aligned}$$

Inequality (3.2.25) enables us to establish the following estimate for the iterations N^n of the operator N:

$$\|N^n(f_2) - N^n(f_1)\|_{t,0,p} \le \frac{C_{11}^n(T_0, M, m^{-1})}{(n-1)!} \int_0^t (t-s)^{n-1} \|f_2 - f_1\|_{s,0,p} ds$$
$$\text{for a.a. } t \in (0, T), \quad \forall n \in \mathbb{N}. \qquad (3.2.26)$$

We now apply the generalized contraction mapping principle (see [30, p. 103]). This ensures equation (3.2.13) to have a unique solution $f \in X(M, T)$ for every $T \in (0, T^*)$, which proves the theorem.

3.3. An Inverse Problem for Electromagnetoelastic Equations with Complete Nonlinear Interaction

In this section, following the original work [31], we present the results of the solution to an inverse problem for a electromagnetoelasticity system in the case of complete nonlinear interaction of electromagnetic and elastic waves. The results of the first initial boundary-value problem solution presented in section 2.3. will be sufficiently used.

Formulation of an Inverse Problem

Let us consider one of possible formulations of inverse problems for the direct problem earlier studied in section 2.3.. In this section we assume that the free member of equation (2.3.7) has representation $f(z, t) = \phi(t) g(z, t)$, where the function ϕ is *unknown*. Let us now formulate inverse problem which will be studied now.

Inverse Problem 3.3.1 *Determine a set of the functions*

$$h : Q_T \to \mathbb{R}, \quad u : Q_T \to \mathbb{R}, \quad \phi : [0, T] \to \mathbb{R}$$

from equations (2.3.6)–(2.3.11) *and*

$$\int_\Omega \rho(z) h h_z dz = -\frac{1}{2} \int_\Omega \rho_z h^2 dz = \psi(t), \quad t \in [0, T], \qquad (3.3.1)$$

where $\rho \in \overset{o}{W}{}_2^1(\Omega)$ and $\psi : [0, T] \to \mathbb{R}$ are giving functions having sufficient smoothness.

The functions r, ν, g, j are supposed to be smooth functions with possible jumps in points $z_m : -l < z_1 < z_2 < \cdots < z_m < l$, $0 < r_0 \leq r(z) \leq r_1 < \infty$, $0 < \nu_0 \leq \nu(z) \leq \nu_0 < \infty$; p is a positive number, and

$$\int_\Omega \rho(z) g(z,t) dz \geq \rho_0 > 0, \quad t \in [0,T]. \tag{3.3.2}$$

At the points of discontinuity we assume the fulfilment of the transmission conditions (2.3.12)–(2.3.13).

We demonstrated in section 2.3. that Direct Problem 2.3.1 has the solution $h \in \mathring{V}_2(Q_T)$, $u \in \mathring{W}_2^{1,1}(Q_T)$ satisfying (2.3.39) and (2.3.43), which are convenient for us to rewrite in the form

$$\frac{1}{2}\int_\Omega (ph^2(z,t_1) + u_t^2(z,t_1) + \nu^2(z)u_z^2(z,t_1))\,dz + \frac{1}{2}\int_{Q_{t_1}} prh_z^2\,dz\,dt$$
$$\leq C_{20}(1 + t_1 \int_{Q_{t_1}} f^2\,dz\,dt), \tag{3.3.3}$$

$$\|u\|_{\mathring{W}_2^{1,1}(Q_T)} \leq C_{20}(1 + \|f\|_{Q_{t_1}}), \qquad \max_{Q_T} |h| + \|h_z\|_{2,Q_T} \leq C_{20}, \tag{3.3.4}$$

where C_{20} is a positive constant. The identities (2.3.16), (2.3.17) can be rewritten in a different form if $u_{tt} \in L_2(0,T; H^{-1}(\Omega))$. In particular, we have for any $\xi(z) \in \mathring{W}_2^1(\Omega)$

$$\langle u_{tt}, \xi\rangle + \int_\Omega \nu^2 u_z \xi\,dz + \int_\Omega phh_z\xi\,dz = \int_\Omega f\xi\,dz, \tag{3.3.5}$$

$$u_t(z,0) = u_1(z), \qquad u(z,0) = u_0(z),$$

valid for a.e. $t \in [0,T]$ (see Corollary 2.3.2). From (3.3.11) we obtain (see [22, Chapter VII, pp. 318–386]

$$\max_{t\in[0,T]} \{\|u\|_{\mathring{W}_2^1(\Omega)} + \|u_t\|_{2,\Omega} + \|u_{tt}\|_{L^2(0,T;H^{-1}(\Omega))}\}$$
$$\leq C_{21}\{\|hh_z\|_{2,Q_T} + \|f\|_{2,Q_T} + \|u_0\|_{\mathring{W}_2^1(\Omega)} + \|u_1\|_{2,\Omega}\}, \tag{3.3.6}$$

where C_{21} is a positive constant. It gives us the existence and uniqueness theorem of the linear initial boundary-value problem, considering $phh_z + f$ as a nonhomogeneous part of the linear equation. Note that in [22] one supposes that ν^2 is a smooth function. In the case when ν^2 is a piecewise smooth function all mentioned in [22] arguments pass without changes.

Consider Direct Problem 2.3.1, for two different functions f_k, $k = 1, 2$. The solutions corresponding to these functions will be mark as h_k, u_k, $k = 1, 2$. To solve Inverse Problem 3.3.1 we need to estimate the differences

$$h_2 - h_1, \qquad u_2 - u_1.$$

Introduce two functions $v(z,t), w(z,t)$ by the formulas

$$v = (h_2 - h_1)\mathrm{e}^{-\lambda t}, \qquad w = (u_2 - u_1)\mathrm{e}^{-\lambda t},$$

where λ is a positive number. The functions v, w are the generalized solution in Q_T of the problem

$$v_t + \lambda v = (rv_z)_z - (h_2 w_t + \lambda h_2 w + u_{1,t} v)_z, \tag{3.3.7}$$

$$w_{tt} + 2\lambda w_t + \lambda^2 w = (\nu^2 w_z)_z - p h_2 v_z - p h_{1,z} v + (f_2 - f_1)\mathrm{e}^{-\lambda t}, \tag{3.3.8}$$

with boundary conditions (2.3.10)–(2.3.11), zero initial data and transmission conditions (2.3.12)–(2.3.13) formulated for the functions v, w. The following estimate is valid for a.e. $t \in [0, t_1]$

$$\begin{aligned}
&\frac{p}{2}\|v(\cdot,t)\|_{2,\Omega}^2 + \frac{1}{2}\|w_t(\cdot,t)\|_{2,\Omega}^2 + \frac{\nu_0^2}{2}\|w_z(\cdot,t)\|_{2,\Omega}^2 + \frac{\lambda^2}{2}\|w(\cdot,t)\|_{2,\Omega}^2 \\
&\quad + 2\lambda\|w_t(\cdot,t)\|_{2,\Omega}^2 + \lambda p\|v\|_{2,Q_t}^2 + pr_0\|v_z\|_{2,Q_t}^2 - \lambda p\|v_z\|_{2,Q_t}^2 \cdot \|w\|_{2,Q_t} \\
&\qquad - pC_{20}\max_{Q_t}|v| \cdot \|v_z\|_{2,Q_t}^2 - pC_{20}\max_{Q_t}|v| \cdot \|w_t\|_{2,Q_t}^2 \\
&\qquad \le \int_{Q_t} |w_t(f_2 - f_1)|\,dz\,dt \le \frac{1}{2\epsilon}\|w_t\|_{2,Q_t}^2 + \epsilon\|f_2 - f_1\|_{2,Q_t}^2.
\end{aligned} \tag{3.3.9}$$

The inequalities $\|w_t\|_{2,Q_t} \le \|w_t\|_{2,Q_{t_1}}$ and $\|f_2 - f_1\|_{2,Q_t} \le \|f_2 - f_1\|_{2,Q_{t_1}}$ are valid for a.e. $t \in [0, t_1]$ too. Using of Theorem 7.1 in [21, Chapter III, p.181] gives us

$$\max_{Q_t}|v| \le C_{21}(\|w_t^2\|_{1,r,Q_t} + \lambda\|w^2\|_{1,r,Q_t}),$$

where $r = 2/(1 - 2\kappa)$, $\kappa \in (0, 1/2)$ and C_{21} is a positive constant. However the first term in the right-hand side of the latter inequality can be estimated as follows

$$\begin{aligned}
\|w_t^2\|_{1,r,Q_t} \le \|w_t^2\|_{1,r,Q_{t_1}} &= \Big(\int_0^{t_1}\Big(\int_\Omega w_t^2\,dz\Big)^r dt\Big)^{1/r} \\
&= \Big(\int_0^{t_1} \|w_t\|_{2,\Omega}^{2r}\,dt\Big)^{1/r} \le t_1^{1/r}\max_{t\in[0,t_1]}\|w_t\|_{2,\Omega}^2.
\end{aligned}$$

In a similar way we obtain

$$\|w^2\|_{1,r,Q_t} \le t_1^{1/r}\max_{t\in[0,t_1]}\|w\|_{2,\Omega}^2,$$

$$\|w_t\|_{2,Q_{t_1}} \le t_1^{1/2}\max_{t\in[0,t_1]}\|w_t\|_{2,\Omega}, \qquad \|v_z\|_{2,Q_{t_1}} \le 2C_{20}.$$

Using the latter inequalities in (3.3.7) gives us

$$\begin{aligned}
&\max_{t\in[0,t_1]}\Big\{\frac{p}{2}\|v\|_{2,\Omega}^2 + \frac{1}{2}\|w_t\|_{2,\Omega}^2 + \frac{\nu_0^2}{2}\|w_z\|_{2,\Omega}^2 + \lambda^2\|w\|_{2,\Omega}^2\Big\} \\
&\quad + 2\lambda\|w_t\|_{2,Q_{t_1}}^2 + pr_0\|v_z\|_{2,Q_{t_1}}^2 + \lambda p\|v\|_{2,Q_{t_1}}^2 - C_{21}t_1^{1/r}p\max_{t\in[0,t_1]}\|w_t\|_{2,\Omega}^2 \\
&\qquad - C_{21}t_1^{1/r}p\max_{t\in[0,t_1]}\|w\|_{2,\Omega}^2 - 2C_8^2 p\lambda t_1^{1/2}\max_{t\in[0,t_1]}\|w\|_{2,\Omega}^2 \\
&\qquad\qquad \le \frac{t_1}{2\epsilon}\max_{t\in[0,t_1]}\|w_t\|_{2,\Omega}^2 + \epsilon\|f_2 - f_1\|_{2,Q_{t_1}}^2.
\end{aligned}$$

Let us choose t_1, λ, ϵ satisfying the inequalities

$$\frac{\lambda^2}{2} > C_{21}pt_1^{1/r} + 2C_8^2pt_1^{1/2}\lambda, \quad \frac{1}{8} > C_{21}pt_1, \quad 2\epsilon = t_1 < \frac{1}{8}.$$

For such values of t_1, λ, ϵ we have

$$\max_{t\in[0,t_1]} \left\{ \frac{p}{2}\,\|v(\cdot,t)\|_{2,\Omega}^2 + \frac{1}{2}\,\|w_t(\cdot,t)\|_{2,\Omega}^2 + \frac{\nu_0^2}{2}\,\|w_z(\cdot,t)\|_{2,\Omega}^2 + \lambda^2\|w(\cdot,t)\|_{2,\Omega}^2 \right\}$$
$$+ 2\lambda\|w_t\|_{2,Q_{t_1}}^2 + pro\|v_z\|_{2,Q_{t_1}}^2 \le t_1\|f_2 - f_1\|_{2,Q_{t_1}}^2.$$

Let us take $t_1 = T$. Therefore

$$\max_{t\in[0,T]} \{\|u_2 - u_1\|_{\mathring{W}_2^1(\Omega)} + \|u_{2,t} - u_{1,t}\|_{2,\Omega} + \sqrt{p}\|h_2 - h_1\|_{2,\Omega} + \sqrt{p}\|h_{2,z} - h_{1,z}\|_{2,\Omega}\}$$
$$\le C_{22}\sqrt{T}\,\|f_2 - f_1\|_{2,Q_T}, \quad (3.3.10)$$

where C_{22} is a positive constant. The following inequality takes place for any function $\xi \in \mathring{W}_2^1(\Omega)$, $\|\xi\|_{\mathring{W}_2^1(\Omega)} = 1$

$$\|u_{2,tt} - u_{1,tt}\|_{H^{-1}(\Omega))} \le \max_{Q_T}|g| \cdot C_{21}T\|\phi_2 - \phi_1\|_2 + \max_{\|\xi\|=1} |\langle f_2 - f_1, \xi\rangle|,$$

where $\|\cdot\|_2$ means the usual $L_2(0,T)$-norm. Using of the latter finally gives us

$$\|u_{2,tt} - u_{1,tt}\|_{L_2(0,T;H^{-1}(\Omega))} \le \max_{Q_T}|g| \cdot \{C_{21}T\|\phi_2 - \phi_1\|_2 + C_{22}\sqrt{T}\,\|\phi_2 - \phi_1\|_2\}$$
$$\le C_{23}\sqrt{T}\,\|\phi_2 - \phi_1\|_2, \quad (3.3.11)$$

where C_{23} is a positive constant. From (3.3.1)–(3.3.2), (3.3.5) we obtain

$$\phi(t) = W(\phi), \quad (3.3.12)$$

where $W : L_2(0,T) \mapsto L_2(0,T)$ is operator defined by formula

$$W(\phi) = \frac{\langle u_{tt}, \rho\rangle + \int_\Omega \nu^2 u_z \rho'\,dz + p\psi}{\int_\Omega \rho g\,dz}. \quad (3.3.13)$$

Here $u = u(z,t;\phi), h = h(z,t;\phi), \phi = \phi(t)$ are the solution of the inverse problem and ψ is the additional information (3.3.1).

Now we are ready to formulate and prove our main results.

Theorem 3.3.1 *Let ϕ be a fixed point of the operator $W(\phi)$, i. e., $\phi = W(\phi)$. Then u, h, ϕ are a solution of Inverse Problem 3.3.1. The reciprocal statement is valid: let u, h, ϕ be a solution of Inverse Problem 3.3.1, then $\phi = W(\phi)$.*

Proof Let ϕ be a fixed point of the operator W. Then u, h are the solution of Direct Problem 2.3.1, and ϕ satisfies equality (3.3.1). Therefore we have

$$p\psi(t) = -\frac{1}{2}\int_\Omega ph^2\rho_z\,dz = \int_\Omega phh_z\rho\,dz,$$

and condition (3.3.1) is valid, i. e., u, h, ϕ are a solution of Inverse Problem3.3.1.

Suppose now that u, h, ϕ are a solution of Inverse Problem 3.3.1, then there is valid equality (3.3.12). Substituting (3.3.1) in (3.3.11) gives us $\phi = W(\phi)$, i. e., ϕ is a fixed point of the W-mapping.

There is valid the following existence and uniqueness theorem.

Theorem 3.3.2 *For sufficiently small values $T > 0$ Inverse Problem 3.3.1 has a unique solution, which can be obtained by the method of successive approximations.*

Proof Let be $M = \{\phi \in L_2(0,T); \quad \|\phi\|_2 \leq R\}$. Show that we can choose constants R, T such that W maps M into itself and is a contractive mapping on this set, i. e.,

$$W(M) \subset M, \tag{3.3.14}$$

$$\|W(\phi_2) - W(\phi_1)\|_2 \leq q\|\phi_2 - \phi_1\|_2 \tag{3.3.15}$$

with some constant $q \in (0,1)$. First, prove inclusion (3.3.14). We have

$$W(\phi) = \frac{\langle u_{tt}, \rho\rangle + \int_\Omega \nu^2 u_z \rho'\,dz + p\psi}{\int_\Omega \rho g\,dz}.$$

It follows from (3.3.4), (3.3.6) that $\|W(0)\|_2 \leq C_{24}$ with some positive constant C_{24}. Using (3.3.2), (3.3.10), (3.3.11) we obtain $\|W(\phi) - W(0)\|_2 \leq C_{25}\|\phi\|_2\sqrt{T}$ or $\|W(\phi)\| \leq C_{24} + C_{25}\|\phi\|_2\sqrt{T}$ with some positive constant C_{25} independent on T. Let be

$$C_{25}\sqrt{T} = q < 1. \tag{3.3.16}$$

Setting C_{24} equalled to $R(1-q)$ we obtain

$$\|W(\phi)\|_2 \leq C_{24} + C_{25}\sqrt{T}\,R = R,$$

i.e., $W(\phi) \subset M$ for any function $\phi \in M$.

Now prove estimate (3.3.15). From (3.3.2), (2.3.44), (3.3.10) and (3.3.11) we have

$$\|W(\phi_2) - W(\phi_1)\|_2 \leq C_{25}\sqrt{T}\,\|\phi_2 - \phi_1\|_2 = q\|\phi_2 - \phi_1\|_2,$$

which proves (3.3.15).

So, application of the contractive principle guaranties that the mapping W has the unique fixed point ϕ in M. By Theorem 3.3.1 u, h, ϕ are the unique solution of Inverse Problem 3.3.1. Theorem 3.3.2 is proved.

References

[1] A. Alvén, *Cosmical Electrodynamics*. Clarendon, Oxford,1950.

[2] M. K. Balakirev, and I. G. Gilinskii, *Waves in Piezocristals*. Nauka, Novosibirsk, 1982 (in Russian).

[3] V. Z. Parton, and V. A. Kudrjavtsev, *Electromagnetoelasticity of Piezoelectric and Electrical-Conducting Bodies*. Nauka, Moscow, 1988 (in Russian).

[4] J. W. Dunkin, and A. C. Eringen, On the propagation of waves in an electromagnetic elastic solid. *Intern. J. Eng. Sci.* **1** (1963), 461–495.

[5] L. Knopoff, The interaction between elastic wave motion and a magnetic field in electrical conductors. *J. Geophys. Res.* **60** (1955), 441–456.

[6] I. E. Tamm, *Foundations of the Theory of Electricity*. Nauka, Moscow, 1976 (in Russian).

[7] A. C. Eringen, and G. A. Maugin, *Electrodynamics of Continua. Vols. I, II*. Springer-Verlag, New York—Berlin, 1990.

[8] S. R. Pride, Governing equations for the coupled electromagnetics and acoustics of porous media. *Phys. Rev.* **B50**, 21 (1994), 15 678–15 698.

[9] L. D. Landau, and E. M. Lifshitz, *Course of Theoretical Physics. Vol. 8. Electrodynamics of Continuous Media*. Pergamon Press, Oxford-Elmsford, New York, 1984.

[10] J. D. Jacson, *Classical Electrodynamics*. John Wiley&Sons Inc., New York, 1999.

[11] A. C. Eringen, *Nonlinear Theory of Continuos Media*. McGraw-Hill, New York, 1962.

[12] L. I. Sedov, *Mechanics of Continua. Vols. I, II*. Nauka, Moscow, 1973 (In Russian).

[13] V. G. Romanov, Structure of a solution to the Cauchy problem for the system of the equations of electrodynamics and elasticity in the case of point sources. *Siberian Math. J.*, **36** (3), 1995, 541–561.

[14] V. I. Priimenko, and M. P. Vishnevskii, The Cauchy problem for a nonlinear model system of electromagnetoelasticity. *56 Seminàrio Brasileiro de Análise, Niteroi, Brasil* (2002), 395–413.

[15] V. Priimenko, and M. Vishnevskii, An initial boundary-value problem for a model electromagnetoelasticity system. *J. Differential Equations* (2007) **235**, 31–55.

[16] M. M. Lavrentiev, V. G. Romanov, and S. P. Shishatskii, *Ill-Posed Problems of Mathematical Physics and Analysis*. Amer. Math. Soc., Providence, RI, 1986.

[17] V. G. Romanov, *Inverse Problems of Mathematical Physics*. The Netherlands, Utrecht: VSP, 1987.

[18] M. M. Lavrentiev, A. V. Avdeev, M. M. Lavrentiev-Jr., and V. I. Priimenko, *Inverse Problems of Mathematical Physics*. The Netherlands, Utrecht: VSP, 2003.

[19] V. G. Romanov, On an inverse problem for a coupled system of equations of electrodynamics and elasticity. *J. Inv. Ill-Posed Problems* (1995) **3**, 321–332.

[20] A. Lorenzi, and V. I. Priimenko, Identification problems related to electro-magneto-elastic interactions. *J. Inv. Ill-Posed Problems* (1996) **4**, 115–143.

[21] O. A. Ladyzhenskaia, V. A. Solonnikov, and M. N. Uralceva, Linear and Quasilinear Equations (Translations of Mathematical Monographs, Vol.23). *Amer. Math. Soc.*, Providence, RI, 1968.

[22] L. C. Evans, Partial Differential Equations (Graduate Studies in Mathematics, Vol. 19). *Amer. Math. Soc.*, Providence, RI, 2002.

[23] M. Necati Özisik, *Heat Conduction*. John Wiley & Sons, Inc., 1993.

[24] J-L. Lions, *Quelques Métodes de Résolution des Problemes aux Limites Nonlinéaires*. Dunod, Paris, 1969.

[25] V. G. Romanov, *Integral Geometry and Inverse Problems for Hyperbolic Equations*. Springer-Verlag, 1974.

[26] Yu. E. Anikonov, *Some Methods for Studying Multidimensional Inverse Problems for Differential Equations*. Nauka, Novosibirsk, 1978. (in Russian).

[27] V. A. Sharafutdinov, *Integral Geometry of Tensor Fields*. The Netherlands, Utrecht: VSP, 1994.

[28] R. G. Mukhometov, The problem of recovering a two-dimensional Riemannian metric and integral geometry. *Dokl. Akad. Nauk SSSR* (1977) **232** (1), 32–35. (in Russian).

[29] D. A. Sheen, A generalized Greens theorem. *Appl. Math. Lett* (1992) **5** (4), 95–98.

[30] L. Schwartz, *Analyse Mathématique. Vols. I, II*. Hermann, Paris, 1967.

[31] V. I. Priimenko, and M. P. Vishnevskii, An inverse problem of electromagnetoelasticity in the case of complete nonlinear interaction. *J. Inv. Ill-Posed Problems* (2005) **13**, 277–301.

In: Nonlinear Analysis Research Trends
Editor: Inès N. Roux, pp. 157-207
ISBN: 978-1-60456-358-0

Chapter 5

NONLINEAR PREDICTIVE CONTROL: STRUCTURE, COMPUTATIONAL CHALLENGES, STOCHASTIC OPTIMIZATION ALGORITHMS

Sorin Olaru [*] ***and Guillaume Sandou*** [**]
SUPELEC, Automatic Control Department, 3 rue Joliot Curie,
91192,Gif-sur-Yvette, France.

Abstract

Predictive control has emerged as a very efficient method to control dynamical systems. The classical approach is to define a linear model of the system to control and then to solve successive on-line optimization problems to compute the best control inputs to apply. The approach is tractable as long as the linear model is sufficient to describe the system behaviour. However, as system become more and more complex, nonlinear models have to be considered. Consequently, the corresponding optimization problems can hardly be exactly solved and there are crucial computational challenges to overcome to use such efficient control methods. That's why a new trend is the use of stochastic optimization so as to get well suited solutions with low computation times. In this chapter, the generic optimal control problem, together with the predictive control principles and their application to linear systems are firstly called up. A brief description of deterministic non linear optimization methods is also provided. Stochastic optimization algorithms are then described, with an analysis of their advantages and drawbacks for the solution of non linear optimization problems. Finally, two examples of the application of non linear predictive control in the area of Power Systems are presented.

I. Introduction

The nonlinear analysis has made large advances in the last years with important contributions to the modeling and control design. Still, there exists a large gap between the theoretical tools

[*] E-mail address: sorin.olaru@supelec.fr
[**] E-mail address: guillaume.sandou@supelec.fr

and the practical nonlinear control methods employed for the industrial processes. The present chapter deals with one of the most popular control techniques encountered in the industry, namely the model predictive control (MPC). This technique can be tailored for dealing with nearly any type of plant model and due to its time-domain formulation can easily handle constraints from the design stage.

Is arguable the measure in which this method is employed in a pure nonlinear manner in the industrial application as long as the identification of linear models is easier than the nonlinear counterparts and operating near a steady state leads generally to good linear approximation. Nevertheless, there are applications (such as batch processes) where the system changes from different operating points and classes of processes with severe nonlinearities for which pure nonlinear control design has to be considered.

Here we review the basic characteristics of the MPC, defined for nonlinear models in the most general case. Then the discussion of the linear versus nonlinear case is done in order to stress the structural properties. In the second part a stochastic optimization based MPC is proposed. Such attempts based on evolutionary optimization are in an early stage on the academic community but promise to tackle in an efficient manner the difficulties raised by what is called in the control theory: "complex systems".

I.1. Modern Control

In order to state the main ideas, consider a multivariable nonlinear process:

$$\begin{cases} x_{k+1} = f(x_k, v_k, u_k, k) \\ y_k = g(x_k, u_k, k); x_0 \in \Re^n; u_k \in \Re^m; y_k \in \Re^p \end{cases} \tag{1}$$

where f, g are continuous vector functions with continuous partial derivatives on the vector arguments x_k, v_k, y_k and u_k representing the state, the disturbance, the output and the control input respectively. One can define the control process by representing the control actions u_k at all the discrete instants, explicitly using an *open loop* control function:

$$u_k = u_k^0 \tag{2}$$

the dependence on the initial state x_0 and the arbitrary initial time in (1) demands the adaptation for each pair (k_0, x_0), which is impractical in a general control system.

The use of a *feedback control* law (figure 1) is the general solution for these problems:

$$u_k = \mu^0(k, x_k) = \mu_k^0(x_k) \tag{3}$$

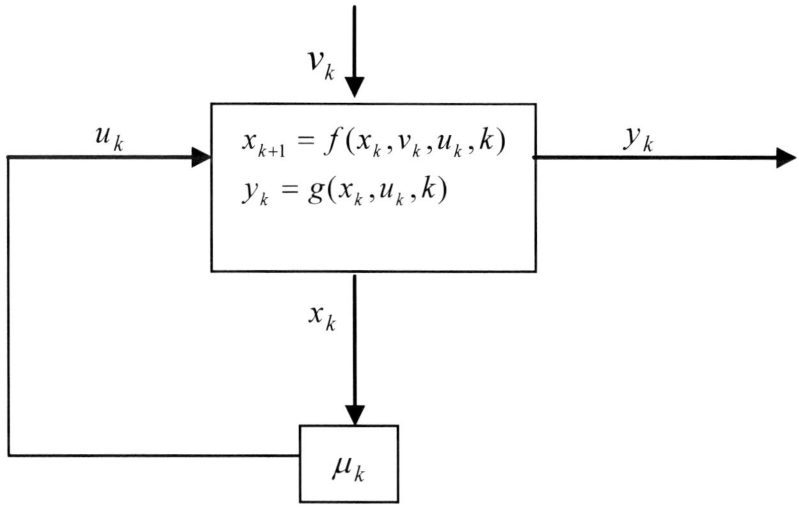

Figure 1. Feedback control law structure.

The modern control theory is marked by the establishment of analytical methods for problems handled only in empirical manner through the classical control (which mostly deals with LTI systems, SISO models, and fix class of compensators). One of the fundamental results of the modern control design is represented by the establishment of the optimal control theory and it is easy to follow the argument which presents the predictive control as one of the straight descents of the optimal control.

I.2. Optimal Control

The optimal control design is principally based on the optimization of a specific performance criterion, the stability and other desirable properties being only consequences of this primer optimality. If the basic ingredient is the performance criterion which gives a measure of the objectives to be achieved, one can observe that the usual manifestation is represented by temporal evaluation of the system trajectory (tracking error / control effort). Such index eludes the controllability or other system realization particularities such that optimal control solvability is not even guaranteed without a prior structural study.

In the general form, the performance index associated with an optimal control problem has the form:

$$J = \sum_{k=0}^{\infty} L(x_k, u_k) \tag{4}$$

and allows a comparative appreciation of the quality of a control law u_k. The real and continuous function L represents the cost per stage (generally positive), enjoying regularity properties (as continuous and partial derivatives of order one and two) with respect to the arguments x_k and u_k. In order to synthesize a control law, the demarcation of the admissible control system by the pair of feasible control function and the associated state trajectory has to be done. The feasibility of the control function is verified by set restriction $u_k \in D_u \in \Re^m$ while the state trajectory has to satisfy constraints $x_k \in D_x \subset \Re^n$.

For each feasible control law one can associate a performance index J_0, the smaller the value of the cost index, the better the quality of the control. The minimal value of the index J corresponds to the optimal control system. With this explanation, the discrete-time optimal control problem can be defined as:

"given the system (Σ), find the feasible control law which minimizes the performance index attached to the system evaluation"

Through the use of the Hamilton-Jacobi equation, the Pontryagin Maximal Principle (Pontryagin et al., 1962), Euler-Lagrange equations and dynamic programming (Bellman, 1957), the optimal control became a well-understood problem at the end of the '60s. Unfortunately, its use in practical control application is rather reduced (even if we can cite direct application in biological systems for example). However, its contribution to the development of the control field in general was tremendous as it represented the departure point for robust control/ predictive control/ adaptive control to name only few mainstreams.

I.3. Predictive Control

Conceiving optimal control policies implies the use of an infinite prediction horizon for accounting the system behavior and coping with all the possible system evolutions with respect to the constraints. However, the infinite horizon will lead in a general case to an infinite number of optimization variables. Adding the presence of constraints and eventually the structural problems inherited from the model used for the prediction, one can have the picture of the difficulties in implementing such an infinite horizon optimal control algorithm.

The Model-based Predictive Control (MPC) is the practical methodology which uses an optimization based philosophy but only emulates the infinite horizon by exploiting through the receding horizon principle the simplicity of an open loop optimal control problems over a finite horizon[1]. For a given discrete state x_k supposed measurable, the optimal control sequence $U = \begin{bmatrix} u_{k|k}^{\mathrm{T}} & \cdots & u_{k+N-1|k}^{\mathrm{T}} \end{bmatrix}^{\mathrm{T}}$ is obtained as solution of an open-loop finite-time control problem:

$$\Pi_k : \begin{cases} U^* = \arg\min\limits_{U} \varphi(x_{k+N}) + \sum\limits_{i=1}^{N-1} L(x_{k+i}, v_{k+i}, u_{k+i}) \\ \text{subject to}: \begin{cases} x_{k+i+1} = f(x_{k+i}, v_{k+i}, u_{k+i}, k+i) \\ h(x_{k+i}, u_{k+i}) \leq 0 \end{cases} \quad \text{for} \quad i = 1, \ldots, N-1 \end{cases} \tag{5}$$

constructed for a finite prediction horizon N, cost per stage $L(.)$, terminal weight $\varphi(.)$ and the system dynamics described by $f(.)$. Figure 2 presents the feedback control system underlining

[1] See (Qin and Badgwell, 2003) for a survey of successful MPC applications in industry.

the fact that the control action is generated by the solution of an optimization problem as in (5).

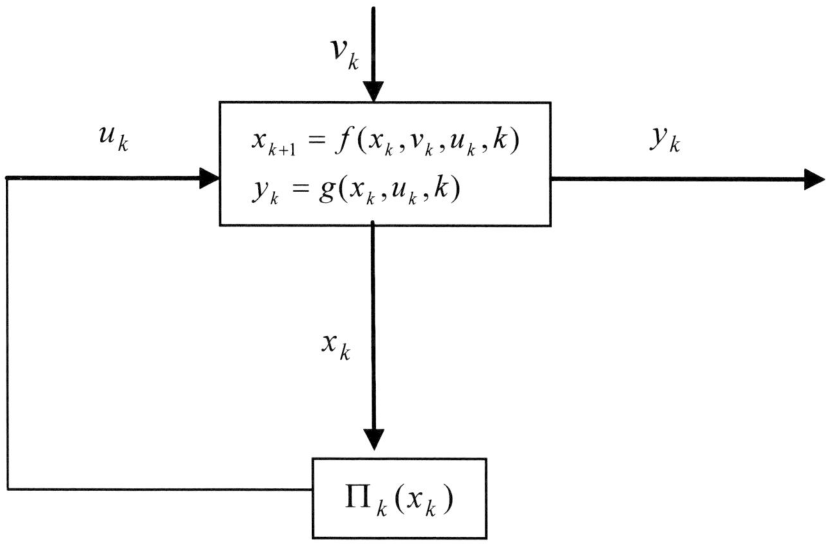

Figure 2. MPC control structure.

The receding horizon principle states that the first component of the sequence U will be effectively applied as control action at the input of the system and the optimization procedure (5) is restarted using the new measurements by receding the prediction window. This leads to a control law described by $\mu_k = U^*(1) = u_k$

Due to the time-domain description, the limitations occurring on physical systems can be easily considered, specifications in terms of constraints on the predicted trajectory can be formulated and auxiliary constraints can be added to the optimization problem to satisfy security limitations or to reinforce stability of the controlled system. In (5), the constraints are expressed in a compact form using element-wise inequalities on functions linking the states and the control actions, *h*(.).

The nonlinear dynamics can accept any description *f*(.) as long as they allow the prediction and the constraints evaluation for a given control scenario.

The presence of constraints alters the global performances of the control system. Consequently, the design of predictive laws requires off-line validation phases, during which the choice of the classical parameters, e.g. prediction horizons and weighting terms, is highly connected to the reliability of the control law. In fact, constraints modify the predictive law, from an initial linear structure to a non linear one with respect to the system state (Bemporad et al, 2002), even if the system itself remains linear. The main consequence is that all tools dedicated to linear systems analysis become ineffective. In this case, new tools (Blanchini, 1999) answering issues such as feasibility, stability and real time implementation of constrained predictive control laws have to be examined.

Linear models together with quadratic objective functions lead to convex formulations which can be solved in the framework of classical QP (Quadratic Programming) and they will be discussed in the section III. This class received a special attention on diverse areas and thus, by the well understood numerical aspects, it has achieved a degree of maturity which offers solutions converging to the optimum in good time, often comparable with the sampling time necessary for the implementation with "fast systems". On the other hand the nonlinear models are leading to complex non-convex optimization problems and the attribute "fast" and "optimal" are hard to be accorded due to the real-time computational constraints. From this point of view, stochastic algorithms may prove to be a good compromise as it will be explained in the second part of this chapter (section V and VI).

II. MPC Ingredients

In this section the main components of the MPC construction are described in order to gain an insight of the degrees of freedom one can dispose at the design stage.

II.1. Prediction Model

The representation of the physical systems is based on experiment and abstraction. The model for a given physical system consists in a set of relations putting in correspondence the output of the system for various inputs realizations. These relations can be obtained either from first principle physical laws or from experimental data and regression analysis. Conceptually, the use of NMPC is independent of the way the prediction model is obtained (is worthy to mention here the models proposed in the literature based on Neural Networks (Henriques et al., 2002) or Volterra series which can describe large scale and complex nonlinear systems).

Nonlinear MPC implies the existence of a nonlinear prediction model. The diversity of the nonlinearities makes a classification merely impossible. In the literature one can find the distinction between *static* (defining the nonlinear dependence between the input and the output in steady state) and *dynamic* nonlinearities (where the nonlinearities involve derivatives or integrals of the inputs, states or outputs). Other delimitation can be made between *deterministic* (fully determined by equations without random components) and *non-deterministic* models.

For the prediction purposes in the MPC framework it is necessary to dispose of a model which assures that together with the initial state, the initial time and the control action over a given interval provides sufficient information to predict both the state evolution and the output of the system. A fundamental difference should be stressed between the continuous time and discrete-time models. In the first case, the value of the inputs, states and outputs are defined over open time intervals while in the discrete-time we are interested in the same quantities at discrete values of the time scale, defined by the *sampling period.*

The MPC design is done in discrete time, and this is not always in concordance with the continuous plant dynamics. The analytical discretization of continuous models is done through the construction of the solution of the differential equations describing the process dynamics. An exact discrete-time model for a continuous plant is feasible for linear systems but for nonlinear continuous time models this turns out to be a difficult task, which appeals at the numerical methods of integration for differential equations.

However, beside the construction, the discrete-time models have to be carefully validated when used for sampled-data nonlinear control design. Indeed, the discretization is a sensitive aspect even for some linear systems. It is known that controllers (parametrized with the sampling period T_e) that stabilize the Euler approximation for all $T_e \geq 0$ but can destabilize the exact discrete-time model. In (Nesic et al., 1999) are derived sufficient conditions for a controller, which stabilizes an approximate discrete-time model, to also stabilize the exact discrete-time model.

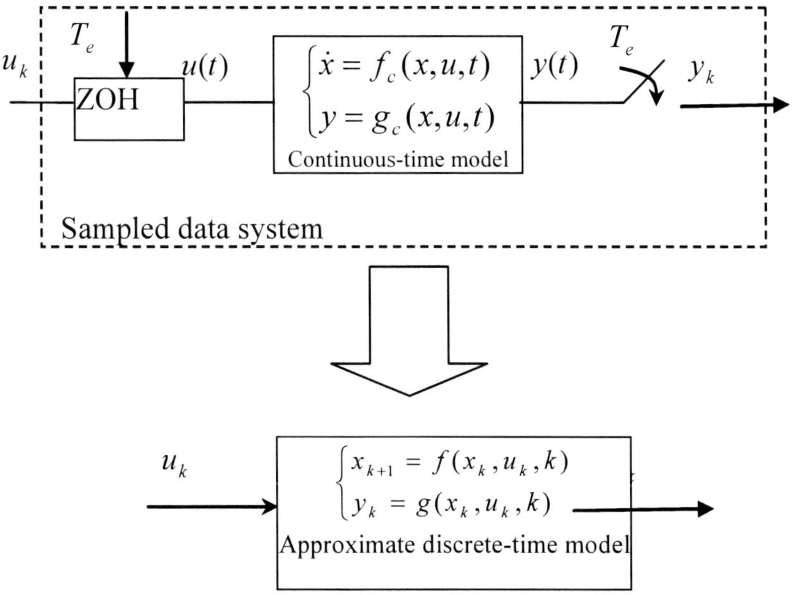

Figure 3. Sampled nonlinear dynamical system and approximate discrete models.

In order to describe the details of the MPC design, we suppose the existence of a discrete-time model (see figure 3):

$$\begin{cases} x_{k+1} = f(k, x_k, u_k, v_k) \\ y_k = g(k, x_k) \end{cases} \tag{6}$$

If $\bar{x}_k$ is a solution of (6) corresponding to an input $\bar{u}_k$, a disturbance $\bar{v}_k$ and delivering the output $\bar{y}_k$, which means :

$$\begin{cases} \bar{x}_{k+1} = f(k, \bar{x}_k, \bar{v}_k, \bar{u}_k) \\ \bar{y}_k = g(k, \bar{x}_k) \end{cases} \tag{7}$$

then considering the variations:

$$\xi_k = x_k - \bar{x}_k ; \mu_k = u_k - \bar{u}_k ; \nu_k = v_k - \bar{v}_k ; \eta_k = y_k - \bar{y}_k \tag{8}$$

One can write in a first approximation:

$$f(k,x_k,u_k,v_k)-f(k,\overline{x}_k,\overline{u}_k,\overline{v}_k)=\left.\frac{\delta f}{\delta x_k}\right|_{k,\overline{x}_k,\overline{u}_k,\overline{v}_k}\xi_k+\left.\frac{\delta f}{\delta u_k}\right|_{k,\overline{x}_k,\overline{u}_k,\overline{v}_k}\mu_k+\left.\frac{\delta f}{\delta v_k}\right|_{k,\overline{x}_k,\overline{u}_k,\overline{v}_k}v_k$$
$$g(k,x_k)-g(k,\overline{x}_k)=\left.\frac{\delta g}{\delta x_k}\right|_{k,\overline{x}_k}\xi_k \tag{9}$$

Subtracting correspondingly (7) from (6) and using (8) and (9) a linear system is obtained:

$$\xi_{k+1}=\left.\frac{\delta f}{\delta x_k}\right|_{k,\overline{x}_k,\overline{u}_k,\overline{v}_k}\xi_k+\left.\frac{\delta f}{\delta u_k}\right|_{k,\overline{x}_k,\overline{u}_k,\overline{v}_k}\mu_k+\left.\frac{\delta f}{\delta v_k}\right|_{k,\overline{x}_k,\overline{u}_k,\overline{v}_k}v_k$$
$$\eta_k=\left.\frac{\delta g}{\delta x_k}\right|_{k,\overline{x}_k}\xi_k \tag{10}$$

Noting

$$A_k=\left.\frac{\delta f}{\delta x_k}\right|_{k,\overline{x}_k,\overline{u}_k,\overline{v}_k},B_k=\left.\frac{\delta f}{\delta u_k}\right|_{k,\overline{x}_k,\overline{u}_k,\overline{v}_k},E_k=\left.\frac{\delta f}{\delta v_k}\right|_{k,\overline{x}_k,\overline{u}_k,\overline{v}_k},C_k=\left.\frac{\delta g}{\delta x_k}\right|_{k,\overline{x}_k} \tag{11}$$

(10) can be rewritten as :

$$\begin{aligned}\xi_{k+1}&=A_k\xi_k+B_k\mu_k+E_kv_k\\ \eta_k&=C_k\xi_k\end{aligned} \tag{12}$$

If the functions f,g,h do not depend explicitly on time, which means that system (6) is *time-invariant*, and the functions $\overline{x}_k,\overline{v}_k,\overline{u}_k$ are constant, then the triplet $(\overline{x},\overline{v},\overline{u})$ represents an operating point (equilibrium) and the matrices (11) are constant. The form (11) is the linearized system around the operating point $(\overline{x},\overline{v},\overline{u})$.

II.2. Cost Functions

The cost function depends on the desired behavior of the system and thus is a subjective matter, often accommodated to the intended implementation scheme. As a general rule the goal is to minimize or maximize some performance criterion. If the control action has to take into account exogenous inputs which deteriorate the performance, the optimization may be translated to min-max or max-min problems. Well known optimal control based on minimization of an infinite time criterion is the LQ control approach.

In equation (5) it was shown that at each sampling time k, a cost function is associated with a control sequence $U=\begin{bmatrix}u_{k|k}^{\mathrm{T}} & \cdots & u_{k+N-1|k}^{\mathrm{T}}\end{bmatrix}^{\mathrm{T}}$ over a prediction horizon *1,..., N*:

$$\min_U \varphi(x_{k+N}) + \sum_{i=1}^{N-1} L(x_{k+i}, u_{k+i}) \tag{13}$$

The cost per stage $L(x_{k+i}, u_{k+i})$ is sometimes called the Lagrangian and together with the function penalizing the terminal state will determine the characteristics of the optimal solution. These scalar functions are often chosen to find a compromise between the tracking performances and the control effort thus declining the cost per stage in the form:

$$\min_U \varphi(x_{k+N}) + \sum_{i=1}^{N-1} \left\{ l(x_{k+i}) + \|u_{k+i}\|_R \right\} \tag{14}$$

where $\|.\|$ expresses an arbitrary Holder norm and R represents a weighting matrix. In view of tracking, the reference trajectory should be known in advance at least for the length of the prediction window. In this case, the performance index becomes:

$$\min_U \varphi(x_{k+N}) + \sum_{i=1}^{N-1} \left\{ \|r_{k+i} - y_{k+i}\|_Q + \|u_{k+i}\|_R \right\} \tag{15}$$

If the cost functions is adapted to the dynamical system models used for the prediction they may allow the construction of exact (analytic) optimal solutions. It is generally the case for linear systems with linear or quadratic cost functions, as it will be seen in section III.

II.3. Constraints

The optimization problems (13-15) are to be solved subject to the dynamical constraints but in the same time, other security or performance properties can be added to the system trajectory. These physical limitations (energy or force) are primary stated in terms of hard constraints on the internal state variables and input control actions: $x_k \in D_x, u_k \in D_u$.

The usual assumptions are: D_x is closed and contains the origin; D_u is compact[2] and contains the origin. Optimization problems (13-15) become:

[2] A notable exception on the compactness on the aforementioned sets can be mentioned however: the case of hybrid systems which allow mode switching or on/off control actions. Supposing $u_k \in \Re^m$ and $x_k \in \Re^n$ the constraints will then be written:

$$u_k \in \{0,1\}^{m_l} \times D_u$$

$$x_k \in \{0,1\}^{n_l} \times D_x$$

with $D_u \subset \Re^{m_c}$, $D_x \subset \Re^{n_c}$ and $m_c + m_l = m$ and $n_c + n_l = n$.

$$\Pi_k : \begin{cases} \min\limits_{U} \varphi(x_{k+N}) + \sum\limits_{i=1}^{N-1} L(x_{k+i}, u_{k+i}) \\ \text{subject to} \quad x_{k+i+1} = f(x_{k+i}, u_{k+i}) \\ \qquad\qquad\quad y_k = g(k, x_k) \\ \qquad\qquad\quad x_{k+i} \in D_x \\ \qquad\qquad\quad u_{k+i} \in D_u \\ \qquad\qquad\quad y_{k+i} \in D_y \end{cases} \tag{16}$$

One can rewrite the restrictions in a compact form by a set of nonlinear inequalities $h(x_k, u_k) \leq 0$ leading to the following optimization problem:

$$\Pi_k : \begin{cases} \min\limits_{U} \phi(x_{k+N}) + \sum\limits_{i=1}^{N-1} L(x_{k+i}, u_{k+i}) \\ \text{subject to} \quad x_{k+i+1} = f(x_{k+i}, u_{k+i}) \\ \qquad\qquad\quad y_k = g(k, x_k) \\ \qquad\qquad\quad h(x_{k+i}, u_{k+i},) \leq 0 \end{cases} \tag{17}$$

By replacing the explicit dependence of the predicted states on the current state and the sequences of future control actions one can observe that the MPC implementation relies on the nonlinear programming problem:

$$\Pi_k : \begin{cases} U^*(x_k) = \arg\min\limits_{U} J(x_k, U) \\ \text{subject to} : c(x_k, U) = 0 \\ \qquad\qquad\quad d(x_k, U) \leq 0 \end{cases} \tag{18}$$

Analysing equation (18), without further information on the structure we can conclude that the computation time is hard to be controlled and in general the complexity is exponential in the number of arguments. More than that, the convexity of both the feasible domain and the cost function can bring important advantages as long as the KKT optimality conditions and the constraints qualification can be exploited.

If disturbances have to be considered over the prediction horizon, the problem is translated into a min-max formulation and the constraints have to be robustly satisfied. An interesting point is the way of handling the prediction as long as the open-loop prediction can be too conservative. If the min-max problem is solved in a dynamic programming manner, the optimal solution will contain control actions as function of the past disturbance realizations $u_{k+i}(x_{k+i|k}, v_k^T \ldots, v_{k+i}^T)$. The conservativeness is reduced with the price of an increased complexity (implied by the nested min-max formulations).

II.4. Feasibility and Stability

The objective of the control law is to obtain a control law $\mu_k = \mu(x_k) = U^*(1)$ such that the origin is asymptotically stable and the region of attraction is large for the close loop system:

$$x_{k+1} = f(k, x_k, \mu(x_k)) \tag{19}$$

A predictive control law is feasible if the related optimization problem Π_k in (18) is feasible.

DEFINITION 1.– A predictive control law is feasible with respect to an initial set D_0, if $\forall x_0 \in D_0$, the system evolution $x_{k+1} = f(k, x_k, \mu(x_k))$ guarantees feasibility of the associated optimization problems for all $k \geq 0$.

Assessing feasibility of a constrained predictive control law is based on the positive invariance concept.

DEFINITION 2.– A domain $D \subset \Re^n$ is positive invariant with respect to $x_{t+1} = \mathrm{f}(x_t)$ if $\forall x_0 \in D$, the solution $x_t \in D$, $\forall t > 0$ (Blanchini, 1999).

Based on this concept, necessary and sufficient feasibility conditions are given below.

PROPOSITION 3.– A MPC law is feasible if and only if there exists a positive invariant set with respect to this MPC law – D_{inv}, satisfying $D_{inv} \subset D_x$ et $D_0 \subset D_{inv}$.

From a practical point of view, positive invariance is guaranteed by adding to the formulation (19) a set of terminal constraints:

$$x_{k+N} \in D_N \tag{20}$$

and an associated local control law:

$$\mu_N(x): D_N \to D_N \tag{21}$$

This modification defines a pseudo-infinite prediction horizon and the tail of the solution $U^*(x_k)$ at time k completed with a suitable terminal control action will provide a feasible solution at time $k+1$. The stability of constrained predictive control is based on the existence of an invariant terminal set (Gilbert and Tan, 1991) and the use of a cost function as a Lyapunov function (Rossiter, 2003), (Vassilaki et al, 1998) for the closed loop system.

Within the predictive framework, this terminal invariant set is defining the 'target' set for the trajectory over the prediction horizon. This approach is known as the 'dual mode' paradigm (Mayne et al, 2000) since the control policy consists in two distinctive parts, the first one ensuring constraints satisfaction and evolution towards the invariant set, the second being equivalent to a local control law which do not violate the constraints and assures the positive invariance of the terminal set with respect to the closed loop dynamic.

The interdependence between feasibility, invariants sets and stability of constrained predictive control is summarized by the following theorem which provides sufficient stability conditions (Mayne et al, 2000).

THEOREM 4.– For a given predictive law $\mu(x_k) = U^*(1)$ obtained by solving :

$$\Pi_k : \begin{cases} U^* = \min_U \max_V \varphi(x_{k+N}) + \sum_{i=1}^{N-1} L(x_{k+i}, u_{k+i}) \\ \text{subject to} \quad x_{k+i+1} = f(x_{k+i}, v_{k+i}, u_{k+1}) \\ \qquad x_{k+i} \in D_x \\ \qquad u_{k+1} \in D_u \\ \qquad x_{k+N} \in D_N \end{cases} \tag{22}$$

the feasibility at time 0 implies the stability at all future times and the stability of the closed loop system $x_{k+1} = f(x_k, v_k, \mu(x_k))$ is guaranteed if:

- H1 $f(\cdot,\cdot,\cdot)$, $L(\cdot,\cdot)$ and $\varphi(\cdot)$ are continuous;
- H2 $f(0,0,0) = 0$, $L(0,0) = 0$ and $\varphi(0) = 0$;
- H3 D_u is compact and contains the origin;
- H4 D_x is closed and contains the origin in its interior;
- H5 $D_N \subset D_x$ is compact and $0 \in \text{int}(D_N)$
- H6 $\mu_N(x) : D_N \to D_u$
- H7 $f_N(D_N) \subset D_N$ (where $f_N(x,v) = f(x, \mu_N(x))$)
- H8 $\varphi(x) \geq L(x, \mu_N(x)) + \varphi(f_N(x))$ for all $x \in D_N$

II.5. MPC Tuning

Several parameters influence the performances and stability of the predictive control law. Due to their mutual interaction, the impact of each of them is difficult to discriminate. The main tuning parameters are (Chen and Allgöwer, 1998; Jadbabaie et al., 1999; Parisini and Zoppoli, 1995; Alamir and Bornard, 1995):

- N, prediction horizon - plays an important role at the control design level, since predictions represent decisive information for a correct performance evaluation. Ideally N should be as large as possible but the structure of the optimization problem becomes more complex, in particular increasing the number of decisional variables (number of future control actions and constraints to be considered). Practically, the influence on the feasibility of the control law is hidden by the fact that the prediction horizon is in direct relationship with the terminal set D_N and the associated control law $\mu_N(x)$. However if all the other parameters are fixed, larger the prediction horizon, the larger the feasible set for the MPC law.

- D_N, terminal set - As previously mentioned, this set is introduced within the concept of 'dual mode' predictive control splitting the control objectives in two parts. The first part aims at driving the system to D_N in finite time, the second starts acting when the terminal set is reached. As a consequence, the existence of this terminal set is directly connected to the existence of a pre-specified feedback law. The important advantage of including such an invariant set in the predictive criterion is the fact that the set of feasible states at time 0 inherits of the positive invariance feature of D_N, and this implies the global feasibility of the predictive law.
- $\mu_N(x)$, the local stabilizing control law over D_N. Classical choice will be the infinite time optimal control law resulting for the unconstrained system. The shape of D_N and the structure of $\mu_N(x)$ are related by the construction.
- $\varphi(x_{k+N})$, terminal cost. This parameter was found to be necessary to ensure stability of the controlled system for unconstrained predictive control, becoming thus equivalent to the pseudo-infinite horizon case. Unfortunately, for constrained predictive control, the terminal cost by itself does not guarantee feasibility or stability. However, it is an important parameter in order to force the desired (asymptotically stable) behavior inside D_N (see the hypothesis H8 in the theorem 4).
- $L(x_{k+i}, u_{k+i})$, the cost per stage - characterizes the structure of the performance index and finally the closed loop characteristics. A high weight on the reference tracking and a low penalty on the control effort will result in a nervous control action more likely to activate the stability constraints.

III. Linear Model Predictive Control

The design of a linear predictive control law is based on the existence of a linear prediction model of the system to be controlled, the use of linear/quadratic const functions and linear constraints on the state and input variables.

III.1. Linear/Quadratic Multiparametric Formulations

Within the general framework of linear systems theory, consider the following state space model:

$$x_{k+1} = Ax_k + Bu_k, \quad k \in Z^+ \tag{23}$$

with $x_k \in \Re^n$ the state vector at time *k*, $u_k \in \Re^m$ the control vector at time *k*, *A* and *B* matrices of adequate dimensions and the pair (A, B) assumed to be stabilisable.

At each sampling time, the current state vector $x_k = x_{k|k}$ (assumed to be measurable) is used to elaborate the open loop optimal control sequence U^* :

$$U^* = \begin{bmatrix} u_{k|k}^{\mathrm{T}} & \cdots & u_{k+N-1|k}^{\mathrm{T}} \end{bmatrix}^{\mathrm{T}} \tag{24}$$

according to the following cost function:

$$U^* = \arg\min_U \left\| P x_{k+N|k} \right\|_p + \sum_{i=1}^{N-1} \left\{ \left\| Q x_{k+i|k} \right\|_p + \left\| R u_{k+i|t} \right\|_p \right\} \tag{25}$$

where $\| \ \|_p$ represents the norm $p = \{1, 2, \infty\}$ and the pair (Q, A) is assumed to be detectable. The prediction horizon N, the weighting terms $Q = Q^{\mathrm{T}} \geq 0$, $R = R^{\mathrm{T}} > 0$ and the final cost defined by $P = P^{\mathrm{T}} \geq 0$ are the tuning knobs of the control law.

The optimization of the cost function is performed subject to constraints imposed by the system dynamic, the physical limitations and by the terminal or stability constraints:

$$\begin{cases} x_{k+i+1|k} = A x_{k+i|k} + B u_{k+i|k} \quad i \geq 0 \\ C x_{k+i|k} + D u_{k+i|k} \leq \gamma, \\ \qquad 0 \leq i \leq N, x_{k+i|k} \in D_x, u_{k+i|k} \in D_u \\ x_{k+N|k} \in D_N, \\ D_N = \left\{ x \in \Re^n \middle| L_{in} x \leq l_{in}; L_{eq} x = l_{eq} \right\} \end{cases} \tag{26}$$

where D_N is an invariant polyhedral domain with respect to the system (24) in close loop with a prespecified stabilizing control law[3].

All constraints previously mentioned are described by systems of linear equalities and inequalities (Olaru and Dumur, 2005). The finite set of constraints (26) can be restructured to obtain a formulation which will be directly tackled by classical optimization routines. Indeed, the predicted state vector can be written in a compact form:

$$X = \begin{bmatrix} x_{k+1/k} \\ x_{k+2/k} \\ \vdots \\ x_{k+N/k} \end{bmatrix} = \begin{bmatrix} A \\ A^2 \\ \vdots \\ A^N \end{bmatrix} x + \begin{bmatrix} B & 0 & \cdots & 0 \\ AB & B & \cdots & 0 \\ \vdots & \vdots & \ddots & \vdots \\ A^{N-1}B & A^{N-2}B & \cdots & B \end{bmatrix} U \tag{27}$$

---

[3] The most popular choice for D_N is the maximal admissible set for the LQ optimal feedback law (Gilbert and Tan, 1991)

$$u_k = K_{LQ} x_k$$

$$K_{LQ} = -(R + B^T P B)^{-1} B^T P A$$

$$P = A^T P A + Q - A^T P^T B (R + B^T P B)^{-1} B^T P A$$

with $x_{k|k} = x$, which enables to structure the optimization problem (25)-(26) by the relations:

- Case $p = 2$:

$$U^* = \arg\min_{U} 0.5U^{\mathrm{T}}HU + x^{\mathrm{T}}FU$$
$$\text{subject to:} \begin{cases} A_{in}U \le B_{in}x + b_{in} \\ A_{eq}U = B_{eq}x + b_{eq} \end{cases} \tag{28a}$$

- Case $p = 1, \infty$:

$$Z^* = \arg\min_{Z} c^{\mathrm{T}}Z$$
$$\text{subject to:} \begin{cases} A_{in}Z \le B_{in}x + b_{in} \\ A_{eq}Z = B_{eq}x + b_{eq} \end{cases} \tag{28b}$$

with $Z = \{U; \xi_1, \ldots, \xi_{N_\xi}\}$ and $\xi_1, \ldots, \xi_{N_\xi}$ auxiliary variables, the number N_ξ of these variables depending on the prediction horizon and on the model (Zadeh and Whalen, 1962).

For both cases (28a)-(28b), the optimal argument includes the control sequence U^*. Only the first part of this sequence is applied to the system, the complete procedure is iterated again at the next sampling time according to the receding horizon principle. Real time implementation is usually performed through on-line optimization procedures (linear or quadratic programming) in order to determine the optimum corresponding to a particular value of the state vector *x*.

III.2. On-Line Optimization Strategies

Linear predictive control is based on on-line strategies if the implementation takes into account current values of the system states and uses iterative linear/quadratic programming procedures to solve:

$$QP: \quad \arg\min_{U} 0{,}5U^{\mathrm{T}}HU + \bar{f}^{\mathrm{T}}U \quad \text{s.t.} \begin{cases} A_{in}U\bar{b}_{in} \\ A_{eq}U = \bar{b}_{eq} \end{cases} \qquad LP: \quad \arg\min_{Z} c^{\mathrm{T}}Z \quad \text{s.t.} \begin{cases} A_{in}Z \le \bar{b}_{in} \\ A_{eq}Z = \bar{b}_{eq} \end{cases} \tag{29}$$

with $\bar{f} = F^T x_k; \bar{b}_{in} = b_{in} + B_{in}x_k; \bar{b}_{eq} = b_{eq} + B_{eq}x_k$, x_t - system state.

Remark: A necessary condition leading to the success of the linear MPC strategy is that the global optimality conditions can be verified in a systematic manner due to the constraints qualification. This is in fact the basic idea behind the most encountered on-line optimization routines.

III.2.1. Active Set Method

This method solving QP problems is one of the most popular strategies, its name coming from the fact that the procedure considers at each iteration the separation between the set of active and inactive constraints (constraints qualification). Successive QP sub-problems with equality constraints must then be solved, which are equivalent by projection to the resolution of an unconstrained problem (Fletcher, 1981). Indeed, the knowledge of an initial feasible point U_0 enables then identifying the set of saturated constraints. The quadratic optimization problem which is obtained replacing these inequalities by equality constraints and ignoring all other constraints:

$$\begin{aligned} U^* &= \arg\min_{U} 0.5U^{\mathrm{T}} HU + \bar{f}^{\mathrm{T}} U \\ &\text{under constraints}: A_{sat}U = b_{sat} \end{aligned} \tag{30}$$

can be solved considering the analytical solution :

$$U_1 = \left(H^{-1}\bar{A}^T (\bar{A}H^{-1}\bar{A}^T)^{-1}\bar{A}H^{-1} - H^{-1} \right)\bar{f} + H^{-1}\bar{A}^T (\bar{A}H^{-1}\bar{A}^T)^{-1}\bar{b} \tag{31}$$

with $\bar{A}U = \bar{b}$ deduced from $A_{sat}U = b_{sat}$ so that the rank of $\bar{A}$ is maximal.

A global optimality test must be performed. If these conditions are not fulfilled, a constraint among those associated to a negative Lagrange coefficient is removed from the active set and the procedure is reiterated. Several possibilities exist choosing the constraint to be cancelled (e.g. remove the one with the Lagrange coefficient of smallest value).

Remark: If the solution U is outside the feasible domain for the initial problem, it should be replaced by the linear combination:

$$U_{10} = \mu_1 U_0 + \mu_2 U_1, \quad 0 \le \mu_1 \le 1, \;\; 0 \le \mu_2 \le 1, \;\; \mu_1 + \mu_2 = 1 \tag{32}$$

which lies on the border. The point U^{10} is feasible but with some saturated constraints which were previously inactive. These ones have to be added to the active set and the procedure is reiterated.

Remark: Finding an initial feasible point is a difficult task. One idea taking into account the optimization structure is to select as initial point at time $k+1$ a sequence which uses again part of the optimal solution of the previous step. This one can however prove to be a non feasible combination; in this case a LP problem has to be solved.

Remark: The exact optimal solution can be found within a finite number of iterations if the initial problem is not degenerated (Fletcher, 1981). The advantage of the method lies in the simplicity of the construction for the particular solutions at each iteration. For sets of constraints of average complexity, this approach is thus one of the most efficient in the MPC framework. However, this is not recommended for large scale problems, for which the number of iterations can be exponential due to the high number of active combinations to be explored.

III.2.2. Pivot Method

Performances in terms of computational load of strategies solving LP problems with pivot methods are well-known (e.g. success of the Simplex method). Their application to LP/QP problems within a MPC framework is possible. In the QP case, transformation into a linear complementary problem (LCP) (Bazaraa et al, 1993) is compulsory.

Consider a vector $q \in \Re^d$ and a matrix $M \in \Re^{d \times d}$. A linear complementary problem (LCP) consists in finding the vectors w and z such that:

$$\begin{aligned} & w - Mz = q \\ & w_i \geq 0 \,; z_i \geq 0 \quad \text{for} \quad i = 1,\ldots,d \\ & w_i z_i = 0 \quad \text{for} \quad i = 1,\ldots,d \end{aligned} \tag{33}$$

or in concluding that such a solution does not exist.

Within the predictive framework, a QP problem similar to (29a) is considered:

$$\begin{aligned} & \hat{U}^* = \arg\min_{\hat{U}} 0.5 \hat{U}^T \hat{H} \hat{U} + \hat{f}^T \hat{U} \\ & \text{under constraints} : \begin{cases} \hat{A}_{in} \hat{U} \leq \hat{b}_{in} \\ \hat{U} \leq 0 \end{cases} \end{aligned} \tag{34}$$

The optimality conditions are given by the following relations ((Bazaraa et al, 1993)):

$$\begin{aligned} & \hat{A}_{in} \hat{U} + y = \hat{b}_{in} \\ & -\hat{H}\hat{U} - \hat{A}_{in}^T \lambda_1 + \lambda_2 = \hat{f} \\ & \hat{U}^T \lambda_2 = 0, \lambda_1^T \rho = 0 \\ & \hat{U}, \rho, \lambda_1, \lambda_2 \geq 0 \end{aligned} \tag{35}$$

with λ_1, λ_2 the vectors of Lagrange multipliers and ρ relaxation variables. Denoting:

$$M = \begin{bmatrix} 0 & -\hat{A}_{in} \\ \hat{A}_{in}^T & \hat{H} \end{bmatrix}, q = \begin{bmatrix} \hat{b}_{in} \\ \hat{f} \end{bmatrix}, w = \begin{bmatrix} \rho \\ \lambda_2 \end{bmatrix}, z = \begin{bmatrix} \lambda_1 \\ \hat{\mathbf{k}}_u \end{bmatrix} \tag{36}$$

drives to the LCP form (34) and in this case pivot algorithms exist, e.g. algorithm developed by Lemke (Lemke, 1968).

Remark: As mentioned in (Camacho and Bordons, 2004), the performances obtained with predictive control depend on the structure of the constraints set, the redundancy and its removal playing an important role in this direction.

III.2.3. Interior Point Method

These algorithms are part of the primal-dual path following methods, which make use of a barrier function and Newton type algorithms to derive polynomial time procedures. Their origins are historically close (Karmarkar, 1984) and their development was remarkable. A state of the art of the research on this topic can be found for example in (Nesterov and Nemirovskii, 1984), (Ye, 1989). Consider as initial formulation (similar to (28a)):

$$\begin{aligned} U^* &= \arg\min_U 0.5U^T HU + \bar{f}^T U \\ &\text{under constraints}: A_{in}U \le \bar{b}_{in} \end{aligned} \tag{37}$$

First step consists in elaborating a barrier function which becomes infinite on the border of the feasible domain. Typical choices of this function are:

$$g(U) = -\sum_i \log(b_i - a_i^T U) \text{ or } g(U) = \frac{1}{\sum_i (b_i - a_i^T U)} \tag{38}$$

with a_i^T the i-th line of A_{in} and b_i the i-th component of vector $\bar{b}_{in}$. Using this function, the optimization (17) can be replaced by:

$$U^*(\gamma) = \arg\min_U \gamma(0.5U^T HU + \bar{f}^T U) + g(U) \tag{39}$$

with γ a scalar which can be considered as a weighting term. Formulation (39) characterizes an unconstrained non linear optimization problem which can be solved by classical Newton type methods ((Bazaraa et al, 1993)). Integrating the barrier function in the optimization criterion avoids going outside the feasible domain during the optimal solution search phase, thus guaranteeing that all intermediary solutions are feasible. If the optimal solution of the initial problem is $\widetilde{U}$, increasing γ makes the solution of (39) evolve so that $\widetilde{U}(\gamma) \to U^*$ when $\gamma \to \infty$, thus improving the quality of the solution up to the limit $\left\| \widetilde{U}^*(\gamma) - \widetilde{U}^* \right\| < \varepsilon$ with ε sufficiently small. The path followed by $\widetilde{U}^*(\gamma)$ is called the 'central path'. The most efficient methods work on the primal and dual problem simultaneously.

Another strategy for the interior point methods (Ben-Tal and Nemirovski, 2001), (Goodwin et al., 2004), (Wright, 1997) starts from the QP problem (28a) with its optimality conditions:

$$\begin{aligned} & HU + \bar{f} + A_{eq}^T v + A_{in}^T \mu = 0 \\ & A_{eq} U = \bar{b}_{eq} \\ & A_{in} U + \xi = \bar{b}_{in} \\ & \mu \geq 0; \xi \geq 0; \mu^T \xi = 0 \end{aligned} \tag{40}$$

with v, μ the vectors of Lagrange multipliers and ξ vector of relaxation variables which allows transformation of inequalities to equalities. These equations are then rewritten in a matrix form, similar to the one appearing in LCP techniques:

$$F(U, v, \mu, \xi) = \begin{bmatrix} HU + A_{eq}^T v + A_{in}^T \mu + \bar{f} \\ -A_{eq} U + \bar{b}_{eq} \\ -A_{in} U - \xi + \bar{b}_{in} \\ \Xi \mathrm{M} e = 0 \end{bmatrix} = 0 \tag{41}$$
$$\mu \geq 0; \xi \geq 0$$

with $\Xi = diag\{\xi_1, \ldots, \xi_r\}; \mathrm{M} = diag\{\mu_1, \ldots, \mu_r\}; e = [1 \ \ldots \ 1]^T$.

The method requires thus solving a system of non linear equations for which Newton type strategies can be considered, using at each iteration a linear approximation of F. A well-known method is the so-called infeasible interior point (IIP) method (Wright, 1997), for which iterative solutions can be non feasible with respect to the initial constraints set but can be found inside the positive quadrant $\mu \geq 0; \xi \geq 0$. A measure of feasibility (and of optimality) of the intermediary solutions considers the duality error:

$$\gamma = \mu^T \xi / r \tag{42}$$

with r the number of inequalities in the constraints set.

Remark: Interior point methods become more and more appreciated, in the detriment of active set methods, due to their polynomial convergence rate. However, the computational load per iteration is more important, which makes them not well adapted to small size problems. Performances are also highly dependent on the care given to avoid numerically ill-conditioned problems.

III.3. Multiparametric Optimization

Starting from formulation (28a) and after transformation, the cost function parameterization can be translated into a constraints parameterization exclusively. As an example, the method given in (Bemporad et al, 2002) starts from the problem:

$$\begin{aligned}&\min_{U} 0.5U^T HU\\&\text{under constraints}: A_{\text{in}}U \leq b_{in} + B_{in}x\end{aligned} \tag{43}$$

The description of the solution for a particular parameter $U^*(x_0)$ can be done using the optimality conditions:

$$\begin{aligned}&HU + A_{in}^T\boldsymbol{\lambda} = 0; \boldsymbol{\lambda} \in \Re^r\\&\boldsymbol{\lambda} \geq 0\\&\boldsymbol{\lambda}^T (A_{in}U - b_{in} - B_{in}x) = 0\\&A_{in}U \leq b_{in} + B_{in}x\end{aligned} \tag{44}$$

the local description of the critical region is deduced in the neighborhood of x_0 :

$$CR_0 : \begin{cases} A_{in}H^{-1}\overline{A}_{in}^T(\overline{A}_{in}H^{-1}\overline{A}_{in}^T)^{-1}(\overline{b}_{in} + \overline{B}_{in}x) \leq b_{in} + B_{in}x \\ (\overline{A}_{in}H^{-1}\overline{A}_{in}^T)^{-1}(\overline{b}_{in} + \overline{B}_{in}x) \leq 0 \end{cases} \tag{45}$$

with $\overline{A}_{in}, \overline{b}_{in}, \overline{B}_{in}$ referring to the inequalities saturated for $\{x_0; U^*(x_0)\}$. Once knowing this non empty critical region, the other parts of the parameters space have to be explored. The exploration technique proposed in (Bemporad et al, 2002) is based on a method introduced in (Dua and Pistikopoulos, 2000) which inverses recursively the inequalities describing CR_0 to define new domains. Then, for each new domain, it is necessary to find an initial point, to solve the related QP problem and recover the active constraints to describe the critical region as in (45).

Once the partition of the state space is obtained:

$$D_x = \bigcup_{i=1}^{N_r} CR_i \tag{46}$$

with N_r the number of regions in the partition, the global explicit solution of the problem (43) will be given by a piecewise affine function of the state:

$$U^*(x) = K_i x + \kappa_i, \quad for \quad x \in CR_i \tag{47}$$

The effective MPC control law applies only the first component of the optimal sequence:

$$\mu_{MPC}(x) = \left[I_n \quad \underbrace{0_n \quad \ldots \quad 0_n}_{N-1} \right] (K_i x + \kappa_i), \quad for \quad x \in CR_i \tag{48}$$

Finally, it can be noticed that strategies based on dynamic programming can be associated to geometric strategies (Goodwin et al., 2004) or to those using the optimality conditions [35] as a mean to decrease computation times. The explicit solutions can be obtained also through geometrical methods based upon the concept of parameterized polyhedra (Olaru and Dumur, 2005).

III.4. Explicit Formulations of Predictive Control Laws

From a practical point of view, the global expression of the predictive law is computed off-line and stocked in a look-up table containing as index the polyhedral domains in the parameters space and as information the linear functions depending on the system state, which correspond to the optimal control signal.

With this table, the on-line implementation of the explicit predictive control law must follow 4 steps:

1. Measure (or estimate) the current parameters vector x
2. Search in the table the critical region which contains x
3. Use the state feedback linear law corresponding to this active region
4. Repeat all the procedure starting from step 1

The identification of the critical region is performed on-line, and the related computational load can be important if the number of partitions is high (the problem is related to the point-location complexity in general). Among all existing positioning methods, the following ones may be interesting:

- Searching in a sequential way to which critical region the current parameters vector belongs. This approach requires checking constraints defining all polyhedral domains, and induces a huge number of elementary operations and in the worst case the examination of all regions.
- A more efficient sequential search method consists in gathering all constraints and imposes that their evaluation is performed only once for the current point. Then, deciding to which region the current point belongs is still sequentially performed, but a multiple check of constraints is avoided. This method also avoids collecting several times the constraints for different inputs to the table. It can still be improved since in the worst case the evaluation of all constraints is required.
- An efficient method for organizing constraints, which also uses the convexity properties of the critical regions, consists in placing the inequalities at the nodes of a binary search tree. The positioning technique investigates this tree starting from the root and looking for an active critical region in a leaf. From the memory point of

view, this method is equivalent to the previous one but the computational load is drastically reduced.

- A sensitive point of the previous method is that in the worst case the number of inequalities to be evaluated equals the depth of the search tree. In these conditions, efficiency can be improved if the search tree is well balanced. In this direction, (Tøndel et al., 2003) proposes a method which elaborates a search tree of minimal depth, by trying to reduce the number of regions that must still be tested when switching from one level of the search tree to another. Performances reported in (Tøndel et al., 2003) show a logarithmic complexity with respect to the number of critical regions.
- Other positioning methods for the evaluation of explicit solutions consider a function based on the performance index to distinguish the active critical region (Borelli, 2003).

The positioning process can use during the design of the binary search tree the preliminary cutting in regions corresponding to regular shapes of the feasible domain (Olaru and Dumur, 2005). Such an approach benefits from the double hierarchical organization of the partitioning to elaborate balanced binary search trees.

IV. Newton-Type Non-linear Programming for NMPC

Returning to the nonlinear framework, one of the direct mutations is to exploit the computational advantages observed in the linear case. Thus, the use of the theory of second variations and quasilinearization lead to NMPC solutions based on a sequence of linear problems. This is the framework of the so-called Succesive Quadratic Programming (SQP)[4] which uses standard numerical tools and expects fast convergence rates upon Newton-type algorithms. The difficulties are also well known mainly in the presence of ill-conditioning, extreme nonlinearities and nonconvexity.

Starting from the general nonlinear optimization to be solved in the NMP case:

$$\Pi_k : \begin{cases} U^* = \arg\min_{U} J(U) \\ \text{subject to} : c(U) = 0 \\ \qquad\qquad\quad d(U) \leq 0 \end{cases} \tag{49}$$

the idea behind SQP is to create iterative solutions U_j and then find new search direction σ by solving :

[4] See (Alkaya et al, 2001) and the references therein for an introduction on the SQP use in process control.

$$\begin{aligned} &\min_{\sigma} \nabla_U J^T (U_j)\sigma + 0.5\sigma^T H_j\sigma \\ &\text{subject to} : c(U_j) + \nabla_U c^T (U_j)\sigma = 0 \\ &\qquad\qquad d(\mathbf{k}_u) + \nabla_U d^T (U_j)\sigma \le 0 \end{aligned} \tag{50}$$

Where H_j is the Hessian of the Lagrange function:

$$\nabla^2 L(U,\lambda) = \nabla^2 J + \sum_i \lambda_i \nabla^2 c_i \tag{51}$$

Good rate of convergence is expected when H_j is asymptotically equivalent with the Hessian at the optimum, $\nabla^2 L^*$. Unfortunately, $\nabla^2 L^*$ is not always positive definite and thus the resolution of the QP (50) is not a direct application of the convex optimization results presented in the previous section. A common way to avoid this difficulties is to consider positive definite approximations for H_j.

Another sensitive aspect is represented by the starting points. The problem (49) may turn to be infeasible if the original admissible domain is nonconvex (the linearization lead to a infeasible combination of constraints). The solution is to add a slack variable and solve the modified optimization problem:

$$\Pi_k : \begin{cases} U^* = \arg\min_U J(U) + \gamma M \\ \text{subject to} : c(U)(1-\gamma) = 0 \\ \qquad\qquad d(U) \le 0 \end{cases} \tag{52}$$

where M is a suitable large scalar such that γ can be forced as small as possible. The solution found either satisfies the original constraints, or $\gamma = 1$ and in this case a different starting point has to be found (if the original problem is consistent).

The SQP methods conduct the solution to (first-order) KKT points that may not represent local minimum, one can examine the second order conditions and follow the negative curvature to assure the convergence to local minima. The algorithms can be improved by combining SQP with trust region and line search methods or in a more general sense by exploiting the structure of the equality constraints which are related to the system dynamics.[5]

To resume the nonlinear programming based MPC versus linear MPC the following important aspects have to be mentioned:

- obtaining discrete-time nonlinear models for the prediction is a hard task comparing with the construction of linear models;

[5] The ideas presented in this section are only brief description on the topic. The field is very active with interesting methodological advances as well as powerful numerical packages available. We refer the reader to references as (Diehl et al., 2005; Kameswaran et al., 2006 ; Diehl et al., 2006) for getting familiar with the latest developments and open problems.

- the optimization problems in Nonlinear MPC might be nonconvex and thus the resolution incomparably more difficult with respect to the linear case where explicit solutions can be constructed even for the constrained case;
- dealing with real-time constraints introduced by small sampling periods is a hard task due to feasibility/ convergence issues of the nonlinear optimization routines;
- for nonlinear MPC is hard to handle strict stability and robustness demands which implies terminal sets with special topology.

Even if SQP methods are providing interesting solutions to the nonlinear MPC, the use of stochastic methods has to be considered as interesting alternative for the tractable on-line optimization. The next section presents some new results on this direction.

V. Stochastic Optimization for non Linear Predictive Control

V.1. Motivations and Principles of Stochastic Optimization

As it has been already exposed in previous sections, predictive control is mainly based on the on-line solution of successive optimization problems. As systems are more and more complex, classical structures of the objective function (linear, quadratic, convex) may not be used anymore to model the system and define the optimal control problem. The following difficulties have often to be dealt with:

- costs and constraints may have no linear expression. They may only be computed by simulators and so have no analytic expressions;
- cost functions may be multimodal and have numerous local minima;
- very large scale cases have to be considered, with numerous variables and constraints. Variables may have to be assigned to discrete values, leading to mixed optimization problems, and the curse of combinatorial complexity.

In such cases, classical optimization methods such as Sequential Quadratic Programming (see section IV) are inefficient because no gradient or hessian can be computed. Further, these methods are based on a local search and may be caught in a local minimum. Finally, there is a computational challenge in the solution of such difficult optimization problems to control non linear complex systems. A new trend is the use of stochastic optimization algorithms among which genetic algorithm may be the most widely known. The goal of such algorithms is not to find *the* global optimum of the problem, but to tackle the harsh difficulties so as to find a "good suboptimal solution" with low computation times.

Consider the following generic optimization problem:

$$\min_{x \in \chi} f(x) \tag{53}$$

The idea of stochastic optimization is to randomly explore the search space so as to detect interesting regions. It can be depicted by the following pseudo-code of figure 4.

- Random Initialisation of $x^{(0)}$
- Initialisation $x^{best} = x^{(0)}$
- **For** $k = 1$ to MaximumNumberOfIterations
  - Randomly choose $x^{(k)} \in \chi$
  - Evaluate $x^{(k)}$
  - **If** $f(x^{(k)}) < f(x^{best})$ **then**
    - $x^{best} = x^{(k)}$
  - **End If**
- **End For**

Figure 4. Pseudo code for stochastic optimization principles.

This pseudo code only supposes that it is possible to evaluate the cost function (for example with a simulator) for all possible values of x. Of course, the pseudo code of figure 4 may be inefficient as it a pure random algorithm. The main point in stochastic optimization is to find a relevant transition rule to "point the random walk" to interesting regions of the search space. The particularities of each stochastic method, also called metaheuristic, will be discussed in section V.3, but there are some common characteristics which can be listed now:

- they are inspired by physical or biological metaphors;
- they are based on a stochastic walk in the search space;
- they only need an evaluation of the cost function, and neither of its gradient nor hessian.

As a result, they can be used to optimize various criterions, as they do not need a particular structure of the objective function and, due to the random walk, are able to escape from local minima.

V.2. Taking into Account Constraints in Stochastic Optimization

As can be seen from the pseudo code of figure 4, metaheuristics walk randomly in the search space. As a result, when optimising an objective function subject to constraints, there is no guarantee that the solution remains in the feasible set. That is why most of the metaheuristics are dedicated to non constrained optimization problems, excepting the case when the method is constructive such as the ant colony algorithm (see section V.3.4). However, it is possible to reformulate constrained optimization problems into non constrained optimization problems, by adding penalty functions. For example, for inequality constraints and equality constraints, the following criterion can be optimised:

$$\min_{x\in\chi} f(x) \quad \text{s.t.} \begin{cases} g(x)\le 0 \\ h(x)=0 \end{cases} \rightarrow \min_{x\in\chi} f(x)+\lambda\left|h(x)\right|+\mu\alpha(g(x)) \quad \text{with} \begin{cases} \alpha(y)=y^2 \text{ si } y>0 \\ \alpha(y)=0 \text{ si } y\le 0 \\ \lambda\ge 0;\mu\ge 0 \end{cases} \tag{54}$$

Numerous strategies exist to define well suited and adaptive penalty functions. The details of the strategies are out of the scope of this chapter. Note that the introduction of penalty function often leads to the introduction of new tuning parameters (such as λ, μ in (54)).

V.3. Stochastic Optimization Methods

V.3.1. Simulated Annealing

Simulated annealing was introduced by Kirkpatrick, Gelatt and Vecchi in 1983 (Kirkpatrick et al., 1983). It is based on the physical annealing process, in which a molten particle is solidified from a very high temperature to a low temperature. The general flowchart of the simulated annealing algorithm is given in figure 5.

The iteration number is equivalent to the temperature level. A candidate solution is generated in the neighbourhood $V(x^{current})$ of the current point. If this solution leads to a decrease of the cost function then it is accepted as the new current point. Otherwise, it is accepted with a probability equal to $\exp(-\Delta E / kT)$, which depends on the current temperature T. As some bad solutions are sometimes accepted, the algorithm may escape from local minima. To implement the algorithm, the user has to define the following strategies:

- Define the topology of the neighbourhood $V(x^{current})$;
- Define the strategy to decrease the temperature parameter T.

Details on these strategies can be found in (Aarts, *et al.*, 1997). The main idea is to set a high temperature in the beginning of the algorithm so as to deeply explore the search space ($\exp(-\Delta E / kT)$ is close to one and so lots of solutions are accepted), and to progressively decrease the temperature to converge to a good suboptimal solution.

- Initialisation $x^{current} = x^{(0)}$
- **For** $k = 1$ to MaximumNumberOfIterations
    - Randomly choose $x^{(k)} \in V(x^{current})$
    - Evaluate $x^{(k)}$
    - **If** $f(x^{(k)}) < f(x^{current})$ **then**
        - $x^{courant} = x^{(k)}$
    - **Else**
        - $\Delta E = f(x^{(k)}) - f(x^{current})$
        - Randomly choose $\rho \in [0,1]$
        - **If** $\rho < \exp^{-\frac{\Delta E}{kT}}$ **then**
            - $x^{courant} = x^{(k)}$
        - **End If**
    - **End If**
    - Decrease T if necessary
- **End For**

Figure 5. General flow-chart of the simulated annealing algorithm.

V.3.2. Genetic Algorithm

Genetic algorithm has emerged as a well-known and efficient metaheuristic method to solve optimization problems. It is based on Darwin theory concerning the evolution of species. The main idea is to make a population of p potential solutions evolve so as to create new potential solutions by using stochastic (or "genetic") operators. For ease of presentation, a binary optimization problem is firstly considered:

$$\min_{(x_1,x_2,\ldots,x_n)\in\{0,1\}^n} f(x_1,x_2,\ldots,x_n) \tag{55}$$

The population of p potential solution, for iteration k is noted: $P^{(k)} = \begin{pmatrix} x_1^{1,(k)} & x_2^{1,(k)} & \cdots & x_n^{1,(k)} \\ x_1^{2,(k)} & x_2^{2,(k)} & \cdots & x_n^{2,(k)} \\ \vdots & \vdots & \ddots & \vdots \\ x_1^{p,(k)} & x_2^{p,(k)} & \cdots & x_n^{p,(k)} \end{pmatrix}$.

The general flow-chart of the genetic algorithm is given in figure 6.

- Create initial population $P^{(0)} = \begin{pmatrix} x_1^{1,(0)} & x_2^{1,(0)} & \cdots & x_n^{1,(0)} \\ x_1^{2,(0)} & x_2^{2,(0)} & \cdots & x_n^{2,(0)} \\ \vdots & \vdots & \ddots & \vdots \\ x_1^{p,(0)} & x_2^{p,(0)} & \cdots & x_n^{p,(0)} \end{pmatrix}$
- **For** $k = 1$ to MaximumNumberOfIterations
  - Create c new solutions by genetic operators: they are children $C^{(k)} = \begin{pmatrix} y_1^{1,(k)} & y_2^{1,(k)} & \cdots & y_n^{1,(k)} \\ y_1^{2,(k)} & y_2^{2,(k)} & \cdots & y_n^{2,(k)} \\ \vdots & \vdots & \ddots & \vdots \\ y_1^{c,(k)} & y_2^{c,(k)} & \cdots & y_n^{c,(k)} \end{pmatrix}$
  - Select p individuals among individuals of $\begin{bmatrix} P^{(k-1)} \\ C^{(k)} \end{bmatrix}$ to form $P^{(k)}$
- **End For**

Figure 6. General flow-chart of the genetic algorithm.

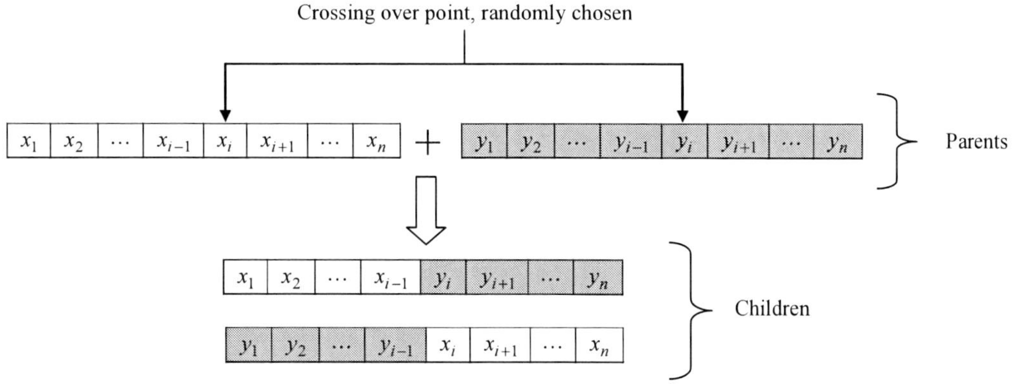

Figure 7. Classical crossing-over operator.

Classical genetic operators (crossing-over and mutation operators) are depicted in figures 7 and 8. For the crossing over operation (figure 7), two potential solutions (labelled "parents") are randomly chosen in the population. They stochastically merge their optimization variables (or "genes") so as to create two new solutions ("children").

The mutation operation (figure 8) deals with the random selection of a potential solution and of one of its genes. This gene is changed to another. The aim of this genetic operator is to keep the population genetic diversity by making new genes appear in the population.

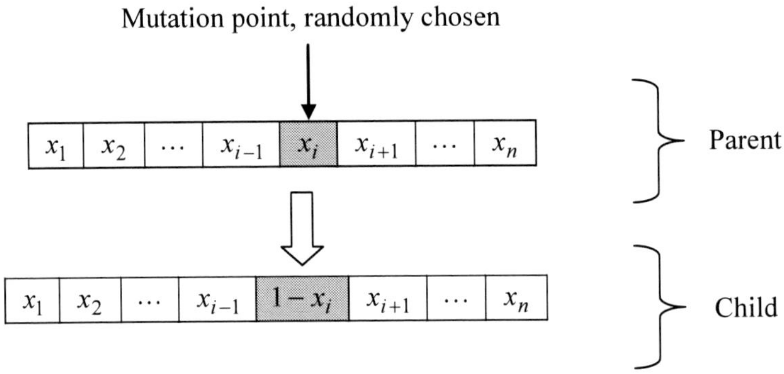

Figure 8. Classical mutation operator.

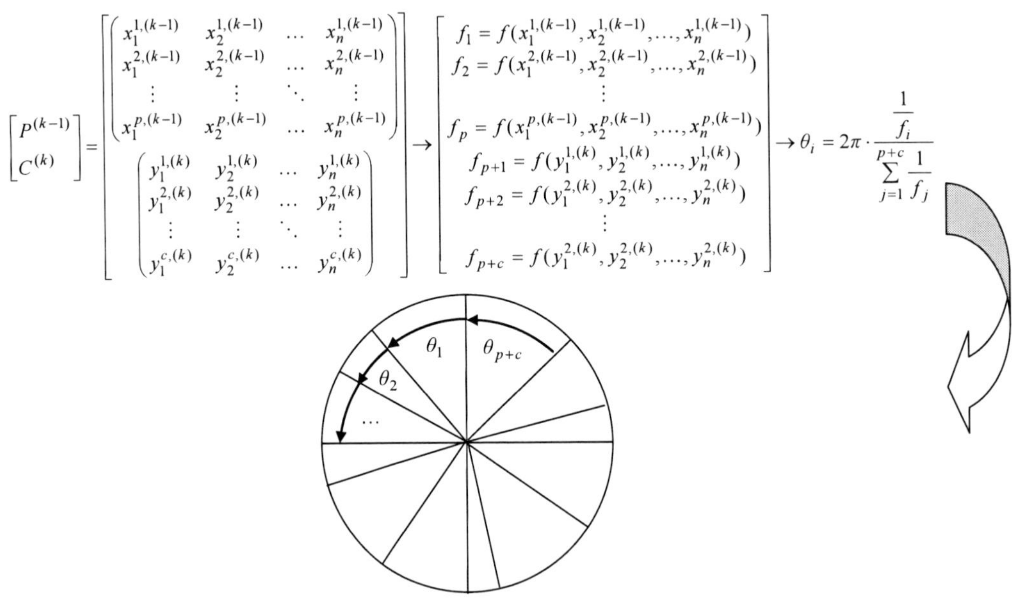

Figure 9. Roulette wheel selection.

Finally, the selection operator is a genetic operator which aims to choose a new population from parents and children. This operation can be made for example using the classical roulette wheel selection, represented in figure 9. After having computed the fitness value of each individual in the population, the probability of selection is proportional to the quality of individuals.

Classically, the size of the population is $p = 100 \cdot n$, where n is the dimension of the problem, $0.5 \cdot p$ to $0.8 \cdot p$ children are created by the crossing-over operator, and $0.01 \cdot p$ to $0.05 \cdot p$ children are created by the mutation operator. The genetic operators represented in figure 7, 8 and 9 are the basic ones. Lots of implementations exist and can be extended to real variable optimization problems. Those improvements of the classical algorithm will not be detailed here, and are fully described in (Koza 1992; Koza 1994; Baeck et al., 2000a; Baeck, *et al.*, 2000b).

V.3.3. Ant Colony Optimization

Ant colony optimization was firstly introduced by Marco Dorigo (Dorigo, 1992; Dorigo et al., 1996). It is based on the way ants are looking for food. Suppose (see figure 10a), that ants have managed to find food. Each particular ant does not know where to go. It only chooses its path depending on the pheromone trail which has been laid on the ground by previous ants. If the path of pheromone is broken because of an obstacle, see figure 10b, first ants randomly choose their path. But, the ants which have chosen the shortest path will arrive first: the trail of pheromone in the shortest path is increasing faster than in the longest path. The positive feedback structure makes all ants finally choose the shortest path at the end of the experience (see figure 10c).

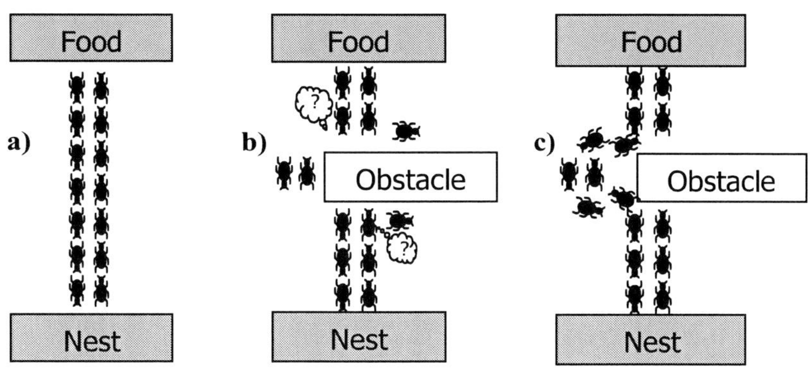

Figure 10. Ants looking for food.

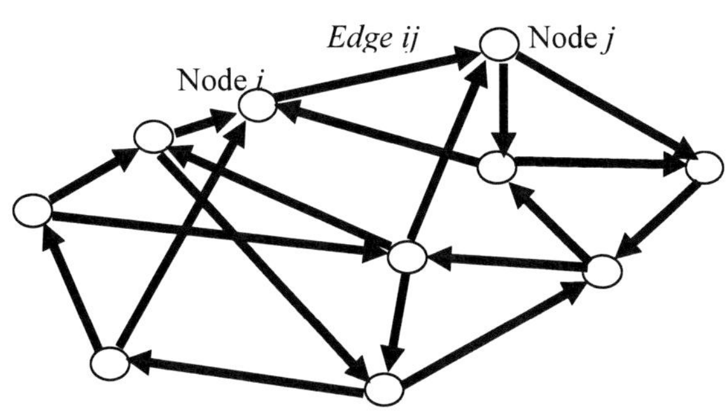

Figure 11. Graph exploration problematic.

The metaphor can be used to perform graph exploration as in figure 11.

Various criterions can be optimised, depending on the graph and the goal of its exploration. For instance, a cost may be associated to each node, and the goal is to minimise the sum of costs while exploring the graph. This is the well known Traveling Salesman Problem, for which ant colony algorithm has been firstly used (Dorigo et al., 1997). During iteration t of the algorithm, F ants walk on the graph of figure 11. If ant f has reached state $i=(u_n^1,\ldots,u_n^K)$, the probability that it chooses the next state $j=(u_{n+1}^1,\ldots,u_{n+1}^K)$ is:

$$p_i^{(f)}(j)=\frac{{\eta_{ij}}^{\alpha}\tau_{ij}(t)^{\beta}}{\sum\limits_{m\in J_f(i)}{\eta_{im}}^{\alpha}\tau_{im}(t)^{\beta}}. \tag{56}$$

- $\tau_{ij}(t)$ is the pheromone trail on edge $i=(u_n^1,\ldots,u_n^K)\rightarrow j=(u_{n+1}^1,\ldots,u_{n+1}^K)$ during iteration t. Its value depends on the results of previous ants;
- η_{ij} is the attractiveness. It refers to the "local choice". For the Traveling Salesman Problem, $\eta_{ij}=1/d_{ij}$, where d_{ij} is the cost associated to the edge ij of the graph;
- α and β are weighting factors;
- $J_f(i)$ is the feasible set (for the travelling Salesman problem, this feasible set contains all nodes connected to node i, but those states which have already been explored by the ant are to be removed).

At the end of iteration t of the algorithm, F ants have computed F potential solutions, which are evaluated. The pheromone trail which laid on the graph of figure 11 is then updated:

$$\tau_{ij}(t+1)=(1-\rho)\,\tau_{ij}(t)+\Delta\tau_{ij}(t) \tag{57}$$

$\Delta\tau_{ij}$ is the update coefficient. Its value depends on the results of ants in iteration t. The more the ants which have walked on edge ij have obtained good results, the higher is $\Delta\tau_{ij}$. Several strategies can be used for this update, see for instance (Dorigo et al., 2005). Finally, the algorithm is based on a positive feedback, as current ants try to influence future ants by enforcing the pheromone trail. As a result, to avoid premature convergence, it is necessary to forget the past mistakes: this is the role of the pheromone evaporation. ρ is the evaporation coefficient. This coefficient is an analogy with physical evaporation of pheromone in nature. Finally, the algorithm can be summed up by the general flow chart of figure 12.

Usually, there are about 20 to 30 ants, while $\alpha\approx 1$ and $\beta\approx 2$. It is often necessary to bound the pheromone trail on each edge ($\sigma\in[\sigma_{\min},\sigma_{\max}]$), to avoid premature convergence; see (Stützle, *et al.*, 2000). Note that ant colony optimization is a constructive algorithm. As a result, it can explicitly take into account all the constraints of the problem, with the help of the feasible sets $J_f(i)$ which are built for each ant. This is one of the most important particularities of this algorithm compared to other well known metaheuristics.

- **For** $k = 1$ to MaximumNumberOfIterations
  - **For** $f = 1$ to F
    - Build a new solution, based on probabilistic law (Eq. 56).
  - **End For**
  - Evaluate the F solutions
  - Update pheromone trail with (Eq. 57)
- **End For**

Figure 12. General flow-chart of the ant colony algorithm.

V.3.4. Particle Swarm Optimization

Particle swarm Optimization (PSO) was firstly introduced by Russel and Eberhart (Eberhart, *et al.*, 1995). This algorithm is mainly dedicated to real variable optimization problems. PSO is inspired by the social behavior of bird flocking or fish schooling.

P particles are moving in the search space. Each of them has its own velocity, and is able to remember where it has found its best performance in the search space. Each particle has some "friends", with which it is able to exchange some pieces of information. The following notations are used:

- x_p^k position of particle p at iteration k;
- v_p^k velocity of particle p at iteration k;
- $b_p^k = \arg\min_{x \in \{b_p^{k-1}, x_p^k\}}(f(x))$ best position found by particle p until iteration k;
- $V(x_p^k) \subset \{1,2,\ldots,P\}$ set of "friend particles" of particle p at iteration k;
- $g_p^k = \underset{x \in \{b_i^k, i \in V(x_p^k)\}}{\arg\min} f(x)$ best position found by the "friend particles" of particle p until iteration k.

The particles move in the search space according to the following transition rule:

$$\begin{aligned} v_p^{k+1} &= w.v_p^{k+1} + c_1 \otimes (b_p^k - x_p^k) + c_2 \otimes (g_p^k - x_p^k) \\ x_p^{k+1} &= x_p^k + v_p^{k+1} \end{aligned} \tag{58}$$

In this equation:

- w is the inertia factor ;

- c_1 (resp. c_2) is a random number in the range $[0, \bar{c}_1]$ (resp. $[0, \bar{c}_2]$);
- $\otimes$ denotes the element wise multiplication of two vectors.

The geometric construction of (58) is represented in figure 13.

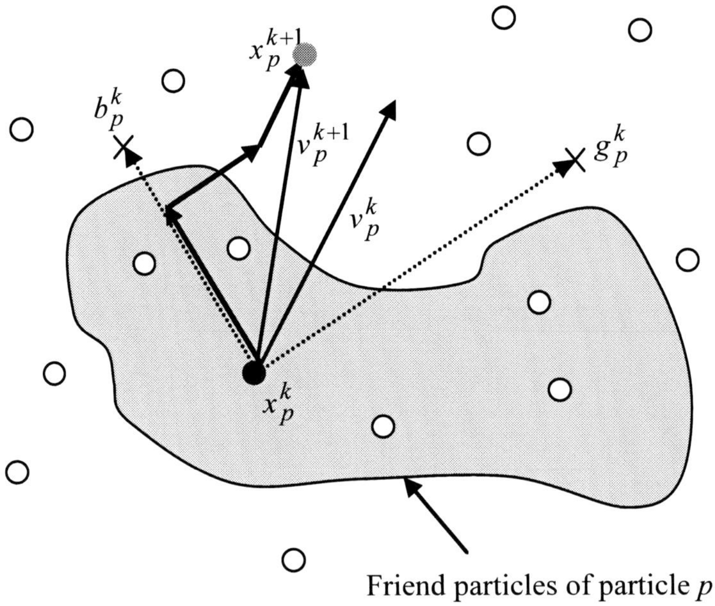

Figure 13. Geometric construction of particle movements.

Finally, PSO can be depicted by the general flow-chart of figure 14.

Randomly choose : $\{x_p^0, v_p^0 ; p = 1, \ldots, P\}$
For $k = 0$ to MaximalNumberOfIterations,
    For particle $p = 1$ to P,
        Make particle p move (58)
        Evaluate new position $f(x_p^{k+1})$
        If $f(x_p^{k+1}) < f(b_p^k)$, **then**
            $b_p^k = x_p^{k+1}$,
        Else,
            $b_p^{k+1} = b_p^k$
        End If
    Fin Pour
    Update g_p^{k+1}
End For

Figure 14. General flow-chart of Particle Swarm Optimization.

The choice of parameters is very important to ensure the satisfying convergence of the algorithm. Numerous studies have treated the topic; see for instance (Shi et al., 1998; Eberhart et al., 2000). Relevant values are for example given in (Eberhart et al., 2006) as "standard" ones:

- Swarm size: $10+\sqrt{n}$, where n is the dimension of the optimization problem;
- $w = 1/(2\ln(2))$;
- $\overline{c_1} = \overline{c_2} = 0{,}5 + \ln(2)$;
- $\dim(V(x_p^k)) \leq 3$.

Several topologies exist for the design of sets $V(x_p^k)$. For a comprehensive study of this topic, see (Kennedy, 1999). In particular, if these sets do not depend on k, neighbourhoods are said to be "social". This choice is the simplest for the implementation of the algorithm.

V.3.5. Some Other Stochastic Optimization Methods

Many other stochastic optimization methods have been developed, but it is not in the scope of this chapter to describe all of them. Here are some references for a description of those methods.

- Tabu search (Glover, 1989; Glover et al., 1997);
- Estimation of Distribution Algorithm (Mühlenbein et al., 1996);
- GRASP ("Greedy Randomized Adaptive Search Procedure") (Feo et al., 1995);
- Cross Entropy (Rubinstein, 1999).

V.4. Conclusion: Advantages and Drawbacks of Stochastic Methods

In this section, the main metaheuristics methods have been reviewed. All these methods have the following characteristics:

- They are based on stochastic movements in the search space, leading to the possibility of avoiding premature convergence in local minima;
- They do not require gradient or hessian computation, leading to the possibility of optimising non linear or non analytic functions.

However, they may present some drawbacks also:

- They require a large number of criterion evaluations, thus it is necessary to have a fast computation or simulation of costs and constraints;
- They require the tuning of the algorithm parameters;
- There is no guarantee on the actual optimality of the solution. In particular, validation of the optimization algorithm can only be performed with statistical data (mean, worst case and standard deviation of the optimization results).

Finally, stochastic optimization methods find very often a suboptimal solution with low computation times for hard optimization problems. Thus, they appear as a new trend in the field of complex non linear predictive control.

VI. Applications in Power Systems

In this section, two applications of non linear predictive control in the field of Power Systems are provided. Indeed, the short term optimal scheduling and control of power systems has emerged as a crucial point for Energy Industries: in a more and more competitive economic context, energy network managers have to control their increasingly complex power systems so as to satisfy consumers' demands keeping the lowest costs.

VI.1. Predictive Control of Supply Temperatures in District Heating Networks

The considered problem is the district heating network of figure 15. It is made of two main distribution networks. The interconnection is controlled by two valves. Both producers are assumed to be heat-only units. Six consumers are installed in the network.

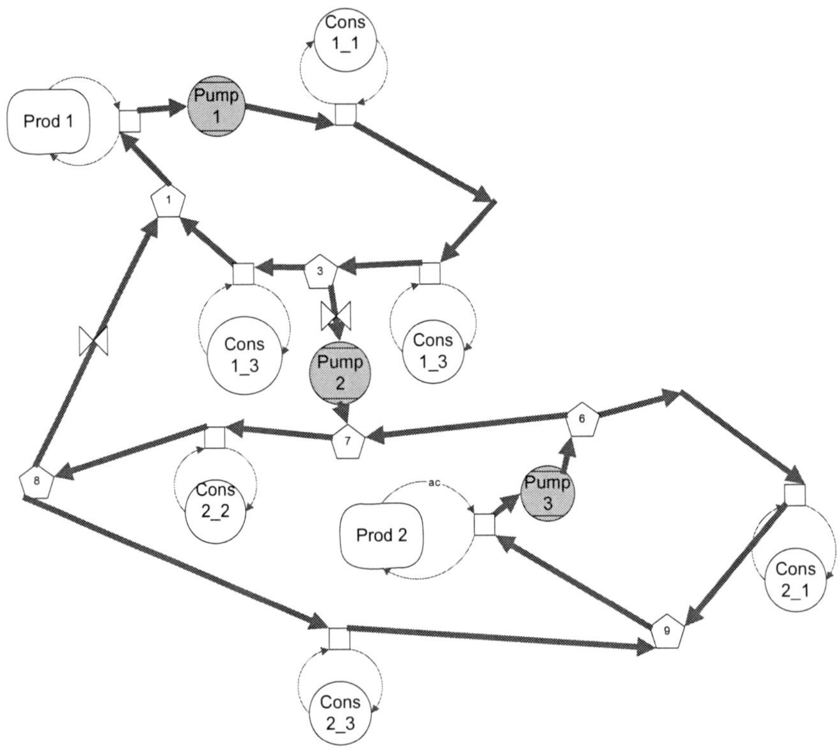

Figure 15. District heating network example.

This district heating network is a part of a network benchmark which has been fully described in (Sandou et al., 2004a).

VI.1.1. Simulation Model of District Heating Network

Inputs of the network (produced powers and opening degrees of valves) are discretized. It can be shown that a sampling period of one hour is a classical and efficient choice for the control of district heating networks and will be used in the following (the subscript for time interval being n).

Production model. Production site k is modelled by a static characteristic, identified from technical data. For the hour n, production costs are computed from produced thermal power Q_n^k [W]:

$$c_{prod}^k(Q_n^k) = a_2(Q_n^k)^2 + a_1 Q_n^k + a_0 \tag{59}$$

Due to this thermal power given to the primary network, an increase in network temperatures occurs:

$$Q_n^k = c_p m_s (T_s^k - T_r^k) \tag{60}$$

m_s [kg.s^{-1}] is the mass flow in primary network, T_s [K] the supply temperature and T_r [K] the return temperature in primary network; c_p [J.kg^{-1}.K^{-1}] is the specific heat of water.

Supply network model - Mass flows and pressures. Mass flows and pressures computation is related to the modelling of pipes, valves, nodes and pumps. Mechanical losses in pipes are derived from:

$$H_{out} = H_{in} - Z_p m_p^2 \tag{61}$$

m_p [kg.s^{-1}] is the mass flow in pipe, H_{in} (resp. H_{out}) [m] the pressure at the input (resp. the output) of the pipe, and Z_p [m.kg^{-2}.s^2] the friction coefficient. For a pipe with a valve, this coefficient is changed into Z_p / d. d is the opening degree of the valve in the range [0,1] (0 for a closed valve to 1 for an open one).

The increase in pressure of pumps compensate for the mechanical losses of (61):

$$\Delta H = a_2\left(m\frac{\omega_0}{\omega}\right)^2 + a_1\left(m\frac{\omega_0}{\omega}\right) + a_0 \tag{62}$$

m is the mass flow through the pump, ω [rad.s^{-1}] its rotation speed and ω_0 its nominal rotation speed. For the mass flow computation, nodes are modelled by mass flow balance equations. Finally, mass flows and pressures have to be computed from a large scale non linear system of algebraic equations obtained from many static equations modelling all pipes, all valves, all pumps and all nodes.

Supply network model - Thermal energy propagation. Let μ_p [J.m^{-2}.s^{-1}.K^{-1}] the thermal loss coefficient, ρ [kg.m^{-3}] the relative density of water, T_0 [K] the external temperature, $T(x,t)$ the temperature in the pipe, R_p [m] its radius, and m_p the mass flow in the pipe. The thermal energy propagation is modelled by a classical partial differential equation (Arvastson, 2001):

$$\frac{\partial T}{\partial t} + \frac{m_p(t)}{\pi \rho R_p^2}\frac{\partial T}{\partial x} + \frac{2\mu_p}{c_p \rho R_p}(T(x,t) - T_0) = 0 \tag{63}$$

For the energy propagation point of view, nodes are modelled by an energy balance equations.

Consumer model. Secondary networks of consumers are connected to the primary network by way of a counter flow heat exchanger, whose notations are those of fig. 16.

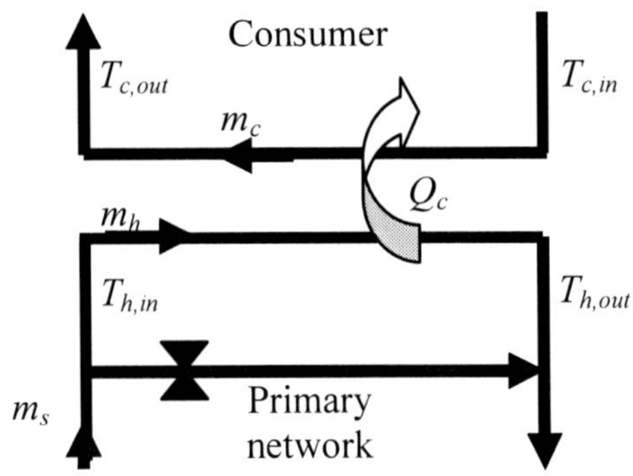

Figure 16. Notations for consumer modelling.

This kind of heat exchanger is classically modelled by:

$$Q_c = eS\frac{(T_{h,in} - T_{c,out}) - (T_{h,out} - T_{c,in})}{\ln(T_{h,in} - T_{c,out}) - \ln(T_{h,out} - T_{c,in})} \tag{64}$$

S [m^2] is the surface of the heat exchanger, and e [W.K^{-1}.m^{-2}] its efficiency. The thermal power given by the primary network is also expressed by:

$$Q_c = c_p m_h (T_{h,in} - T_{h,out}) \tag{65}$$

Finally, the thermal power received by the consumer is:

$$Q_c = c_p m_c (T_{c,out} - T_{c,in}) \tag{66}$$

Assuming that there is no thermal losses between primary and consumer networks, and assuming that T_c and $T_{c,out}$, and mass flow m_h is determined by the valve opening degree, $T_{c,in}$, Q_c and $T_{h,out}$ can be computed as a function of $T_{h,in}$. Of course, Q_c is an increasing function of m_h and thus the maximal power which can be given to a consumer is obtained for $m_h = m_s$. Consequently, the given power is finally:

$$Q_c = \min(Q_{dem}, Q_{\max}) \tag{67}$$

Q_{dem} is the consumer's demand and $Q_{\max}$ is the maximum power which can be given by the network. $Q_{\max}$ is computed by solving the non linear algebraic system (64), (65) and (66), in the particular case $m_h = m_s$.

VI.1.2. Model for Optimization

Model for optimization. An ideal open loop control can be computed from the minimisation of production costs:

$$\min_{\substack{Q_n^k, d_n^v \\ n\in[1:N], k\in[1:K], v\in[1:V]}} \left\{ \sum_{k=1}^{K} \sum_{n=1}^{N} \left(c_{prod}^k (Q_n^k)\right) \right\} \tag{68}$$

Q_n^k is the thermal power produced by site *k* during hour *n*, and d_n^v the opening degree of valve *v*. The constraints are those of the district heating network (mainly bounds on temperatures, pressures and mass flows) and the satisfaction of consumers' demands. Note that consumers' demands are predicted variables.

The reference model, described in previous section VI.1.1, is fully representative of a district heating network: the objective function and constraints could be computed using this model. However, as the simulation of the network is about a few seconds (solution of non linear algebraic equations and simulation of partial differential equations), it is not tractable for optimization and control purposes. Thus, an adapted model, well suited to optimization issues, has to be developed.

Production and consumer parts of the reference model which have been developed for simulation purposes are highly tractable: they will be used for optimization. Contrariwise, the computation of varying time delays is time consuming: constant time delays are considered for optimization purposes. So, (64) leads to:

$$T_{out}(t) \approx (1-\rho_{th})T_{in}(t-\tau_p) \tag{69}$$

τ_p is the time delay of the pipe, supposed to be constant.

This assumption allows globally modelling thermal propagation as a non linear dynamic system, which can be quickly solved. A sampling period of 1/6 hour has been chosen for the optimization model. Note however that costs and constraints are still highly non linear and can only be computed from a simulator.

Regularisation of the optimization problem. However, thermal losses in pipes are very low for a district heating network: thus, the network globally acts as a storage tank. As a result, lots of solutions are feasible solutions with quite similar costs. Thus, the optimization problem has to be regularized with a penalisation criterion for non smooth solutions:

$$c_{pen}^k = \lambda \sum_{n=2}^{N-1} (Q_{n+1}^k - 2Q_n^k + Q_{n-1}^k)^2 \tag{70}$$

The penalisation term can be interpreted as a way to take into account dynamics and attrition costs of production sites.

VI.1.3. Control Procedure of District Heating Networks

Control procedure. The solution of (68) is an open loop control of the network. This solution can not be applied to the real system as:

- the system can be globally viewed as an integrator and by consequence is unstable;
- the solution has been computed from a simplified model;
- load prediction errors are disturbances to be rejected in order to get a satisfying behavior.

Finally, a closed loop framework is required. Predictive control is a natural way to extend optimization results in a close loop structure. The control procedure is depicted on figure 17 upon the receding horizon principle which has been defined in previous sections.

The scheduling algorithm computes produced powers. However:

- the system may lead to instability if more power than required is produced, because of the storage effect of the distribution network;
- temperature supply variables are the actual outputs of district controllers in real networks.

Thus, to assure stability of the control law and to get convenient control inputs, a change of variables is performed: produced powers are replaced by supply temperatures as control inputs. This change of variables is implicit and performed by simulating the optimization model. Note that the optimization has to be updated. This may be difficult as these state variables are different from the simulation model state variables for which time varying propagation delays are considered. However an estimator can be deduced from the simulation model.

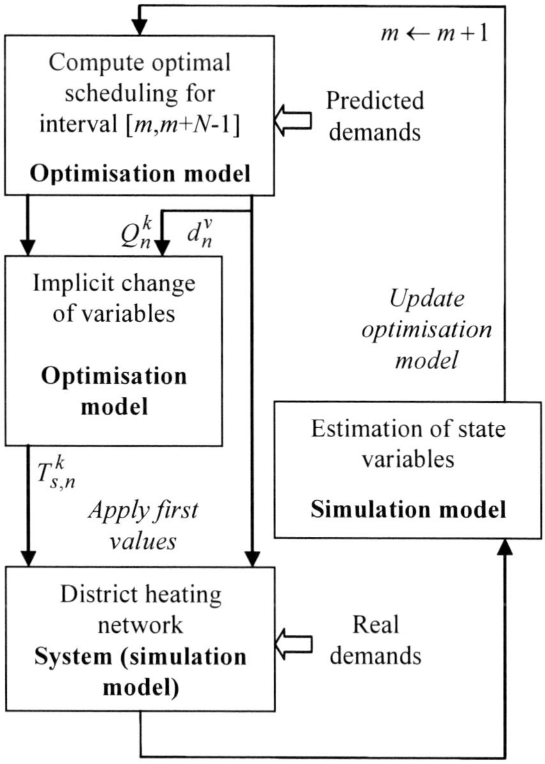

Figure 17. Control strategy of district heating networks.

Robustification of the control law. As a consequence of optimality, the open loop control try to produce as less power as possible, while satisfying predicted consumers' demands. It may lead to unfeasibility of next step optimization problems, especially for underestimated consumers' demands. Then, in this approach, the classical automatic control notion of robustness appears to be strongly related to the notion of unfeasibility. The robustness property can be obtained by using the thermal storage effect of the network. Indeed, as previously mentioned, if supply temperature controls are increased, most of the extra power supply will remain in the network and will be usable later. Thus, the increase in production costs over the whole time interval remains very slight, and a robust behavior is also achieved.

VI.1.4. Numerical Results

The control procedure has been validated on a district heating network benchmark of figure 15. The consumers' demand shape is depicted in figure 18. The worst case is considered: it is assumed that all demands have been underestimated with an error of 5%.

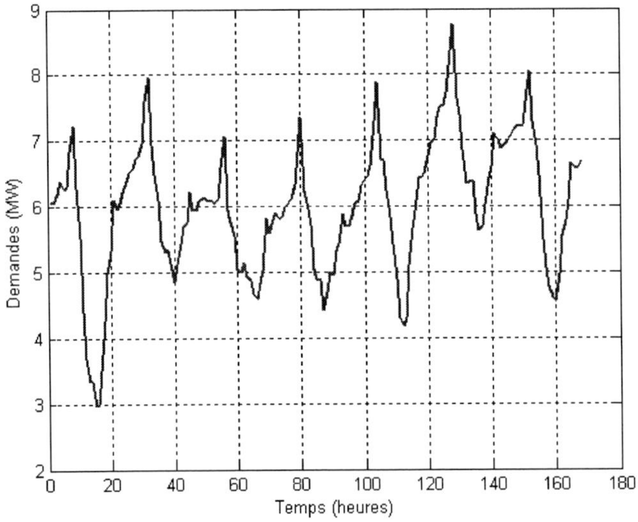

Figure 18. Consumers' demand shape.

The control law has been implemented with Matlab 6.1 and its optimization toolbox 2.1.1 on a Pentium IV, 2.5 GHz (optimization with a classical Successive Quadratic Procedure). The time horizon for optimization solutions is N = 12 hours. The scheduling algorithm is performed in 4 minutes. Supply temperatures for both producers are given in figure 19.

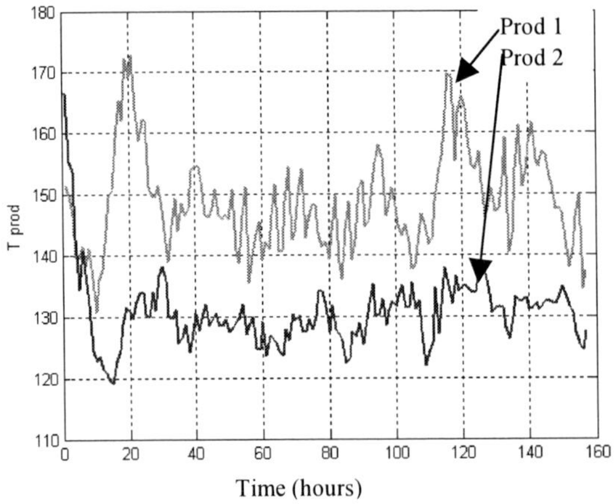

Figure 19. Temperature control with load prediction errors (5%).

The global shape of supply temperature controls is quite similar to consumers' demands shape. Due to propagation times and the anticipative effect of predictive control, temperature peaks occur several hours before consumers' demands peaks. Supply temperatures are not as smooth as consumers' demands because of modelling uncertainties and load prediction errors. Remember that producers are heat-only units and so there is no potential economic benefit from producing more than required and using the distribution network as a heat storage tank. The optimal strategy is then to produce as later as possible, to reduce thermal losses, while taking into account propagation delays. As consumers' demands have been underestimated, temperatures are relatively low between hours 40 and 100. However, because of the robustification procedure, the successive optimization problems remain feasible. At the end of the simulation, the remaining energy in the network is quite low, and a high peak is observed (hour 120).

VI.2. Predictive Control of Power Production

VI.2.1. Open Loop Control and Unit Commitment

Unit Commitment is a classical large scale mixed integer problem in power systems. The goal is to compute the optimal scheduling of several production units while satisfying consumer demand and technical constraints. Integer variables are the on/off status of production units, and real variables are produced powers. Finally, the Unit Commitment problem is:

$$\min_{\{u_n^k, Q_n^k\}} \sum_{n=m}^{m+N-1} \left(\sum_{k=1}^{K} (c_{prod}^k(Q_n^k, u_n^k) + c_{on/off}^k(u_n^k, u_{n-1}^k)) \right). \tag{71}$$

Where:

- N is the length of time horizon
- K is the number of production unit,
- u_n^k is the on/off status of production unit k during time interval n.
- Q_n^k is the produced power of production unit k during time interval n.

Production costs and start up and shut down costs are:

$$\begin{cases} c_{prod}^k(Q_n^k, u_n^k) = \alpha_1^k Q_n^k + \alpha_0^k u_n^k \\ c_{on/off}^k(u_n^k, u_{n-1}^k) = c_{on}^k u_n^k(1 - u_{n-1}^k) + c_{off}^k u_{n-1}^k(1 - u_n^k) \end{cases}. \tag{72}$$

The constraints of the problem are:

- capacity constraints:

$$Q_{\min}^k u_n^k \le Q_n^k \le Q_{\max}^k u_n^k, \; \forall n \in \{m, \ldots, m+N-1\}, \forall k \in \{1, \ldots, K\}, \tag{73}$$

- consumer demand satisfaction:

$$\sum_{k=1}^{K} Q_n^k \ge \hat{Q}_n^{dem}, \; \forall n \in \{m, \ldots, m+N-1\}, \tag{74}$$

- time up and time down constraints:

$$\begin{cases} \left(u_{n-1}^k = 0, u_n^k = 1\right) \Rightarrow \left(u_{n+1}^k = 1, u_{n+2}^k = 1, \cdots, u_{n+T_{up}^k-1}^k = 1\right) \\ \left(u_{n-1}^k = 1, u_n^k = 0\right) \Rightarrow \left(u_{n+1}^k = 0, u_{n+2}^k = 0, \cdots, u_{n+T_{down}^k-1}^k = 0\right) \end{cases}, \qquad (75)$$

- ramp constraints:

$$\left|Q_n^k - Q_{n-1}^k\right| \le \Delta Q^k \quad \forall n \in \{m, \ldots, m+N-1\}, \forall k \in \{1, \ldots, K\}. \qquad (76)$$

Discrete dynamics on the system are expressed through logical equations (75) and continuous dynamics are stated as power increment limitations (76).

VI.2.2. Closed Loop Control

As in section VI.1, from the optimization point of view, consumer's demand is supposed to be perfectly known over the whole time horizon, leading to the computation of the optimal scheduling, which is a reference trajectory for integer and real control inputs. However, prediction errors of the consumer load may lead to a deficient behavior of the system. Thus, a closed loop control is required, and the problem refers to the control of a hybrid system. Predictive control is used and depicted in figure 20.

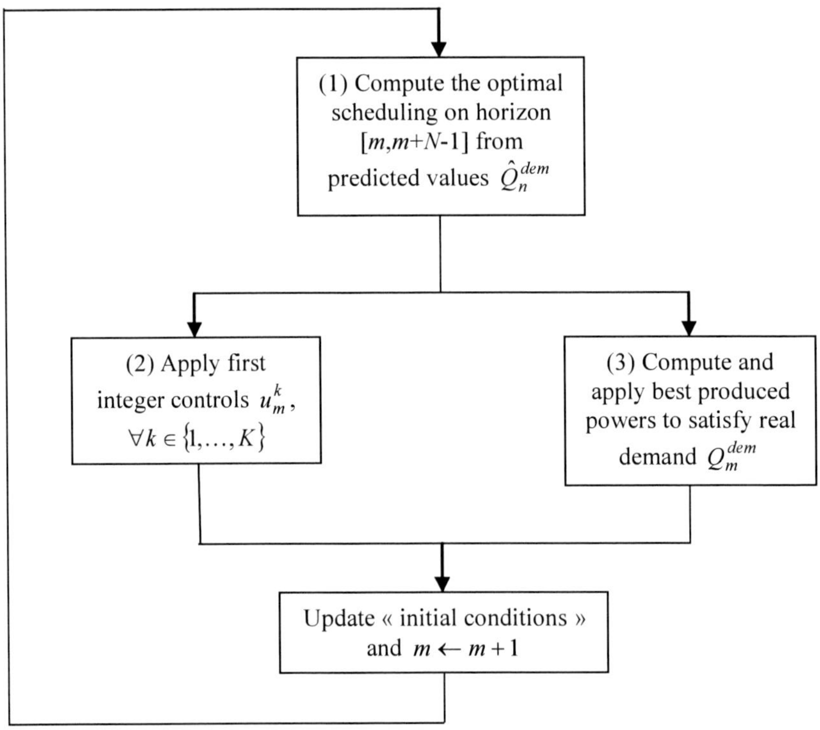

Figure 20. Closed loop control synopsis.

The optimal scheduling, considering predicted consumer load $\hat{Q}_n^{dem}$, is computed on time interval $[m, m+N-1]$. The first values of integer scheduling u_m^k are applied to the system. Simultaneously, real variables are slightly updated so as to fulfil, if possible, the real demand at time *m*, Q_m^{dem}. Production unit local regulations are perfect: if production unit *k* has produced power Q_{m-1}^k during time interval $m-1$, then it is able to produce, if still switched on, whatever power in the power range $[\max(Q_{m-1}^k - \Delta Q^k, Q_{\min}^k), \min(Q_{m-1}^k + \Delta Q^k, Q_{\max}^k]$ during time interval *m*.

VI.2.3. Optimization Procedure

General synopsis of the optimization procedure. Many optimization methods have been tested to solve this problem and are listed in the survey paper (Sen et al., 1998). A sequential stochastic optimization procedure will be used and is given in figure 21.

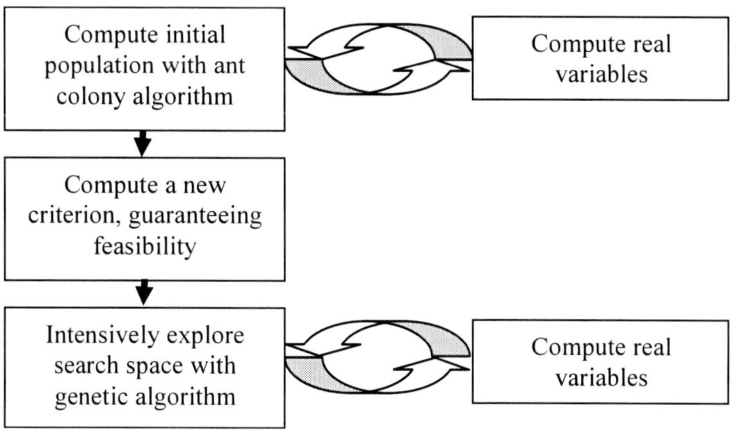

Figure 21. Cooperative algorithm synopsis.

The idea is to firstly sample the feasible space with an ant colony algorithm (feasible solutions are constructed) to feed a genetic algorithm. The algorithm supposes that real variables can be quickly computed, and so, it is hybridized with an exact solution algorithm for real variables.

Ant colony optimization. The Unit Commitment problem can be formulated as a graph exploration problem as shown in figure 22, see also previous work (Sandou et al., 2004b).

Production costs are associated to each node, and start up and shut down costs are associated to each edge. Finally, optimising cost function (72) refers to a graph exploration and can be performed by ant colony optimization (see section V.3.3).

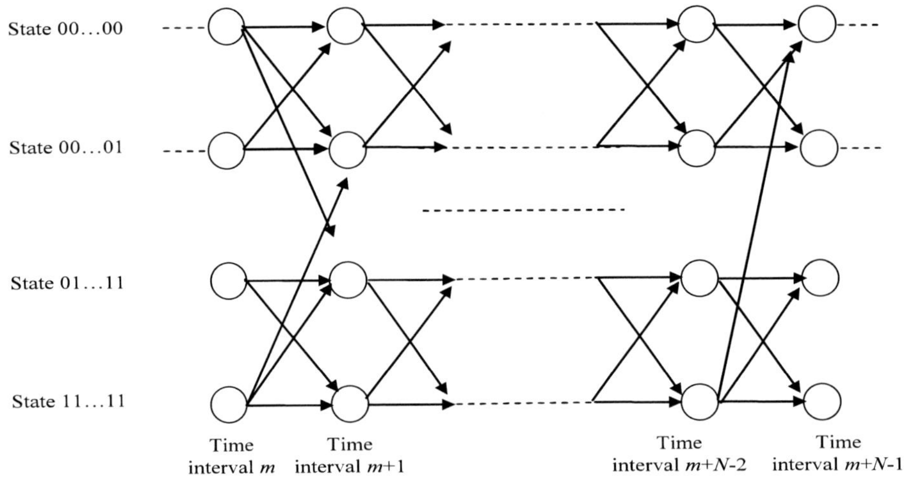

Figure 22. Graph structure for Unit Commitment.

Computation of real variables. Binary variables are computed for instance by ant colony optimization. For each feasible scheduling sequence $(u_n^k; k=1,\ldots,K; n=m,\ldots,m+N-1)$, it corresponding optimal real variables Q_n^k can be computed by:

$$\begin{aligned}
&\arg\min_{\{Q_n^k\}} \sum_{n=m}^{m+N-1}\left(\sum_{k=1}^{K}(c_{prod}^k(Q_n^k,u_n^k)+c_{on/off}^k(u_n^k,u_{n-1}^k))\right)\\
&=\arg\min_{\{Q_n^k\}}\left(\sum_{n=m}^{m+N-1}\sum_{k=1}^{K}\alpha_1^k Q_n^k u_n^k+\alpha_0^k u_n^k+c_{on}^k u_n^k(1-u_{n-1}^k)+c_{off}^k u_{n-1}^k(1-u_n^k)\right)\\
&=\arg\min_{\{Q_n^k\}}\left(\sum_{n=m}^{m+N-1}\sum_{k=1}^{K}\alpha_1^k Q_n^k u_n^k\right)
\end{aligned} \tag{77}$$

The problem can be divided into N optimization problems:

$$\min_{\{Q_n^k,k=1,\ldots,K\}}\left(\sum_{k=1}^{K}\alpha_1^k Q_n^k u_n^k\right)$$
$$\text{subject to}\begin{cases}\displaystyle\sum_{k=1}^{K}Q_n^k\geq\hat{Q}_n^{dem}\\ Q_{\min}^k u_n^k\leq Q_n^k\leq\underbrace{\begin{pmatrix}Q_{\min}^k(1-u_{n-1}^k)+\\ \left(\min(Q_{\max}^k,Q_{n-1}^k+\Delta Q^k)\right)u_{n-1}^k\end{pmatrix}}_{=Q_{\max}^k(n)}u_n^k\end{cases} \tag{78}$$

$\hat{Q}_n^{dem}$ is the predicted consumer's demand. Without loss of generality, consider that:

$$\alpha_1^1 \le \alpha_1^2 \le \ldots \le \alpha_1^K \tag{79}$$

Then, the optimal solution of problem (Eq. 25) is to produce as much as possible with low-cost units, while satisfying capacity constraints:

$$\begin{cases} Q_n^1 = \min\left(\max\left(\hat{Q}_n^{dem} - \sum_{i=2}^{K} Q_{\min}^i u_n^i, Q_{\min}^1\right), Q_{\max}^1(n)\right) u_n^1 \\ \vdots \\ Q_n^k = \min\left(\max\left(\hat{Q}_n^{dem} - \sum_{i=1}^{k-1} Q_n^i - \sum_{i=k+1}^{K} Q_{\min}^i u_n^i, Q_{\min}^k\right), Q_{\max}^k(n)\right) u_n^k \\ \vdots \\ Q_n^K = \min\left(\max\left(\hat{Q}_n^{dem} - \sum_{i=1}^{K-1} Q_n^i, Q_{\min}^K\right), Q_{\max}^K(n)\right) u_n^K \end{cases} \tag{80}$$

$$Q_{\max}^k(n+1) = Q_{\min}^k(1-u_n^k) + \left(\min(Q_{\max}^k, Q_n^k + \Delta Q^k)\right) u_n^k$$

Feasibility criterion. If a feasible solution is known (and it is case the case after the ant colony optimization), a feasibility criterion can be defined:

$$\min_{\substack{\{u_n^k, Q_n^k\} \\ n=1,\cdots,N \\ k=1,\ldots,K}} \left(\begin{array}{l} \sum_{n=1}^{N} \sum_{k=1}^{K} \left(c_{prod}^k(Q_n^k, u_n^k) + c_{on/off}^k(u_n^k, u_{n-1}^k)\right) \\ + ((1+\varepsilon)\, c^f + h(\{u_n^k, Q_n^k\})) \cdot B(\{u_n^k, Q_n^k\}) \end{array} \right) \tag{81}$$

- c^f is the cost of this feasible known solution
- ε is a small positive real,
- $h(\{u_n^k, Q_n^k\})$ is a penalty function for non feasible solutions $\{u_n^k, Q_n^k\}$.
- $B(\{u_n^k, Q_n^k\})$ is a boolean function with value 1 for non feasible solutions and 0 for feasible ones.

Finally, any infeasible solution has a higher cost than the cost of the feasible known solution. This criterion can be optimised by genetic algorithm, and constraints are implicitly taken into account.

Knowledge based genetic algorithm. The classical crossing-over and mutation operators are depicted in figure 23 for the Unit Commitment problem

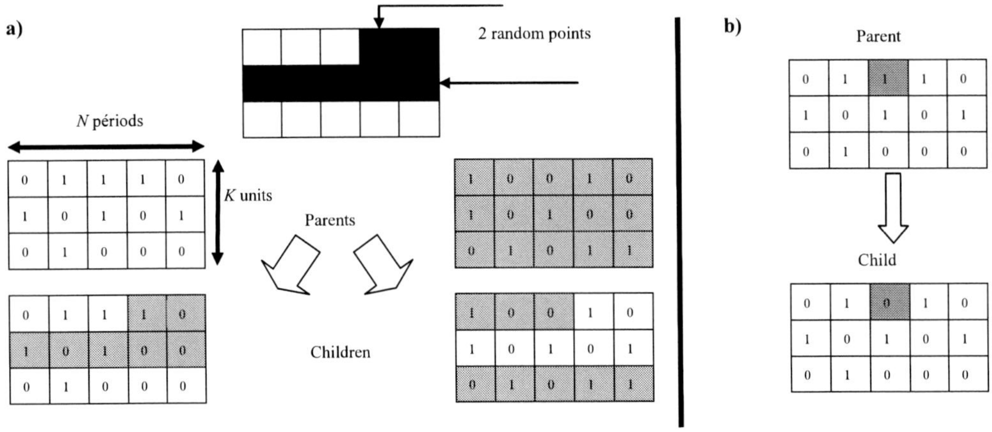

Figure 23. Classical crossing-over (a) and mutation (b) operators.

The classical genetic algorithm does converge, but slowly as classical operators may not be well suited to Unit Commitment problem. Consider for instance the planning of figure 24.

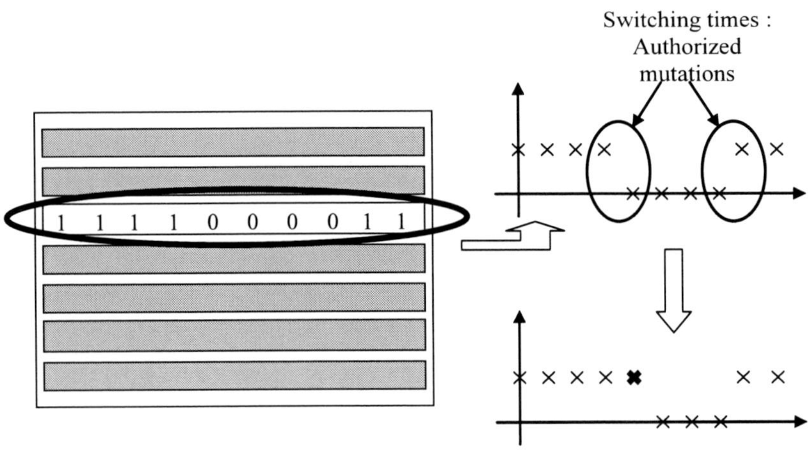

Figure 24. Selective mutation operator.

Because of time down and time up constraints, a random mutation will very often lead to an infeasible solution. However, switching times are particular points where a mutation has a high probability to create a new feasible solution. Thus, a selective mutation operator has been designed. The operator detects switching times, and allows a mutation only for these particular points. There is no guarantee that the feasibility is achieved, but the "probability of feasibility" is higher.

VI.2.4. Simulation Results

Real time updating of produced powers. As already explained, the "optimal" scheduling of the power system is computed from predicted values $\hat{Q}_n^{dem}$ on time interval $[m, m+N-1]$.

u_m^k are applied to the system. Real variables are updated at time *m*, when real value Q_m^{dem} is known, using (81) considering Q_m^{dem} instead of $\hat{Q}_n^{dem}$.

Case study. A "four unit" academic case is considered, whose characteristics are given in table 1. As in section VI.1, a worst case is considered: it is assumed that the consumer load is always underestimated. The prediction error is a random value in the range [-5%,0%]. The time horizon is $N = 24$ hours. The simulation is performed on a 4 day total horizon. Results, obtained with Matlab 6.1 on a PIV 2.5 GHz, are given in figure 25.

Table 1. Characteristics of the benchmark example

Unit	Qmin (MW)	Qmax (MW)	ΔQ (MW)	α0 (€)	α1 (€/MWh)	con (€)	coff (€)	Tdown (h)	Tup (h)
1	10	40	10	25	2.6	10	2	2	4
2	10	40	10	25	7.9	10	2	2	4
3	10	40	10	25	13.1	10	2	3	3
4	10	40	10	25	18.3	10	2	3	3

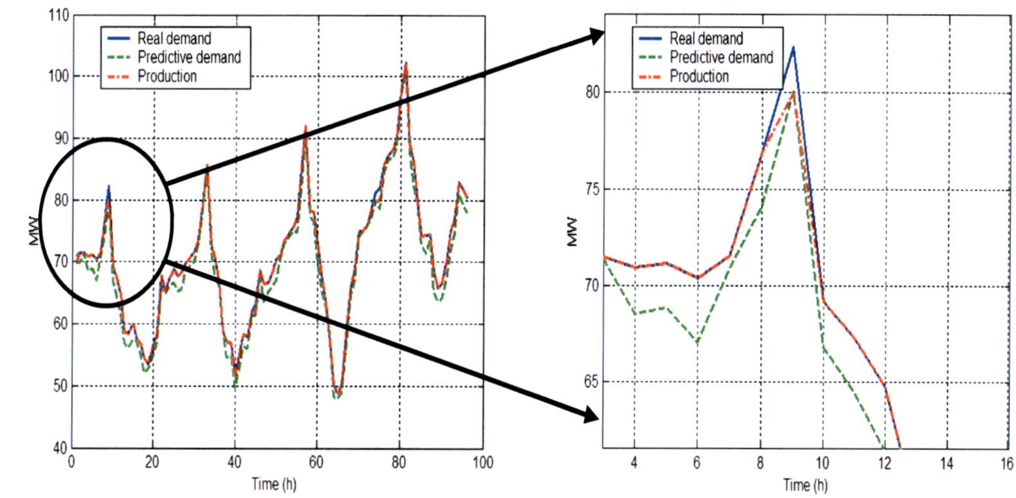

Figure 25. Simulation results.

The optimization of the 96 binary variable problem is performed in just 40 seconds. The production is very close to the real demand, except for some peaks which have been underestimated. Due to the computation of successive economical optimal solutions and real time slight updates, global costs are very close to global optimal costs.

References

Aarts, E. H., Korst, J. and van Laarhoven, P. J., "Simulated Annealing". In E. H. Aarts and J. K. Lenstra, editors, *Local Search in Combinatorial Optimization, chapter 4. John Wiley and Sons*, 1997.

Alamir M. and Bornard G., "Stability of a truncated infinite constrained receding horizon scheme : the general nonlinear case", *Automatica*, vol. 31, n°9, pp. 1353-1356, 1995.

Alkaya, D., Vasantharajan S. and Biegler L. T., "Successive Quadratic Programming: Applications in the Process Industry", *Encyclopedia of Optimization,* vol. 5., pp. 400-413, C. Floudas and P. Pardalos (eds.), Kluwer, 2001.

Arvastson, L., "Stochastic modelling and operational optimization in district heating systems", PhD. Thesis, Lund Institute of Technology, Mathematical Statistics, Centre for Mathematical Sciences, ISBN 91-628-4855-0, 2001.

Baeck, T., Fogel, D. B. and Michalewicz, Z.. "Evolutionary computation 1: basic algorithms and operators", *Institute of physics publishing*, 2000a.

Baeck, T., Fogel, D. B. and Michalewicz, Z., "Evolutionary computation 2: advanced algorithms and operators", *Institute of physics publishing*, 2000b.

Bazaraa M.S., Sherali H.D. and Shetty C.M., "Nonlinear programming. Theory and algorithms", Second edition, *John Wiley and Sons*, 1993.

Bellman R., "Dynamic Programming", *Princeton University Press*, Princeton, 1957.

Bemporad A., Morari M., Dua V. and Pistikopoulos E., "The Explicit Linear Quadratic Regulator for Constrained Systems", *Automatica*, vol. 38, pp. 3-20, 2002.

Ben-Tal A. and Nemirovski A., "Lectures on Modern Convex Optimization. Analysis, Algorithms, and Engineering Applications", *SIAM*, 2001.

Blanchini F., "Set invariance in control", *Automatica*, vol. 35, pp. 1747-1767, 1999.

Borelli F., "Constrained Optimal Control of Linear and Hybrid Systems", *Springer-Verlag*, Berlin, 2003.

Camacho E.F. and Bordons C., "Model Predictive Control", *Springer*, 2004.

Chen H. and Allgöwer F., "A quasi infinite horizon nonlinear model predictive control scheme with guaranteed stability", *Automatica*, vol. 14, n°10, p. 1205-1217, 1998.

Diehl M., Jarre F., and Vogelbusch C., "Loss of Superlinear Convergence for an SQP-type method with Conic Constraints". *SIAM Journal on Optimization*, vol. 16, pp. 1201–1210, 2006.

Diehl M., Bock H.G., and Schlöder J.P.: "A Real-Time Iteration Scheme for Nonlinear Optimization in Optimal Feedback Control", *SIAM Journal on Control and Optimization*, vol 43, n°5, pp. 1714-1736, 2005.

Dorigo M., "Optimization, learning and natural algorithms". PhD Thesis, Politecnico di Milano, Italy, 1992.

Dorigo M., Maniezzo V. and Colorni A., "The Ant System: Optimization by a Colony of Cooperating Agents", *IEEE Transactions on Systems, Man and Cybernetics-Part B*, vol. 26, n°1, pp. 1-13, 1996.

Dorigo M. and Gambardella, L. M., "Ant Colony System : a Cooperative Learning Approach to the Traveling Salesman Problem", *IEEE Transactions on Evolutionary Computation*, tome 1, pp. 53-66, 1997.

Dorigo, M. and Blum, C., "Ant colony optimization theory: A survey", *Theoretical Computer Science*, vol. 344, n°2-3, pp.243-278, 2005.

Dua V. and Pistikopoulos E.N., "An algorithm for the solution of multiparametric mixed integer linear programming problems", *Annals of Operations Research*, pp. 123-139, 2000.

Eberhart, R. C. and Shi, Y., "Comparing inertia weigthts and constriction factors in particle swarm optimization", *Proceedings of the IEEE Congress on Evolutionary Computation (CEC 2000)*, San Diego, CA, USA, pp. 84-88, 2000.

Eberhart, R. C and Kennedy, J., "A new optimizer using particle swarm theory", *Proceedings of the Sixth International Symposium on Micromachine and Human Science*, Nagoya, Japan. pp. 39-43, 1995.

Feo, T. and Resende, M., "Greedy randomized adaptive search procedure", *Journal of global optimization*, tome 42, pp. 860-878, 1995.

Fletcher R., "Practical Methods of Optimization 2: Constrained Optimization", *John Wiley and Sons*, Chichester, 1981.

Gal T. , "Postoptimal analyses, Parametric Program-ming and Related Topics", Second edition, *De Gruyter,* Berlin, 1994.

Gilbert E.G. and Tan K.T., "Linear systems with state and control constraints: The theory and application of maximal output admissible sets", *IEEE Transactions on Automatic Control*, vol. 36, n°9, pp. 1008-1020, 1991.

Glover, F., "Tabu search – part I", *ORSA journal on Computing*, tome 1, pp. 190-206, 1989.

Glover, F. and Laguna, M., "Tabu search". *Kluwer academic publishers*, 1997.

Goodwin G.C., Seron M.M. and De Dona J.A., "Constrained Control and Estimation", *Springer-Verlag*, London, 2004.

Henriques J., Gil P., Dourado A., and Duarte-Ramos H., "Nonlinear Multivariable Predictive Control: Neural versus First Principle Modelling Approach", *IASTED Control and Applications,* 2002.

Jadbabaie A., Yu J. and Hauser J., "Unconstrained receding-horizon control of nonlinear systems", *IEEE transactions on automatic control*, vol. 45, n°5, pp. 776-783, 1999.

Kameswaran S. and Biegler, L. T., "A Survey of Simultaneous Dynamic Optimization Strategies", *Computers and Chemical Engineering*, vol. 30, pp. 1560-1575, 2006.

Karmarkar N., "A new polynomial-time algorithm for linear programming", *Combinatorics,* vol. 4, pp. 373-395, 1984.

Kennedy, J. and M. Clerc, "Standard PSO", http://www.particleswarm.info/Standard_PSO_2006.c, 2006.

Kennedy, J., "Small worlds and mega-minds: effects of neighborhood topology on particle swarm performance", *Proceedings of IEEE Congress on Evolutionary Computation (CEC 1999)*, Piscataway, NJ, USA, pp. 1931-1938, 1999.

Kirkpatrick, S., Gelatt, C. D. and Vecchi, M. P., "Optimization by Simulated Annealing", *Science*, vol 220, n°4598, pp. 671-680, 1983.

Koza, J. R., "Genetic programming: on the programming of computers by means of natural selection", *MIT Press*, 1992.

Koza, J. R., "Genetic programming II: automatic discovery of reusable programs", MIT Press, 1994.

Lemke C.E., "Mathematics of the decision science, chapter on Complementary Pivot Theory", *Editors G.B. Dantzing and A.F. Veinott*, 1968.

Mayne D.Q., Rawlings J.B., Rao C.V. and Scockaert P.O.M., "Constrained model predictive control: Stability and optimality", *Automatica*, vol. 36, pp. 789-814, 2000.

Mühlenbein, H. and Paaβ, G., "From recombination of genes to the estimation of distributions I, binary parameters". *Lecture Notes in Computer Science 1411: Parallel problemm solving from nature, tome PPSN IV*, pp. 178-187, 1996.

Nesic , D., Teel, A. R., and Kokotovic, P. V., "Sufficient conditions for stabilization of sampled-data nonlinear systems via discrete-time approximations", *Systems and Control Letters,* vol. 38, pp. 259 – 270, 1999.

Nesterov Y. and Nemirovskii A., "Interior-Point Polynomial Algorithms in Convex Programming", *SIAM*, 1994.

Olaru S. and Dumur D., "Avoiding Constraints Redundancy in Predictive Control Optimization Routines", *IEEE Transaction on Automatic Control*, vol. 50, n°9, pp. 1459-1466, 2005.

Parisini T. and Zoppoli R., "A receding horizon reguator for nonlinear systems and a neural approximation", *Automatica*, vol. 3, n°10, p. 1443-1451, 1995.

Pontryagin L. S., Boltyanskii V. G., Gamkrelidze R. V. and Mishchenko E. F., "The mathematical theory of optimal processes", *Interscience publisher*, 1962.

Qin, SJ. and Badgwell, TA., "A survey of industrial model predictive control technology", *Control Engineering Practice,* vol. 11, n°7, pp. 733-764, 2003.

Rossiter J.A., "Model-based predictive control. A practical approach", *CRC Press*, 2003.

Rubinstein, R., "The simulated entropy method for combinatorial and continuous optimization", *Methodology and computing in applied probability*, tome 2, pp. 127-190, 1999.

Sandou, G., Font, S., Tebbani, S., Hiret, A. and Mondon, C., "Global modelling and simulation of a district heating network", *Proceedings of the 9th International Symposium on District Heating and Cooling*, Espoo, Finland, pp. 117-124, 2004a.

Sandou G., Font S., Tebbani S., Hiret A. and Mondon C., "Optimisation par colonies de fourmis d'un site de génération d'énergie", *Journal Européen des Systèmes Automatisés, Numéro spécial Métaheuristiques pour l'optimisation difficile*, vol. 38, n°9/10, pp. 1097-1119, 2004b.

Sen, S. and Kothari, D. P., "Optimal Thermal Generating Unit Commitment: a Review", *Electrical Power and Energy Systems*, vol. 20, n°7, pp. 443-451, 1998.

Seron M.M., Goodwin G.C. and De Dona J.A., "Characterisation of Receding Horizon Control for Constrained Linear Systems", *Asian Journal of Control*, vol. 5, n°2, pp. 271-286, 2002.

Shi, Y. and Eberhart, R. C., "Parameter selection in particle swarm optimization", *Evolutionary Programming VII: Proceedings of the Seventh Annual Conference on Evolutionary Programming*, New York, USA, pp. 591-600, 1998.

Stützle T. and Hoos, H. H., "MAX-MIN Ant System", *Future Generation Computer Systems*, vol. 16, pp. 889-914, 2000.

Tøndel P., Johansen T. and Bemporad A., "An algorithm for multiparametric quadratic programming and explicit MPC solutions", *Automatica*, vol. 39, pp. 489-497, 2003.

Vassilaki M.C., Hennet C. and Bitsoris G., "Feedback control of linear discrete-time systems under state and control constraints", *International Journal of Control*, vol. 47, pp. 1727-1735, 1988.

Wright S.J., "Primal-Dual Interior Point Methods", *SIAM Publications*, 1997.

Ye Y., "Interior Point Algorithms for quadratic programming", *Paper Series No 89-29*, Department of Management Sciences, University of Iowa, Iowa, 1989.

Zadeh L. and Whalen L., "On optimal Control and Linear Programming", *IEEE Transaction on Automatic Control*, vol. 7, pp. 45-46, 1962.

In: Nonlinear Analysis Research Trends
Editor: Inès N. Roux, pp. 209-233
ISBN 978-1-60456-358-0

Chapter 6

NON-FRAGILE OBSERVER-BASED STABILIZATION OF A CLASS OF SWITCHED NONLINEAR SYSTEMS UNDER PARTIAL STATE MEASUREMENTS

Salim Ibrir*
Department of Mechanical & Industrial Engineering,
1455 de Maisonneuve Blvd. West, Montreal,
Quebec, H3G 1M8, Canada

Abstract

In most cases, the conflict between stability and filtering prevents the search of the appropriate solution, and the stability/fitering problem may still unsolvable when the desired performances are high. This paper deals with robust observer-based stabilization of multiple-models systems subject to Lipschitzian nonlinearities. First, new sufficient and necessary linear matrix inequality conditions are provided for designing switching nonlinear observers. After the decomposition of the observer-based problem into two convex optimization problems, the observer-based control of systems under consideration is totally solved in efficient numerically way. In case of presence of system and output uncertainties, a robust nonlinear observer-based scheme is developed to enfeeble the effects of uncertainties. In this case, a dynamic output feedback stabilization issue with a prescribed degree of uncertainty attenuation is considered. The design of the observer and the controller gains is conditioned by the solutions of three linear matrix inequalities. A numerical example is provided to demonstrate the potential of our results.

Key-Words: Nonlinear Observers; Robust filtering; Switched Systems; Separation principle; Linear Matrix Inequalities (LMIs).

1. Introduction

In the last decades, research in switched nonlinear dynamical systems has been mainly focused on stabilization and observer design problems rather than observer-based stabilization issues, see e.g., (Cuzzola & Morari, 2002), (Pettersson & Lennartson, 2002), (Savkin,

*E-mail address: sibrir@encs.concordia.ca

Petersen, Skafidas, & Evans, 1996), (Miller & Rubinovich, 2003), (Liu, 1997). With few exceptions (Farra, Mhaskar, & Christofides, 2005), robustness by means of nonlinear observers has not been fully discussed and still one of the major problems that necessitates a particular attention. Robust stabilization of single and multiple models nonlinear systems subject to system and output uncertainties have been the subject of numerous papers, see e.g., (Cuzzola & Morari, 2002), (Balluchiy, Benvenuti, Benedetto, & Vincentelli, 2002), (Pettersson & Lennartson, 2002), (Savkin et al., 1996), (Li, Wen, & Soh, 2003). However, these contributions have been achieved under the assumption that the state vector of the considered system is available for feedback. This assumption which simplifies the design of robust controllers becomes, however, unrealistic when the state vector is not totally measured. As a result, new fundamental issues arise and call for a complete redesign of the feedback guaranteeing the desired performances. Among fundamental issues that usually arise are the choice of the feedback and the observer gains that guarantee the stability of the nonlinear system with a certain degree of performance. These problems come essentially from the fact that a high-gain observer necessitates, in general, a high-gain output injection that could defeat the inherently nonlinearities, see for example (Thau, 1973), (Ciccarella, Mora, & Germani, 1993), (Raghavan & Hedrick, 1994), (Rajamani, 1998). But, in the same time, this output injection increases considerably the amount of uncertainties, and as consequence, the performance of the closed-loop system will degrade through stabilizing feedbacks that use the noisy estimates. In addition, selecting the right observer gain is also a difficult and a challenging task, especially when the values of Lipschitz constants of nonlinearities are notably high. We refer the reader to the references (Raghavan & Hedrick, 1994), (Rajamani & Cho, 1998), (Aboky, Sallet, & Vivalda, 2002) for further details on how to characterize the relation between the distance to unobservability and the Lipschitz constants of nonlinearities. The reader is referred to the reference (Ibrir, 2005) for new results on observer design for discrete-time and continuous-time nonlinear systems.

When addressing the observer-based control problem, the complexity of selecting the observer and the feedback gain becomes more accentuated due to the nonlinear coupling of variables to be determined. Optimality issues by means of nonlinear observers make also other constraints on selecting the right gains. Thanks to recent developments in convex optimization, there has been much interest in the LMI approach due to its numerical capability in handling more general type of systems and solving multi-objective problems. However, the available results presented in a linear matrix inequality setting are generally sufficient, and sometimes the solutions of the convex optimization problems are not given in terms of linear matrix inequalities, and hence, iterative steps are required to arrive to the desired solution. Evidently, when the considered system involves multiple nominal models with different nonlinearities, the search of observer-based feedback becomes more difficult and call for generic algorithms that can be numerically tractable and easy to implement.

Switching systems with Lipschitzian nonlinearities are generally encountered in robotics and many other disciplines as biology and electrical engineering. Besides all these arguments that motivate the research in this area, there are at least two other cases where the Lipschitz property is of crucial help in design of observer-based controllers. The first case generally happens when the state space region is limited. Hence, non-Lipschitzian nonlinearities can be viewed as Lipschitz ones. The second case is related to many practical requirements and lack of suitable analysis tools. For example, in aeronautics, the requirement of building linear stabilizing feedbacks for aircraft systems with multiple models can also be seen as sufficient argument to conduct the analysis to the linear well-known established theory.

Referring to the existing literature, the stability analysis of switching systems can be handled by a common or multiple Lyapunov functions, see for example (Daafouz, Riedinger, & Iung, 2002) and the references therein. Sometimes the link between the construction of multiple Lyapunov functions and conditions of switching, that are quite dependent to the Lyapunov functions themselves, poses, in general, more constraints on the design of output stabilizing feedbacks (Farra et al., 2005). For this reason, in this paper, we try to lay down a systematic approach to robust observer-based stabilization of multiple-models systems with arbitrary switching and subject to Lipschitzian nonlinearities. The proposed technique reduces the complexity of construction of observer-based stabilizing feedbacks, in the sense that, a common Lyapunov function is used to get switching controllers *without* any constraint on the manner how the system switches between the different modes. However, in our design, the observer-controller switches between different values depending on the current mode. Although the controller is mode-dependent, the analysis of stability is carried out with a common Lyapunov function, which consequently allows the switching to be arbitrary. The second main focus of the paper is to propose less conservative conditions that are numerically tractable and free from any equality constraints. Observer-based stabilization with uncertainty attenuation is also discussed in convex optimization setting. The plan of the paper is as follows. First, we begin by giving *sufficient and necessary* linear matrix inequality conditions for observer design of switched systems subject to Lipschitzian nonlinearities. The proposed LMI conditions can be seen as an extension of LMI-based observer design techniques proposed for nonlinear single mode systems. The proposed LMI conditions are less *restrictive* since they enjoy the properties to be both necessary and sufficient. Moreover, the numerical tractability of these LMIs makes the design of observer gains straightforward and easy to obtain using commercial softwares. Subsequently, the problem of stabilization with dynamic output feedbacks is considered. The stabilization issue under partial state measurements is known to be a non convex problem due to nonlinear coupling of unknown variables that determine the controller and the observer gains. In this paper we show how to decouple the problem of the observer-based stabilization into two separate convex optimization problems. It will be shown that the solution of the first convex optimization problem implies the feasibility of stabilization of the switched system by means of full state feedbacks, while the computation of the observer gain is given by the solution of the second convex optimization issue. It is important to highlight that our design is neither iterative nor restrictive, in the sense that, by the use of the developed results, the construction of multiple observer-based controllers for multiple models becomes straightforward and numerically tractable.

When uncertainties are present, a mode-dependent robust nonlinear observer is used to estimate the missing states and feeding back the switched system. The developed observer enjoys the property of rejecting output and system uncertainties, and facilitates the design of a robust mode-dependent feedback that steers the system to the origin with a certain level of performance. It will be demonstrated that the robust observer-based control can be also decomposed into two linear matrix inequalities that are related linearly by another convex optimization problem.

The notation $A > 0$ stands for the positive definiteness of the matrix A. A' is the transpose matrix of A. $\|\cdot\|$ is used to notify the usual Euclidean norm. We note by $\boldsymbol{I}$, and $\mathbf{0}$, the identity matrix and the null matrix of appropriate dimensions, respectively. In symmetric block matrices, we use $\star$ as an ellipsis for terms that are induced by symmetry. The following lemmas and facts are frequently used in the proof of the main statement of this paper. We prefer herein to recall them.

Fact 1. *For given matrices $\mathscr{A}_1$, and $\mathscr{A}_2$ with appropriate dimensions, we have*

$$\mathscr{A}_1'\mathscr{A}_2 + \mathscr{A}_2'\mathscr{A}_1 \leq \varepsilon \mathscr{A}_1'\mathscr{A}_1 + \varepsilon^{-1}\mathscr{A}_2'\mathscr{A}_2.$$

and

$$\mathscr{A}_1'\mathscr{A}_2 + \mathscr{A}_2'\mathscr{A}_1 \leq \mathscr{A}_1'\mathscr{P}^{-1}\mathscr{A}_1 + \mathscr{A}_2'\mathscr{P}\mathscr{A}_2$$

where μ is any positive constant and $\mathscr{P}$ is an arbitrary symmetric positive definite matrix of appropriate dimensions.

The next lemma is known by the Schur complement. The proof of this lemma is given in (Boyd, Ghaoui, Feron, & Balakrishnan, 1994).

Lemma 1. *Given constant matrices $\mathcal{M}$, $\mathcal{N}$, $\mathcal{Q}$ of appropriate dimensions where $\mathcal{M}$ and $\mathcal{Q}$ are symmetric, then $\mathcal{Q} > 0$ and $\mathcal{M} + \mathcal{N}'\mathcal{Q}^{-1}\mathcal{N} < 0$ if and only if*

$$\begin{bmatrix} \mathcal{M} & \mathcal{N}' \\ \mathcal{N} & -\mathcal{Q} \end{bmatrix} < 0,$$

or equivalently

$$\begin{bmatrix} -\mathcal{Q} & \mathcal{N} \\ \mathcal{N}' & \mathcal{M} \end{bmatrix} < 0,$$

Lemma 2. *Let S, Y and Z be given $k \times k$ symmetric matrices such that $S \geq 0$, $Y < 0$, and $Z \geq 0$. Furthermore, assume that*

$$\left(\eta' Y \eta\right)^2 - 4\left(\eta' Y \eta\right)\left(\eta' Z \eta\right) > 0 \tag{1}$$

for all nonzero $\eta \in \mathrm{IR}^k$. Then, there exists a constant λ such that the matrix $\lambda^2 S + \lambda Y + Z$ is negative definite.

Proof. For the proof see (Petersen, 1987).

Lemma 3. *Given any $x \in \mathrm{IR}^n$ and four matrices P, D, F and G of appropriate dimensions, we have*

$$\max\{(x'PDFGx)^2 : F'F \leq \boldsymbol{I}\} = x'PDD'Px\, x'G'Gx. \tag{2}$$

Proof. See (Petersen, 1987) for the proof.

2. Nonlinear Observer Design

Consider the switched nonlinear system

$$\begin{cases} \dot{x}(t) = A_i\, x(t) + f_i(x(t)) + B_i\, u(t) + E_i\, d_1(t), \\ y_i(t) = C_i\, x(t) + G_i\, d_2(t); \quad i \in \mathscr{Q} = \{1,\, 2,\, \cdots,\, q\}, \end{cases} \tag{3}$$

where q is a positive integer that describes the number of the system modes. $x(t)$ is n-dimensional vector, $u(t)$ is an m dimensional control input, and $(y_i(t))_{1\leq i\leq q}$ are the switching outputs of dimension p. The matrices $(A_i)_{1\leq i\leq q}$, $(B_i)_{1\leq i\leq q}$, $(C_i)_{1\leq i\leq q}$, $(E_i)_{1\leq i\leq q}$, $(G_i)_{1\leq i\leq q}$ are constant matrices of appropriate dimensions. $d_1(t)$ and $d_2(t)$ stand for the system and the output disturbances. We assume that for $t \geq 0$, $\sup \|d_1(t)\| \leq \bar{d}_1$, and $\sup \|d_2(t)\| \leq \bar{d}_2$ where $\bar{d}_1$ and $\bar{d}_2$ are some positive constants. Throughout this paper, nonlinearities $(f_i(x(t)))_{1\leq i\leq q}$ depend only on the state $x(t)$. For system (3), the following hypotheses are considered.

Hypothesis 1. *We assume that for each i, the pair $(A_i,\, C_i)$ is detectable.*

Hypothesis 2. *The vectors $(f_i(x(t)))_{1\leq i\leq q} \; : \mathbb{R}^n \longmapsto \mathbb{R}^n$ are Lipschitz nonlinearities satisfying*

$$\mathscr{G}_i(x(t)) = \frac{\partial f_i(x(t))}{\partial x(t)} = M_i^{(f)} F_i(x(t)) N_i^{(f)}, \; \forall x(t);\; 1 \leq i \leq q, \tag{4}$$

where $\left(M_i^{(f)}\right)_{1\leq i\leq q}$, $\left(N_i^{(f)}\right)_{1\leq i\leq q}$ are constants matrices, and $(F_i(x(t)))_{1\leq i\leq q}$ are norm-bounded matrices satisfying $F_i^{'}(x(t))F_i(x(t)) \leq \boldsymbol{I}$ for all $i \in \mathscr{Q}$.

Hypothesis 3. *For each mode $i \in \mathscr{Q}$, the condition of smoothness $f_i(0) = 0$ is verified.*

Hypothesis 4. *The switch between the different modes of the system is arbitrary, and switching instants are assumed to be known.*

System (3) under the above hypotheses may describe many dynamics of robotic systems. This class of systems appears also in chaotic synchronization problems, see e.g., (Huijberts, Lilge, & Nijmeijer, 2001), (Pogromsky & Nijmeijer, 1998) and the references therein. System (3) under hypotheses 1-4 may also describe parameter varying nonlinear dynamics where the system parameters switch between different constants values.

Remark 1. *When $f(\cdot)$ is differentiable, hypothesis 2 is exactly the Lipschitz condition written in matrix form. Remark also that this representation of the Lipschitz property does not involve any approximation of nonlinear terms by their norms. Therefore, this key formulation will help later to give sufficient and necessary conditions that are less conservative than those proposed in the literature.*

Remark 2. *The Lipschitz property of nonlinearities as defined in (4) is a* flexible representation *since for a given nonlinearity $f_i(x(t))$, the matrices $M_i^{(f)}$, $N_i^{(f)}$ and $F_i(x(t))$ admit infinity of representations.*

In this section we show that the observer design issue for the switched nonlinear system (3) is equivalent to quadratic stability of multiple models linear systems with *known* uncertainties. First, we consider nonlinear observer design for switched systems (3) with $d_1(t) = d_2(t) = 0$. To compute the observer gains, a mode-dependent linear matrix inequality condition is given. The proposed conditions are sufficient and necessary for quadratic stability of the observation errors. The result is given by the following statement.

Theorem 1. *Consider system (3) with $d_1(t) = d_2(t) = 0$ and let*

$$\dot{\hat{x}}(t) = A_i\,\hat{x}(t) + f_i(\hat{x}(t)) + B_i\,u(t) + P^{-1}Y(i)(C_i\,\hat{x}(t) - y_i(t)), \tag{5}$$

be the corresponding mode-dependent nonlinear observer where P is a symmetric and positive definite matrix and $(Y(i))_{1\leq i\leq q} \in \mathrm{IR}^{n\times p}$ are some constant matrices. Then, the derivative of the common Lyapunov function $V(e(t)) = (\hat{x}(t) - x(t))'P(\hat{x}(t) - x(t))$ is negative or equal to zero, if and only if, there exist a set of positive constants $(\varepsilon_i)_{1\leq i\leq q}$ such that the following linear matrix inequalities hold for each mode i

$$\begin{bmatrix} A_i'P + PA_i + Y(i)C_i + C_i'Y'(i) + \varepsilon_i\,{N'}_i^{(f)}N_i^{(f)} & P\,M_i^{(f)} \\ {M'}_i^{(f)}P & -\varepsilon_i\,\boldsymbol{I} \end{bmatrix} < 0. \tag{6}$$

Proof.

$a)$ **Sufficiency** The dynamics of the observation error is

$$\begin{aligned} \dot{e}(t) &= \big(A_i + P^{-1}Y(i)C_i\big)e(t) + f_i(\hat{x}) - f_i(x), \\ &= \int_0^1 \big(A_i + P^{-1}Y(i)C_i\big)e(t) \\ &\quad + M_i^{(f)}F_i(\hat{x}(t) - \lambda\,e(t))N_i^{(f)}\,e(t)\;\mathrm{d}\lambda. \end{aligned} \tag{7}$$

By associating the common Lyapunov function $V(e(t)) = e'(t)Pe(t)$ to the last dynamics. Then, we have

$$\begin{aligned} \dot{V}(e(t)) = \int_0^1 & e'(t)\big(A_i + P^{-1}Y(i)C_i\big)'Pe(t) \\ & + e'(t)P\big(A_i + P^{-1}Y(i)C_i\big)e(t) \\ & + e'(t)\Big(M_i^{(f)}F_i(\hat{x}(t) - \lambda\,e(t))\,N_i^{(f)}\Big)'Pe(t) \\ & + e'(t)P\Big(M_i^{(f)}F_i(\hat{x}(t) - \lambda\,e(t))N_i^{(f)}\Big)e(t)\;\mathrm{d}\lambda. \end{aligned} \tag{8}$$

A sufficient condition to fulfill the negativity of $\dot{V}(e(t))$ is

$$\begin{aligned} & \big(A_i + P^{-1}Y(i)C_i\big)'P + P\big(A_i + P^{-1}Y(i)C_i\big) \\ & + \Big(M_i^{(f)}F_i(\hat{x}(t) - \lambda\,e(t))\,N_i^{(f)}\Big)'P \\ & + P\Big(M_i^{(f)}F_i(\hat{x}(t) - \lambda\,e(t))N_i^{(f)}\Big) < 0. \end{aligned} \tag{9}$$

Using result of Fact 1 and $F_i'(s)F_i(s) \leq \boldsymbol{I}$ for all $s \in \mathbb{R}^n$, we conclude that the observation error decays to zero when time elapses, if and only if there exist a set of positive constants $(\varepsilon_i)_{1\leq i\leq q}$ such that the mode-dependent linear matrix inequality holds

$$\begin{aligned} & A_i'P + PA_i + Y(i)C_i + C_i'Y'(i) \\ & + \varepsilon_i N_i'^{(f)} N_i^{(f)} + \varepsilon_i^{-1} P M_i^{(f)} M_i'^{(f)} P < 0, \end{aligned} \tag{10}$$

which is equivalent by the Schur complement to inequality (6).

$b)$ **Necessity** To prove the necessity of the conditions (6), let $\eta \in \mathbb{R}^n$. Then, $\dot{V}(e(t)) \leq 0$, if for all $\eta \in \mathbb{R}^n$, we have

$$\begin{aligned} & \int_0^1 \eta'(t)\big(A_i + P^{-1}Y(i)C_i\big)'P\eta(t) \\ & + \eta'(t)P\big(A_i + P^{-1}Y(i)C_i\big)\eta(t) \\ & + \eta'(t)\Big(M_i^{(f)}F_i(\hat{x}(t) - \lambda\,\eta(t))\,N_i^{(f)}\Big)'P\eta(t) \\ & + \eta'(t)P\Big(M_i^{(f)}F_i(\hat{x}(t) - \lambda\,\eta(t))N_i^{(f)}\Big)\eta(t)\;\mathrm{d}\lambda < 0. \end{aligned} \tag{11}$$

or

$$\begin{aligned} & \eta'(t)\big(A_i'P + PA_i + Y(i)C_i + C_i'Y'(i)\big)\eta(t) \\ & \leq \int_0^1 -\eta'(t)\Bigg(\Big(M_i^{(f)}F_i(\hat{x}(t) - \lambda\,\eta(t))\,N_i^{(f)}\Big)'P \\ & + P\Big(M_i^{(f)}F_i(\hat{x}(t) - \lambda\,\eta(t))N_i^{(f)}\Big)\Bigg)\eta(t)\;\mathrm{d}\lambda. \end{aligned} \tag{12}$$

Since $F_i'(s)F_i(s) \leq \boldsymbol{I}$ for all $s \in \mathbb{R}^n$, then, we obtain from the last inequality that

$$\begin{aligned} & \eta'(t)\big(A_i'P + PA_i + Y(i)C_i + C_i'Y'(i)\big)\eta(t) \\ & \leq \int_0^1 -2\max\Big\{\eta'(t)\Big(PM_i^{(f)}F_i(\hat{x}(t) - \lambda\,\eta(t))N_i^{(f)}\Big)\eta(t):\; F_i'(s)F_i(s) \leq \boldsymbol{I}\Big\}\;\mathrm{d}\lambda. \end{aligned} \tag{13}$$

This immediately implies that

$$\begin{aligned} & \Big(\eta'(t)\big(A_i'P + PA_i + Y(i)C_i + C_i'Y'(i)\big)\eta(t)\Big)^2 \\ & > \left(\int_0^1 2\max\Big\{\eta'(t)\Big(PM_i^{(f)}F_i(\hat{x}(t) - \lambda\,\eta(t))N_i^{(f)}\Big)\eta(t):\; F_i'(s)F_i(s) \leq \boldsymbol{I}\Big\}\;\mathrm{d}\lambda\right)^2. \end{aligned} \tag{14}$$

Using result of Lemma 3, we could write that

$$\begin{aligned}&\Big(\eta'(t)\big(A_i'P + PA_i + Y(i)C_i + C_i'Y'(i)\big)\eta(t)\Big)^2\\&> 4\,\eta'(t)\Big(PM_i^{(f)}M_i'^{(f)}P\Big)\eta(t)\,\eta'(t)\Big(N_i'^{(f)}N_i^{(f)}\Big)\eta(t).\end{aligned} \tag{15}$$

Finally, by the use of Lemma 2, we conclude that the last inequality holds if there exist a set of $\lambda_i > 0$, $1 \leq i \leq q$ such that

$$\lambda_i^2\Big(PM_i^{(f)}M_i'^{(f)}P\Big) + \lambda_i(A_i'P + PA_i + Y(i)C_i + C_i'Y'(i)) + N_i'^{(f)}N_i^{(f)} < 0. \tag{16}$$

By putting $\lambda_i = \varepsilon_i^{-1}$, we get by the Schur complement the linear matrix inequality (6). This ends the proof.

Remark 3. *From the last development, we see that nonlinearities act as well-known perturbations which gives us a good insight to use the concept of quadratic stability of linear systems subject to structured uncertainties.*

Notice that LMI (6) is not conservative since it is a sufficient and necessary condition for quadratic stability of the observation error. In addition, hypothesis 2 is a *necessary condition* that can be always verified for globally Lipschitz nonlinearities. In the next sections, we will show how to make the high-gain observer robust against measurements errors and system uncertainties and how to use this observer to achieve asymptotic stability of the switched system with prescribed degree of performance.

3. Observer-Based Stabilization

In this section we shall develop multiple-gain nonlinear observer-controllers that steer system (3) to the origin in the particular case where $d_1(t) = d_2(t) = 0$. The existence of the observer-based controllers is conditioned by the solution of a set of linear matrix inequalities. To this end, we propose for each mode i the following observer-controller

$$\begin{cases}\dot{\hat{x}}(t) = A_i\,\hat{x}(t) + f_i(\hat{x}(t)) + B_i\,u(t) + L(i)(C_i\,\hat{x}(t) - y_i(t)),\\ u_i(t) = K(i)\,\hat{x}(t),\end{cases} \tag{17}$$

where the observer-gain $L(i)$ and the controller-gain $K(i)$ are some constant vectors of appropriate dimensions to be determined. Let $K(i) = Y_1(i)\,P_1^{-1}$, and $L(i) = P_2^{-1}\,Y_2(i)$ be the feedback and the observer gains, respectively, where $Y_1(i) \in \mathbb{R}^{m\times n}$ and $Y_2(i) \in \mathbb{R}^{n\times p}$, are constant mode-dependent matrices, and $P_1 \in \mathbb{R}^{n\times n}$, $P_2 \in \mathbb{R}^{n\times n}$ are symmetric, positive and definite matrices. Under the action of the above feedback (17), the dynamics of system (3) and the dynamics of the observation error $e(t) = \hat{x}(t) - x(t)$ becomes

$$\begin{cases}\dot{x}(t) = (A_i + B_iK(i))\,x(t) + B_iK(i)\,e(t) + f_i(x(t)),\\ \dot{e}(t) = (A_i + L(i)\,C_i)\,e(t) + f_i(\hat{x}(t)) - f_i(x(t)).\end{cases} \tag{18}$$

Under hypothesis 3, the last system can be rewritten as

$$\begin{cases} \dot{x}(t) = (A_i + B_i\, K(i))\, x(t) + B_i\, K(i)\, e(t) \\ \qquad + \displaystyle\int_0^1 \mathscr{G}_i((1-\lambda)x(t))\, x(t)\, \mathrm{d}\,\lambda, \\ \dot{e}(t) = (A_i + L(i)\, C_i)\, e(t) \\ \qquad + \displaystyle\int_0^1 \mathscr{G}_i(\hat{x}(t) - \lambda e(t))\, e(t)\, \mathrm{d}\,\lambda. \end{cases} \tag{19}$$

The last system can be rewritten in matrix form as

$$\begin{bmatrix} \dot{x}(t) \\ \dot{e}(t) \end{bmatrix} = \int_0^1 \begin{bmatrix} \mathscr{A}_{1,1}(K(i),\, \lambda,\, x(t)) & \mathscr{A}_{1,2}(K(i)) \\ \mathbf{0} & \mathscr{A}_{2,2}(L(i),\, \lambda,\, x(t),\, e(t)) \end{bmatrix} \times \begin{bmatrix} x(t) \\ e(t) \end{bmatrix} \mathrm{d}\,\lambda, \tag{20}$$

where

$$\begin{aligned} &\mathscr{A}_{1,1}(K(i),\, \lambda,\, x(t)) = A_i + B_i\, Y_1(i)\, P_1^{-1} + \mathscr{G}_i((1-\lambda)x(t)), \\ &\mathscr{A}_{1,2}(K(i)) = B_i\, Y_1(i)\, P_1^{-1}, \\ &\mathscr{A}_{2,2}(L(i),\, \lambda,\, x(t),\, e(t)) = A_i + P_2^{-1} Y(i)\, C_i + \mathscr{G}_i(\hat{x}(t) - \lambda e(t)). \end{aligned} \tag{21}$$

For each mode, we consider the Lyapunov function candidate

$$V(x(t), e(t)) = x'(t) P_1^{-1} x(t) + e'(t) P_2 e(t) \tag{22}$$

where P_1, and P_2 are n by n symmetric, and positive definite matrices. The first derivative of the Lyapunov function (22) along the trajectories of (20) is

$$\begin{bmatrix} x(t) \\ e(t) \end{bmatrix}' \left(\int_0^1 \begin{bmatrix} \mathscr{A}_{1,1}(K(i),\, \lambda,\, x(t)) & \mathscr{A}_{1,2}(K(i)) \\ \mathbf{0} & \mathscr{A}_{2,2}(L(i),\, \lambda,\, x(t),\, e(t)) \end{bmatrix}' \times \begin{bmatrix} P_1^{-1} & \mathbf{0} \\ \mathbf{0} & P_2 \end{bmatrix} \mathrm{d}\,\lambda + \begin{bmatrix} P_1^{-1} & \mathbf{0} \\ \mathbf{0} & P_2 \end{bmatrix} \times \int_0^1 \begin{bmatrix} \mathscr{A}_{1,1}(K(i),\, \lambda,\, x(t)) & \mathscr{A}_{1,2}(K(i)) \\ \mathbf{0} & \mathscr{A}_{2,2}(L(i),\, \lambda,\, x(t),\, e(t)) \end{bmatrix} \mathrm{d}\,\lambda \right) \begin{bmatrix} x(t) \\ e(t) \end{bmatrix}.$$

This implies that $\dot{V}(x(t), e(t)) < 0$ if the following matrix inequality holds

$$\int_0^1 \begin{bmatrix} \mathscr{S}_{1,1}(Y_1(i),\, P_1,\, \lambda,\, x(t)) & \mathscr{S}_{1,2}(Y_1(i),\, P_1) \\ \mathscr{S}_{1,2}'(Y_1(i),\, P_1) & \mathscr{S}_{2,2}(P_2,\, Y_2(i),\, \lambda,\, x(t),\, e(t)) \end{bmatrix} \mathrm{d}\,\lambda < 0, \tag{23}$$

where

$$\begin{aligned} \mathscr{S}_{1,1}(Y_1(i),\, P_1,\, \lambda,\, x(t)) &= A_i'\, P_1^{-1} + P_1^{-1}\, A_i \\ &\quad + P_1^{-1}\Big(B_i Y_1(i) + Y_1'(i) B_i'\Big) P_1^{-1} \\ &\quad + \mathscr{G}_i'((1-\lambda)x(t)) P_1^{-1} + P_1^{-1} \mathscr{G}_i((1-\lambda)x(t)), \\ \mathscr{S}_{1,2}(Y_1(i),\, P_1) &= P_1^{-1} B_i Y_1(i) P_1^{-1}, \\ \mathscr{S}_{2,2}(P_2,\, Y_2(i),\, \lambda,\, x(t),\, e(t)) &= A_i' P_2 + P_2 A_i + Y_2(i) C_i + C_i' Y_2'(i) \\ &\quad + \mathscr{G}_i'(\hat{x}(t) - \lambda e(t)) P_2 + P_2 \mathscr{G}_i(\hat{x}(t) - \lambda e(t)). \end{aligned} \tag{24}$$

Pre- and post multiplying inequality (23) by the matrix

$$\begin{bmatrix} P_1 & \mathbf{0} \\ \star & \boldsymbol{I} \end{bmatrix}, \tag{25}$$

we obtain the following matrix inequality

$$\int_0^1 \begin{bmatrix} \mathscr{L}_{1,1}(Y_1(i),\, P_1,\, \lambda,\, x(t)) & \mathscr{L}_{1,2}(Y_1(i),\, P_1) \\ \mathscr{L}_{1,2}^{'}(Y_1(i),\, P_1) & \mathscr{S}_{2,2}(P_2,\, Y_2(i),\, \lambda,\, x(t),\, e(t)) \end{bmatrix} \mathrm{d}\,\lambda < 0, \tag{26}$$

where

$$\begin{aligned} \mathscr{L}_{1,1}(Y_1(i),\, P_1,\, \lambda,\, x(t)) &= P_1\, A_i^{'} + A_i\, P_1 + B_i Y_1(i) + Y_1^{'}(i) B_i^{'} \\ &\quad + P_1 \mathscr{G}_i^{'}((1-\lambda)x(t)) + \mathscr{G}_i((1-\lambda)x(t))P_1, \\ \mathscr{L}_{1,2}(Y_1(i),\, P_1) &= B_i Y_1(i)\, P_1^{-1}. \end{aligned} \tag{27}$$

Let $(P_3(i))_{1\le i\le q}$ be a set of symmetric and positive definite matrices of dimensions n by n. Then, the linear matrix inequality (26) can be rewritten as follows

$$\begin{aligned} &\int_0^1 \begin{bmatrix} \boldsymbol{I} & \mathbf{0} & \mathbf{0} \\ \mathbf{0} & P_1^{-1} & \boldsymbol{I} \end{bmatrix} \times \\ &\begin{bmatrix} \mathscr{L}_{1,1}(Y_1(i),\, P_1,\, \lambda,\, x(t)) & B_i Y_1(i) & \mathbf{0} \\ Y_1^{'}(i) B_i^{'} & -P_3(i) & \mathbf{0} \\ \mathbf{0} & \mathbf{0} & \mathscr{S}_{2,2}(P_2,\, Y_2(i),\, \lambda,\, x(t),\, e(t)) + P_1^{-1} P_3(i) P_1^{-1} \end{bmatrix} \\ &\times \begin{bmatrix} \boldsymbol{I} & \mathbf{0} \\ \mathbf{0} & P_1^{-1} \\ \mathbf{0} & \boldsymbol{I} \end{bmatrix} \mathrm{d}\,\lambda < 0. \end{aligned} \tag{28}$$

From inequality (28), we conclude that $V(x(t),\, e(t)) < 0$ if the following hold simultaneously

$$\int_0^1 \begin{bmatrix} \mathscr{L}_{1,1}(Y_1(i),\, P_1,\, \lambda,\, x(t)) & B_i Y_1(i) \\ Y_1^{'}(i) B_i^{'} & -P_3(i) \end{bmatrix} \mathrm{d}\,\lambda < 0, \tag{29}$$

and

$$\int_0^1 \mathscr{S}_{2,2}(P_2,\, Y_2(i),\, \lambda,\, x(t),\, e(t)) + P_1^{-1} P_3(i) P_1^{-1} \mathrm{d}\,\lambda < 0. \tag{30}$$

The matrix inequality (29) implies that the state feedbacks $u_i = K(i)\,x(t)$, $1 \le i \le q$ are stabilizing controllers, while inequality (30) stands for the condition under which the observation error is quadratically stable. Now let us work out equivalent linear matrix inequalities to the matrix inequalities (29), and (30). Using fact 1 and taking into account the definitions of the nonlinearities, we have for any $\varepsilon_1(i) > 0$

$$\begin{aligned}\int_0^1 \mathscr{L}_{1,1}(Y_1(i), P_1, \lambda, x(t)) \leq P_1 A_i + A_i' P_1 + Y_1(i) B_i + B_i' Y_1'(i) \\ + \varepsilon_1(i) M_i^{(f)} M_i'^{(f)} \\ + \varepsilon_1^{-1}(i) P_1 N_i'^{(f)} N_i^{(f)} P_1,\end{aligned} \tag{31}$$

From (31) and by the use of the Schur complement, we conclude that the matrix inequality (29) hold if the following *linear matrix inequality* holds

$$\begin{bmatrix} P_1 A_i + A_i' P_1 + Y_1(i) B_i + B_i' Y_1'(i) + \varepsilon_1(i) M_i^{(f)} M_i'^{(f)} & Y_1(i) B_i & P_1 N_i'^{(f)} \\ \star & -P_3(i) & \mathbf{0} \\ \star & \star & -\varepsilon_1(i) \boldsymbol{I} \end{bmatrix} < 0. \tag{32}$$

In order to make the term $P_1^{-1}\, P_3(i)\, P_1^{-1}$ linear with respect to P_1^{-1}, let $(\alpha_i)_{1 \leq i \leq q}$ be a set of positive constants such that

$$P_3(i) \leq \alpha_i^2 P_1 \;\forall i. \tag{33}$$

Then, we conclude that

$$P_1^{-1} P_3(i) P_1^{-1} \leq \alpha_i^2 P_1^{-1} \;\forall i. \tag{34}$$

Inequality (33) is equivalent by the Schur complement to the following matrix inequality

$$\begin{bmatrix} -P_1 & \boldsymbol{I} \\ \boldsymbol{I} & -\alpha_i^2 P_3^{-1}(i) \end{bmatrix} < 0. \tag{35}$$

Taking into account fact 1, we have for each mode i

$$-\alpha_i^2 P_3^{-1}(i) \leq -2\alpha_i \boldsymbol{I} + P_3(i). \tag{36}$$

This implies that (35) holds if the following linear matrix inequality holds

$$\begin{bmatrix} -P_1 & \boldsymbol{I} \\ \boldsymbol{I} & -2\alpha_i \boldsymbol{I} + P_3(i) \end{bmatrix} < 0. \tag{37}$$

From (34), we can write that (30) holds if the following is verified

$$\int_0^1 \mathscr{S}_{2,2}(P_2, Y_2(i), \lambda, x(t), e(t)) + \alpha_i^2 P_1^{-1} \mathrm{d}\lambda < 0 \tag{38}$$

Using again fact 1, we have

$$\begin{aligned}&\int_0^1 \mathscr{S}_{2,2}(P_2, Y_2(i), \lambda, x(t), e(t)) + \alpha_i^2 P_1^{-1} \mathrm{d}\lambda \\ &\leq A_i' P_2 + P_2 A_i + C_i Y_2(i) + Y_2'(i) C_i' \\ &+ \varepsilon_2(i) N_i'^{(f)} N_i^{(f)} \\ &+ \varepsilon_2^{-1}(i) P_2 M_i^{(f)} M_i'^{(f)} P_2 + \alpha_i^2 P_1^{-1}.\end{aligned} \tag{39}$$

Consequently, inequality (30) holds if the following linear matrix inequality holds for each mode i

$$\begin{bmatrix} A_i' P_2 + P_2 A_i + C_i Y_2(i) + Y_2'(i) C_i' + \varepsilon_2(i) N_i'^{(f)} N_i^{(f)} & \alpha_i \boldsymbol{I} & P_2 M_i^{(f)} \\ \star & -P_1 & \boldsymbol{0} \\ \star & \star & -\varepsilon_2(i) \boldsymbol{I} \end{bmatrix} < 0. \tag{40}$$

We have proved the following statement.

Theorem 2. *Consider system (3) satisfying hypotheses 1-4 with $d_1(t) = d_2(t) = 0$. Then, system (3) is globally asymptotically stable under the action of the observer-controller (17) provided that there exist sets of symmetric and positive definite matrices $P_1 \in \mathbb{R}^{n\times n}$, $P_2 \in \mathbb{R}^{n\times n}$, and $(P_3(i))_{1\le i\le q} \in \mathbb{R}^{n\times n}$, a set of real matrices $(Y_1(i))_{1\le i\le q} \in \mathbb{R}^{m\times n}$, $(Y_2(i))_{1\le i\le q} \in \mathbb{R}^{n\times p}$, and sets of positive constants $(\varepsilon_1(i))_{1\le i\le q}$, $(\varepsilon_2(i))_{1\le i\le q}$ and $(\alpha_i)_{1\le i\le q}$ such that the linear matrix inequalities (32), (37), and (40) hold for each mode i; $1 \le i \le q$.*

Remark 4. *The LMI (32) ensures the existence of a stabilizing controller and LMI (40) guarantees the existence of a converging observer. LMI (37) describes the link between P_1 and $P_3(i)$. The mode dependent matrix $P_3(i)$ has been introduced to decouple the original observer-based control issue into two separate problems: robust controller design issue and observer design problem. The proposed LMIs are neither subject to equality constraints nor iterative which make the design of observer-controllers quite straightforward and easy to implement.*

Remark 5. *It is always possible to solve the linear matrix inequalities (32), (37) and (40) with respect to mode-independent matrices P_1, P_2, Y_1, Y_2, ε_1, ε_2 and α. In this case, if a solution exists, a constant-gain observer-based controller can be built to ensure the stability of the system whatever the switch is.*

In the next section we shall consider the observer-based stabilization issue with guaranteed cost control.

4. Robustness with Respect to System and Output Uncertainties

To overcome the effects of uncertainties, we shall not conserve the same structure of the high-gain observer discussed in the last section. Consider the augmented system

$$\begin{cases} \dot{x}(t) = A_i\, x(t) + f_i(x(t)) + B_i\, u(t) + E_i\, d_1(t), \\ \dot{z}(t) = C_i\, x(t) + A_z(i)\, z(t) + G_i\, d_2(t), \\ y_s(t) = z(t), \end{cases} \tag{41}$$

where $y_s(t)$ is the new smooth output of the system such that $z(0) = 0$ and $A_z(i)$ is a mode-dependent Hurwitz matrix such that the subsystems $\dot{z}(t) = A_z(i)\, z(t)$ are asymptotically stable for all $i \in \mathcal{Q}$. In fact, we plan to feedback the observer with the new output $y_s(t)$ instead of the noisy output $y(t)$ so as to enfeeble the effect of the system and the output uncertainties. To this end, let us define the following matrices $\boldsymbol{A}_i$, $\boldsymbol{B}_i$, $\boldsymbol{C}$, $\boldsymbol{D}_i$, and D

of dimensions $(n+p) \times (n+p)$, $(n+p) \times m$, $p \times (n+p)$, $(n+p) \times (n+p)$, and $2(n+p) \times 2(n+p)$, respectively, where

$$\xi(t) = \begin{bmatrix} z(t) \\ x(t) \end{bmatrix}, \ \boldsymbol{A}_i = \begin{bmatrix} A_z(i) & C_i \\ \mathbf{0} & A_i \end{bmatrix}, \ \boldsymbol{B}_i = \begin{bmatrix} \mathbf{0} \\ B_i \end{bmatrix}, \boldsymbol{f}_i(\xi(t)) = \begin{bmatrix} \mathbf{0} \\ f_i(x(t)) \end{bmatrix}, \tag{42}$$

$$\boldsymbol{C} = \begin{bmatrix} \boldsymbol{I} & \mathbf{0} \end{bmatrix}, \boldsymbol{D}_i = \begin{bmatrix} G_i & \mathbf{0} \\ \mathbf{0} & E_i \end{bmatrix}, \ D_i = \begin{bmatrix} \boldsymbol{D}_i & \mathbf{0} \\ \mathbf{0} & -\boldsymbol{D}_i \end{bmatrix}, \ \omega(t) = \begin{bmatrix} d_2(t) \\ d_1(t) \end{bmatrix}. \tag{43}$$

Remark 6. *The matrices $A_z(1), \cdots, A_z(q)$ must be chosen so as the pairs $(\boldsymbol{A}_i, \boldsymbol{B}_i)$, and $(\boldsymbol{A}_i, \boldsymbol{C})$ are controllable and observable, respectively for each $i \in \mathscr{Q}$.*

The new nonlinearity $\boldsymbol{f}_i(\xi(t))$ preserves the conditions of hypothesis 2. Consequently, we can write that

$$\frac{\partial \boldsymbol{f}_i(\xi(t))}{\partial \xi(t)} = \boldsymbol{M}_i \boldsymbol{F}_i(\xi(t)) \boldsymbol{N}_i; \ \ \forall i \in \mathscr{Q}, \tag{44}$$

where $\boldsymbol{M}_i$, $\boldsymbol{N}_i$, and $\boldsymbol{F}_i(\xi(t))$ are $(n+p)$ by $(n+p)$ matrices defined as

$$\boldsymbol{M}_i = \begin{bmatrix} \mathbf{0} & \mathbf{0} \\ \mathbf{0} & M_i^{(f)} \end{bmatrix}, \ \boldsymbol{N}_i = \begin{bmatrix} \mathbf{0} & \mathbf{0} \\ \mathbf{0} & N_i^{(f)} \end{bmatrix}, \ \boldsymbol{F}_i(\xi(t)) = \begin{bmatrix} \mathbf{0} & \mathbf{0} \\ \mathbf{0} & F_i(x(t)) \end{bmatrix}.$$

System (41) takes then the form

$$\begin{cases} \dot{\xi}(t) = \boldsymbol{A}_i \, \xi(t) + \boldsymbol{f}_i(\xi(t)) + \boldsymbol{B}_i \, u(t) + \boldsymbol{D}_i \, \omega(t), \\ y_s(t) = \boldsymbol{C} \, \xi(t). \end{cases} \tag{45}$$

Recall that our objective is to design a high-gain observer for system (45) such that system (45) under the feedback $u_i(t) = K_\xi(i) \, \hat{\xi}(t)$ is asymptotically stable when $\omega(t) \equiv 0$ where $\hat{\xi}(t)$ is the estimate of $\xi(t)$ given by the high-gain observer

$$\begin{cases} \dot{\hat{\xi}}(t) = \boldsymbol{A}_i \, \hat{\xi}(t) + \boldsymbol{f}_i(\hat{\xi}(t)) + \boldsymbol{B}_i \, K_\xi(i) \, \hat{\xi}(t) + L_\xi(i)(\boldsymbol{C} \, \hat{\xi}(t) - \boldsymbol{C} \, \xi(t)), \\ \hat{y}_s(t) = \boldsymbol{C} \, \hat{\xi}(t), \end{cases} \tag{46}$$

where $K_\xi(i)$ and $L_\xi(i)$ are constant matrices of appropriate dimensions. For the null initial conditions, i.e, $\hat{\xi}(0) = \xi(0) = 0$ and non-null $\omega(t)$, the controller and the observer gains $K_\xi(i)$ and $L_\xi(i)$ must be chosen to satisfy the following criterion

$$\sup_{t \geq 0} \|\boldsymbol{C}\hat{\xi}(t) - \boldsymbol{C}\xi(t)\| \leq \gamma; \ \ \forall i \in \mathscr{Q}, \tag{47}$$

where γ is a positive constant.

Remark 7. *With this representation, we see clearly that $\omega(t)$ appears only on the state dynamics of system (45), and the new output $y_s(t)$ behaves as noise free. Constructing, a high-gain observer for system (45) will reduce the effect of $\omega(t)$ and the observer gain shall be decoupled from the effects of uncertainties.*

Let $S_1 \in \mathbb{R}^{(n+p)\times(n+p)}$, and $S_2 \in \mathbb{R}^{(n+p)\times(n+p)}$ be two symmetric and positive definite matrices, and let $(W_1(i))_{1\le i\le q} \in \mathbb{R}^{m\times(n+p)}$ and $(W_2(i))_{1\le i\le q} \in \mathbb{R}^{(n+p)\times p}$ be sets of constant matrices such that $K_\xi(i) = W_1(i)S_1^{-1}$ and $L_\xi(i) = S_2^{-1}W_2(i)$. Let $e(t) = \hat{\xi}(t) - \xi(t)$ be the observation error, and define $v(\lambda, t) = (1-\lambda)\xi(t)$, $w(\lambda, t) = \hat{\xi}(t) - \lambda e(t)$ then, we have

$$\begin{bmatrix} \dot{\xi}(t) \\ \dot{e}(t) \end{bmatrix} = \int_0^1 \begin{bmatrix} \boldsymbol{A}_i + \boldsymbol{B}_i W_1(i)S_1^{-1} + \boldsymbol{M}_i\boldsymbol{F}_i(v)\boldsymbol{N}_i & \boldsymbol{B}_i W_1(i)S_1^{-1} \\ \boldsymbol{0} & \boldsymbol{A}_i + S_2^{-1}W_2(i)\boldsymbol{C} + \boldsymbol{M}_i\boldsymbol{F}_i(w)\boldsymbol{N}_i \end{bmatrix} \times \begin{bmatrix} \xi(t) \\ e(t) \end{bmatrix} + D_i\,\omega(t)\mathrm{d}\lambda. \tag{48}$$

Define the Lyapunov function $\mathscr{W}(e(t), \xi(t)) = \eta(t)'S\eta(t)$ associated to the dynamics (48) where

$$S = \begin{bmatrix} S_1^{-1} & \boldsymbol{0} \\ \boldsymbol{0} & S_2 \end{bmatrix},\ \eta(t) = \begin{bmatrix} \xi(t) \\ e(t) \end{bmatrix}. \tag{49}$$

Since $\mathscr{W}(0) = 0$, then the optimality condition (47) is satisfied for the null initial conditions if the following holds

$$\int_0^t \|\widetilde{\boldsymbol{C}}\eta(s)\|^2 - \gamma^2\|\omega(s)\|^2 \mathrm{d}s < 0 \tag{50}$$

where $\widetilde{\boldsymbol{C}} = \begin{bmatrix} \boldsymbol{C} & \boldsymbol{0} \end{bmatrix} \in \mathbb{R}^{p\times 2(n+p)}$. The stability of the dynamics (48) is guaranteed along with the optimality condition (50) if the following holds

$$\mathscr{C} = \int_0^t \|\widetilde{\boldsymbol{C}}\eta(s)\|^2 - \gamma^2\|\omega(t)\|^2 + \dot{\mathscr{W}}(e(s), \xi(s))\mathrm{d}s < 0. \tag{51}$$

We have

$$\begin{aligned}
\mathscr{C} \le & \int_0^t \int_0^1 \|\widetilde{\boldsymbol{C}}\eta(s)\|^2 - \gamma^2\|\omega(s)\|^2 \\
& +\eta'(s)\left(\begin{bmatrix} \boldsymbol{A}_i + \boldsymbol{B}_i W_1(i)S_1^{-1} + \boldsymbol{M}_i\boldsymbol{F}_i(v)\boldsymbol{N}_i & \boldsymbol{B}_i W_1(i)S_1^{-1} \\ \boldsymbol{0} & \boldsymbol{A}_i + S_2^{-1}W_2(i)\boldsymbol{C} + \boldsymbol{M}_i\boldsymbol{F}_i(w)\boldsymbol{N}_i \end{bmatrix}' S \right. \\
& \left. +S \begin{bmatrix} \boldsymbol{A}_i + \boldsymbol{B}_i W_1(i)S_1^{-1} + \boldsymbol{M}_i\boldsymbol{F}_i(v)\boldsymbol{N}_i & \boldsymbol{B}_i W_1(i)S_1^{-1} \\ \boldsymbol{0} & \boldsymbol{A}_i + S_2^{-1}W_2(i)\boldsymbol{C} + \boldsymbol{M}_i\boldsymbol{F}_i(w)\boldsymbol{N}_i \end{bmatrix} \right)\eta(s) \\
& +\eta'(s)S\,D_i\omega(s) + \omega'(s)D'_iS'\eta(s)\ \ \mathrm{d}s\,\mathrm{d}\lambda \\
= & \int_0^t \int_0^1 \begin{bmatrix} \eta(s) \\ \omega(s) \end{bmatrix}' \begin{bmatrix} X_{1,1}(\lambda, s) & X_{1,2} & X_{1,3} & X_{1,4} \\ \star & X_{2,2}(\lambda, s) & X_{2,3} & X_{2,4} \\ \star & \star & X_{3,3} & X_{3,4} \\ \star & \star & \star & X_{4,4} \end{bmatrix} \begin{bmatrix} \eta(s) \\ \omega(s) \end{bmatrix} \ \mathrm{d}s\,\mathrm{d}\lambda
\end{aligned}$$

where all the elements $X_{i,j}$ are $(n+p)$ by $(n+p)$ matrices defined as follows

$$\begin{aligned}
X_{1,1}(\lambda, t) &= \boldsymbol{A}_i' S_1^{-1} + S_1^{-1}\boldsymbol{A}_i \\
&\quad + S_1^{-1}\Big(\boldsymbol{B}_i W_1(i) + W_1'(i)\boldsymbol{B}_i'\Big) S_1^{-1} \\
&\quad + S_1^{-1}\boldsymbol{M}_i\boldsymbol{F}_i(v)\boldsymbol{N}_i + (\boldsymbol{M}_i\boldsymbol{F}_i(v)\boldsymbol{N}_i)' S_1^{-1} + \boldsymbol{C}'\boldsymbol{C}, \\
X_{1,2} &= S_1^{-1}\boldsymbol{B}_i W_1(i) S_1^{-1}, \\
X_{1,3} &= S_1^{-1}\boldsymbol{D}_i, \\
X_{1,4} &= \boldsymbol{0}, \\
X_{2,2}(\lambda, t) &= \boldsymbol{A}_i' S_2 + S_2\boldsymbol{A}_i + (\boldsymbol{M}_i\boldsymbol{F}_i(w)\boldsymbol{N}_i)' S_2 + S_2\boldsymbol{M}_i\boldsymbol{F}_i(w)\boldsymbol{N}_i \\
&\quad + W_2(i)\boldsymbol{C} + \boldsymbol{C}' W_2'(i), \\
X_{2,3} &= \boldsymbol{0}, \\
X_{2,4} &= -S_2\boldsymbol{D}_i, \\
X_{3,3} &= -\gamma^2\boldsymbol{I}, \\
X_{3,4} &= \boldsymbol{0}, \\
X_{4,4} &= -\gamma^2\boldsymbol{I}.
\end{aligned}$$

This implies that $\mathscr{C} < 0$ if the following holds

$$\int_0^t\int_0^1 \begin{bmatrix} X_{1,1}(\lambda, s) & X_{1,2} & X_{1,3} & X_{1,4} \\ \star & X_{2,2}(\lambda, s) & X_{2,3} & X_{2,4} \\ \star & \star & X_{3,3} & X_{3,4} \\ \star & \star & \star & X_{4,4} \end{bmatrix} \mathrm{d}\lambda\,\mathrm{d}s < 0. \tag{52}$$

Pre- and post multiplying the last matrix inequality by the matrix

$$\begin{bmatrix} S_1 & \boldsymbol{0} & \boldsymbol{0} & \boldsymbol{0} \\ \star & \boldsymbol{I} & \boldsymbol{0} & \boldsymbol{0} \\ \star & \star & \boldsymbol{I} & \boldsymbol{0} \\ \star & \star & \star & \boldsymbol{I} \end{bmatrix}, \tag{53}$$

we obtain the following matrix inequality

$$\int_0^t\int_0^1 \begin{bmatrix} \mathscr{M}_{1,1}(\lambda, s) & \boldsymbol{B}_i W_1(i) S_1^{-1} & \boldsymbol{D}_i & \boldsymbol{0} \\ \star & \mathscr{M}_{2,2}(\lambda, s) & X_{2,3} & X_{2,4} \\ \star & \star & X_{3,3} & X_{3,4} \\ \star & \star & \star & X_{4,4} \end{bmatrix} \mathrm{d}\lambda\,\mathrm{d}s < 0, \tag{54}$$

where

$$\begin{aligned}
\mathscr{M}_{1,1}(\lambda, s) &= S_1\boldsymbol{A}_i + \boldsymbol{A}_i' S_1 + \boldsymbol{B}_i W_1(i) + W_1'(i)\boldsymbol{B}_i' \\
&\quad + S_1\boldsymbol{C}'\boldsymbol{C}S_1 + S_1\boldsymbol{N}_i'\boldsymbol{F}_i(v)'\boldsymbol{M}_i' + \boldsymbol{M}_i\boldsymbol{F}_i(v)\boldsymbol{N}_i S_1, \\
\mathscr{M}_{2,2}(\lambda, s) &= X_{2,2}(\lambda, s).
\end{aligned} \tag{55}$$

For any $S_3(i) = S_3'(i) > 0$, $i \in \mathscr{Q}$, we rewrite inequality (54) as follows

$$\int_0^t \int_0^1 \begin{bmatrix} \boldsymbol{I} & 0 & 0 & 0 & 0 \\ 0 & 0 & S_1^{-1} & \boldsymbol{I} & 0 \\ 0 & \boldsymbol{I} & 0 & 0 & 0 \\ 0 & 0 & 0 & 0 & 0 \end{bmatrix} \times$$

$$\begin{bmatrix} \mathscr{M}_{1,1}(\lambda, s) & \boldsymbol{D}_i & \boldsymbol{B}_i W_1(i) & 0 & 0 \\ \star & -\gamma^2 \boldsymbol{I} & 0 & 0 & 0 \\ \star & \star & -S_3(i) & 0 & 0 \\ \star & \star & \star & \mathscr{M}_{2,2}(\lambda, s) + S_1^{-1} S_3(i) S_1^{-1} & -S_2 \boldsymbol{D}_i \\ \star & \star & \star & \star & -\gamma^2 \boldsymbol{I} \end{bmatrix} \tag{56}$$

$$\times \begin{bmatrix} \boldsymbol{I} & 0 & 0 & 0 \\ 0 & 0 & \boldsymbol{I} & 0 \\ 0 & S_1^{-1} & 0 & 0 \\ 0 & \boldsymbol{I} & 0 & 0 \\ 0 & 0 & 0 & \boldsymbol{I} \end{bmatrix} \mathrm{d}\lambda\, \mathrm{d}s < 0.$$

This implies that (54) holds if the following holds for each $i \in \mathscr{Q}$

$$\int_0^t \int_0^1 \begin{bmatrix} \mathscr{M}_{1,1}(\lambda, s) & \boldsymbol{D}_i & \boldsymbol{B}_i W_1(i) \\ \star & -\gamma^2 \boldsymbol{I} & 0 \\ \star & \star & -S_3(i) \end{bmatrix} \mathrm{d}\lambda\, \mathrm{d}s < 0, \tag{57}$$

and

$$\int_0^t \int_0^1 \begin{bmatrix} \mathscr{M}_{2,2}(\lambda, s) + S_1^{-1} S_3(i) S_1^{-1} & -S_2 \boldsymbol{D}_i \\ \star & -\gamma^2 \boldsymbol{I} \end{bmatrix} \mathrm{d}\lambda\, \mathrm{d}s < 0. \tag{58}$$

If we choose the i-th matrix $S_3(i)$ such that

$$\frac{1}{\alpha_i^2} S_3(i) - S_1 < 0, \quad \alpha_i > 0 \tag{59}$$

and using the fact that for any $\varepsilon_1(i) > 0$ and $\varepsilon_2(i) > 0$, we have

$$\begin{aligned} \int_0^1 \mathscr{M}_{1,1}(\lambda, s) \mathrm{d}\lambda \leq\; & S_1 \boldsymbol{A}_i' + \boldsymbol{A}_i S_1 + W_1(i) \boldsymbol{B}_i + \boldsymbol{B}_i' W_1'(i) \\ & + \varepsilon_1^{-1}(i) S_1 \boldsymbol{N}_i' \boldsymbol{N}_i S_1 + \varepsilon_1(i) \boldsymbol{M}_i \boldsymbol{M}_i' \\ & + S_1 \boldsymbol{C}' \boldsymbol{C} S_1, \end{aligned} \tag{60}$$

and

$$\begin{aligned} \int_0^1 \mathscr{M}_{2,2}(\lambda, s) \mathrm{d}\lambda \leq\; & \boldsymbol{A}_i' S_2 + S_2 \boldsymbol{A}_i + W_2(i) \boldsymbol{C} + \boldsymbol{C}' W_2'(i) \\ & + \varepsilon_2^{-1}(i) S_2 \boldsymbol{M}_i \boldsymbol{M}_i' S_2(i) + \varepsilon_2(i) \boldsymbol{N}_i' \boldsymbol{N}_i. \end{aligned} \tag{61}$$

Based upon these results inequalities (59), (57), (58) are equivalent by the Schur complement to the following linear matrix inequalities, respectively

$$\begin{bmatrix} -S_1 & \boldsymbol{I} \\ \boldsymbol{I} & -2\alpha_i \boldsymbol{I} + S_3(i) \end{bmatrix} < 0; \quad i \in \mathscr{Q}, \tag{62}$$

$$\begin{bmatrix} S_1\boldsymbol{A}_i' + \boldsymbol{A}_i S_1 + W_1(i)\boldsymbol{B}_i + \boldsymbol{B}_i' W_1'(i) + \varepsilon_1(i)\boldsymbol{M}_i\boldsymbol{M}_i' & \boldsymbol{B}_i W_1(i) & S_1\boldsymbol{N}_i' & \boldsymbol{D}_i & S_1\boldsymbol{C}' \\ \star & -S_3(i) & \boldsymbol{0} & \boldsymbol{0} & \boldsymbol{0} \\ \star & \star & & -\varepsilon_1(i)\boldsymbol{I} & \boldsymbol{0} \\ \star & \star & \star & -\gamma^2\boldsymbol{I} & \boldsymbol{0} \\ \star & \star & \star & \star & -\boldsymbol{I} \end{bmatrix} < 0 \;\; i \in \mathscr{Q}, \tag{63}$$

$$\begin{bmatrix} \boldsymbol{A}_i' S_2 + S_2\boldsymbol{A}_i + W_2(i)\boldsymbol{C} + \boldsymbol{C}' W_2'(i) + \varepsilon_2(i)\boldsymbol{N}_i'\boldsymbol{N}_i & \alpha_i\boldsymbol{I} & S_2\boldsymbol{M}_i & -S_2\boldsymbol{D}_i \\ \star & -S_1 & \boldsymbol{0} & \boldsymbol{0} \\ \star & \star & -\varepsilon_2(i)\boldsymbol{I} & \boldsymbol{0} \\ \star & \star & \star & -\gamma^2\boldsymbol{I} \end{bmatrix} < 0 \;\; i \in \mathscr{Q}. \tag{64}$$

We summarize the design of the the robust observer-based controller in the following statement.

Theorem 3. *Consider the switched nonlinear system (45) and the nonlinear observer (46). If there exist sets of symmetric and positive definite matrices* $S_1 \in \mathbb{R}^{(n+p)\times(n+p)}$, $S_2 \in \mathbb{R}^{(n+p)\times(n+p)}$, $(S_3(i))_{1\leq i\leq q} \in \mathbb{R}^{(n+p)\times(n+p)}$, *sets of real matrices* $(W_1(i))_{1\leq i\leq q} \in \mathbb{R}^{m\times(n+p)}$, $(W_2(i))_{1\leq i\leq q} \in \mathbb{R}^{(n+p)\times p}$, *and sets of positive constants* $(\varepsilon_1(i))_{1\leq i\leq q}$, $(\varepsilon_2(i))_{1\leq i\leq q}$, $(\alpha_i)_{1\leq i\leq q}$, *and a positive constant* γ *such that the linear matrix inequalities (62), (63) and (64) hold for each* $i \in \mathscr{Q}$. *Then, system (45) under the action of the mode dependent observer-based controller* $u_i(t) = K_\xi(i)\hat{\xi}(t)$ *is globally asymptotically stable when* $\omega(t) \equiv 0$, *and for the null initial condition and non-null* $\omega(t)$, *we have*

$$\sup_{t\geq 0} \|y_s(t) - \hat{y}_s(t)\| \leq \gamma. \tag{65}$$

5. Simulations

5.1. Application to Multiple Models Robot

The single-link flexible joint robot with two configurations is described by the following dynamics

$$\begin{aligned}
&\dot{\theta}_m(t) = \omega_m(t), \\
&\dot{\omega}_m(t) = \frac{k}{J_m}\left(\theta_\ell(t) - \theta_m(t)\right) - \frac{B}{J_m}\omega_m(t) + \frac{K_\tau}{J_m}u(t), \\
&\dot{\theta}_\ell(t) = \omega_\ell(t), \\
&\dot{\omega}_\ell(t) = -\frac{k}{J_\ell(i)}\left(\theta_\ell(t) - \theta_m(t)\right) - \frac{m(i)gh}{J_\ell(i)}\sin(\theta_\ell(t)),\ 1 \le i \le 2, \\
&y_1(t) = \theta_m(t) + \alpha_2\, d_2(t), \\
&y_2(t) = \omega_m(t) + \alpha_3\, d_2(t).
\end{aligned} \tag{66}$$

where J_m represents the inertia of the actuator (d.c. motor), and $(J_\ell(i))_{1\le i\le 2}$ stands for the values of the link inertia. θ_m and θ_ℓ are the angles of rotations of the motor and the link, respectively. $\dot{\theta}_m$ and $\dot{\theta}_\ell$ are their angular velocities. k, K_τ, $m(i), 1 \le i \le 2$ are positive constants with $g = 9.81\,Kg/s^2$, and $h = 0.1\,b$, see table 5.1.. The output $y_1(t)$ and $y_2(t)$ are subject to measurement errors with the following coefficients: $\alpha_1 = 0.1$, $\alpha_2 = 0.2$. This multiple models system (66) results from the variation of the system load. Hence there are two parameters $m(i)$ and $J_\ell(i)$ that are subject to change at any moment. According to (66), we have fixed the nominal matrices as

$$\begin{aligned}
&A_1 = \begin{bmatrix} 0 & 1 & 0 & 0 \\ -\frac{k}{J_m} & -\frac{B}{J_m} & \frac{k}{J_m} & 0 \\ 0 & 0 & 0 & 1 \\ \frac{k}{J_\ell(1)} & 0 & -\frac{k}{J_\ell(1)} & 0 \end{bmatrix}, A_2 = \begin{bmatrix} 0 & 1 & 0 & 0 \\ -\frac{k}{J_m} & -\frac{B}{J_m} & \frac{k}{J_m} & 0 \\ 0 & 0 & 0 & 1 \\ \frac{k}{J_\ell(2)} & 0 & -\frac{k}{J_\ell(2)} & 0 \end{bmatrix}, \\
&B(1) = B(2) = \begin{bmatrix} 0 \\ \frac{K_\tau}{J_m} \\ 0 \\ 0 \end{bmatrix}, C_1 = C_2 = \begin{bmatrix} 1 & 0 & 0 & 0 \\ 0 & 1 & 0 & 0 \end{bmatrix}, \\
&G_1 = G_2 = \mathrm{diag}\Big(0.1, 0.2\Big), \\
&F_{1,2}(x(t)) = \mathrm{diag}\Big(0, 0, -\cos(\theta_\ell(t)), 0\Big), \\
&N_i^{(f)} = \begin{bmatrix} 0 & 0 & 0 & 0 \\ 0 & 0 & 0 & 0 \\ 0 & 0 & \sqrt{\frac{m(i)gh}{J_\ell(i)}} & 0 \\ 0 & 0 & 0 & 0 \end{bmatrix}, M_i^{(f)} = \begin{bmatrix} 0 & 0 & 0 & 0 \\ 0 & 0 & 0 & 0 \\ 0 & 0 & 0 & 0 \\ 0 & 0 & \sqrt{\frac{m(i)gh}{J_\ell(i)}} & 0 \end{bmatrix}, \\
&A_z(1) = A_z(2) = -\boldsymbol{I}.
\end{aligned} \tag{67}$$

By solving the set of LMIs of Theorem 3, we get

$$S_1 = \begin{bmatrix} 1.4118 & 0.0503 & -0.0776 & -0.7118 & -0.2027 & 0.0991 \\ 0.0503 & 1.6641 & 0.6140 & -0.2409 & 0.3269 & -2.6087 \\ -0.0776 & 0.6140 & 1.2027 & -1.1051 & 0.7860 & -3.8620 \\ -0.7118 & -0.2409 & -1.1051 & 19.0979 & 1.4442 & -0.8191 \\ -0.2027 & 0.3269 & 0.7860 & 1.4442 & 1.5819 & -2.5779 \\ 0.0991 & -2.6087 & -3.8620 & -0.8191 & -2.5779 & 23.8692 \end{bmatrix},$$

$$S_2 = \begin{bmatrix} 60.1463 & 25.4578 & -25.3728 & -0.0737 & 1.6619 & -1.8954 \\ 25.4578 & 18.0129 & -7.0995 & 0.1890 & -3.1774 & -1.0378 \\ -25.3728 & -7.0995 & 30.0404 & 1.0813 & -16.6696 & 2.2450 \\ -0.0737 & 0.1890 & 1.0813 & 0.2885 & -0.9733 & 0.0409 \\ 1.6619 & -3.1774 & -16.6696 & -0.9733 & 13.7316 & -1.3573 \\ -1.8954 & -1.0378 & 2.2450 & 0.0409 & -1.3573 & 0.6002 \end{bmatrix},$$

$$S_3(1) = \begin{bmatrix} 2.9626 & 0.0627 & -0.0412 & -0.0620 & 0.0141 & 0.0170 \\ 0.0627 & 3.0588 & 0.3905 & -0.1528 & 0.3170 & 0.1343 \\ -0.0412 & 0.3905 & 1.5705 & -0.5304 & 1.4181 & -0.0026 \\ -0.0620 & -0.1528 & -0.5304 & 3.6292 & -0.2399 & -0.2251 \\ 0.0141 & 0.3170 & 1.4181 & -0.2399 & 3.2193 & 0.4494 \\ 0.0170 & 0.1343 & -0.0026 & -0.2251 & 0.4494 & 3.5939 \end{bmatrix},$$

$$S_3(2) = \begin{bmatrix} 2.7745 & 0.0530 & -0.0548 & -0.0219 & -0.0087 & 0.0150 \\ 0.0530 & 2.8862 & 0.4565 & 0.0284 & 0.3977 & 0.1526 \\ -0.0548 & 0.4565 & 1.6551 & -0.1805 & 1.6050 & 0.1095 \\ -0.0219 & 0.0284 & -0.1805 & 3.2565 & 0.2256 & 0.0021 \\ -0.0087 & 0.3977 & 1.6050 & 0.2256 & 3.3911 & 0.5584 \\ 0.0150 & 0.1526 & 0.1095 & 0.0021 & 0.5584 & 3.5004 \end{bmatrix},$$

$$Y_1(1) = \begin{bmatrix} -0.0672 & -0.3293 & -0.5436 & 0.4334 & -0.7770 & -0.4042 \end{bmatrix},$$

$$Y_1(2) = \begin{bmatrix} -0.0333 & -0.4081 & -0.7156 & -0.0672 & -1.0766 & -0.5262 \end{bmatrix},$$

$$Y_2(1) = \begin{bmatrix} 16.0232 & 0.0000 \\ 11.9652 & -1.3661 \\ -27.8248 & 13.5332 \\ -1.7039 & -8.4052 \\ -30.5856 & -31.6035 \\ -0.4883 & 3.4829 \end{bmatrix}, \; Y_2(2) = \begin{bmatrix} 11.3597 & 0 \\ 6.2723 & -2.3915 \\ -16.4289 & 19.3407 \\ -1.8288 & -8.6032 \\ -37.9692 & -35.7914 \\ 2.5235 & 5.1785 \end{bmatrix},$$

$$\varepsilon_1(1) = 4.3531,\ \varepsilon_2(1) = 2.6369,$$
$$\varepsilon_1(2) = 3.6381,\ \varepsilon_2(2) = 1.8009,$$
$$\alpha_1 = 3.0209,\ \alpha_2 = 2.8674,\ \gamma = 0.9660.$$

5.2. Second Example

Here, we introduce another example in order to show that the proposed LMIs are not restrictive when nonlinearities are of high Lipschitz constants. Let us consider the following

Table 1. Robot parameters.

System parameters	Values
Motor inertia, J_m $(Kg\,m^2)$	$3.7\ 10^{-3}$
Link inertia, $J_\ell(i)$ $(Kg\,m^2)$	$9.3\ 10^{-3}, 9.1\ 10^{-3}$
Pointer Mass, $m(i)$ Kg	$2.1\ 10^{-1}, 4\ 10^{-1}$
Link length, $2b$ (m)	$3.1\ 10^{-1}$
Torsional spring constant, k $(Nm\,rad^{-1})$	$1.8\ 10^{-1}$
Viscous friction coefficient, B $(Nm\,V^{-1})$	$4.6\ 10^{-2}$
Amplifier gain, K_τ $(Nm\,V^{-1})$	$8\ 10^{-2}$

switched systems with two configurations

$$\begin{aligned}
A_1 &= \begin{bmatrix} 1 & -1 & 1 \\ 0 & 1 & 1 \\ 0 & -1 & -1 \end{bmatrix}, \; B_1 = \begin{bmatrix} 1 & 1 \\ -1 & 2 \\ 1 & 0 \end{bmatrix}, \; C_1 = \begin{bmatrix} 1 & 0 & 0 \\ 0 & 0 & 1 \end{bmatrix}, \\
f_1(x(t)) &= \begin{bmatrix} 0 \\ 0 \\ \frac{5}{2}\Big(\exp(-2\,x_1^2) - 1\Big) \end{bmatrix}, \; E_1 = \begin{bmatrix} 0 & 0 & 0 \\ 0 & 0 & 0 \\ 0 & 10^{-2} & 0 \end{bmatrix}, \; D_1 = 5\ 10^{-2}\boldsymbol{I}, \\
A_2 &= \begin{bmatrix} 1 & -1 & 1 \\ 0 & 1 & 1 \\ 1 & 1 & 0 \end{bmatrix}, \; B_2 = \begin{bmatrix} 0 & 1 \\ -1 & 2 \\ 1 & 1 \end{bmatrix}, \\
f_2(x(t)) &= \begin{bmatrix} 0 \\ 0 \\ \frac{1}{3}\Big(5\,x_1(t) - \ln(x_1^2(t) + 1)\Big) + 4\arctan(x_3(t)/2) \end{bmatrix}, \\
C_2 &= C_1\ E_2 = E_1,\ D_2 = D_1.
\end{aligned} \tag{68}$$

We suppose that for $t < 1$ $[S]$ the system admits the first representation, while for $t \geq 1$ $[S]$ the system dynamics is described by the second representation. The jacobian of the nonlinearities $f_1(x(t))$ and $f_2(x(t))$ can be rewritten as

$$\mathscr{G}_1(x(t)) = \underbrace{\begin{bmatrix} 0 & 0 & 0 \\ 0 & 0 & 0 \\ 0 & \sqrt{2} & 0 \end{bmatrix}}_{M_1^{(f)}} \underbrace{\begin{bmatrix} 0 & 0 & 0 \\ 0 & -\frac{659}{200}\,x_1(t)\,\exp(-2\,x_1^2(t)) & 0 \\ 0 & 0 & 0 \end{bmatrix}}_{F_1(x(t))} \underbrace{\begin{bmatrix} 0 & 0 & 0 \\ \frac{\sqrt{2}\,10^3}{659} & 0 & 0 \\ 0 & 0 & 0 \end{bmatrix}}_{N_1^{(f)}}, \tag{69}$$

$$\mathscr{G}_2(x(t)) = \underbrace{\begin{bmatrix} 0 & 0 & 0 \\ 0 & 0 & 0 \\ \sqrt{2} & 0 & \sqrt{2} \end{bmatrix}}_{M_2^{(f)}} \underbrace{\begin{bmatrix} 0 & 0 & \frac{1}{6}\frac{5x_1^2(t) - 2x_1(t) + 5}{x_1^2(t) + 1} \\ 0 & 0 & 0 \\ \frac{4}{4 + x_3^2} & 0 & 0 \end{bmatrix}}_{F_2(x(t))} \underbrace{\begin{bmatrix} 0 & 0 & \sqrt{2} \\ 0 & 0 & 0 \\ \sqrt{2} & 0 & 0 \end{bmatrix}}_{N_2^{(f)}}. \tag{70}$$

For $A_z(1) = A_z(2) = -\boldsymbol{I}$, the solutions of LMIs (62), (63) and (64) with respect to their variables are

$$S_1 = \begin{bmatrix} 1.6721 & -0.0356 & -0.0440 & 0.1285 & -0.0826 \\ -0.0356 & 1.2747 & 0.1798 & 0.0344 & 0.1163 \\ -0.0440 & 0.1798 & 0.5286 & 0.6667 & -0.2075 \\ 0.1285 & 0.0344 & 0.6667 & 1.9241 & -0.8644 \\ -0.0826 & 0.1163 & -0.2075 & -0.8644 & 1.2009 \end{bmatrix},$$

$$S_2 = \begin{bmatrix} 909.4242 & 45.7848 & -161.9243 & -4.3528 & 0.3320 \\ 45.7848 & 425.7645 & -149.5910 & -103.0807 & -13.1510 \\ -161.9243 & -149.5910 & 102.7835 & 54.0712 & -1.4200 \\ -4.3528 & -103.0807 & 54.0712 & 38.5075 & -0.8927 \\ 0.3320 & -13.1510 & -1.4200 & -0.8927 & 1.7962 \end{bmatrix},$$

$$S_3(1) = \begin{bmatrix} 4.3749 & -0.0056 & -0.0999 & 0.2545 & -0.0696 \\ -0.0056 & 4.2696 & 0.4417 & -0.0815 & 0.1546 \\ -0.0999 & 0.4417 & 2.8035 & 1.2905 & 0.4206 \\ 0.2545 & -0.0815 & 1.2905 & 5.7367 & -1.0462 \\ -0.0696 & 0.1546 & 0.4206 & -1.0462 & 4.3000 \end{bmatrix},$$

$$Y_1(1) = \begin{bmatrix} 0.1643 & -0.5213 & -1.2340 & 1.0955 & -1.3358 \\ -0.1919 & -0.3182 & -1.3178 & -2.3071 & 0.0836 \end{bmatrix},$$

$$Y_2(1) = \begin{bmatrix} 471.1609 & 0 \\ 29.4573 & 261.7321 \\ -753.4220 & 19.3812 \\ -151.4178 & -127.5968 \\ 120.9710 & -180.6999 \end{bmatrix},$$

$\varepsilon_1(1) = 0.6210,\ \varepsilon_2(1) = 3.3965,\ \alpha_1 = 3.6294,\ \gamma = 1.5777.$

$$S_3(2) = \begin{bmatrix} 4.5510 & -0.0355 & -0.1451 & 0.2810 & -0.0588 \\ -0.0355 & 4.5300 & 0.4580 & -0.2215 & 0.5065 \\ -0.1451 & 0.4580 & 2.7121 & 0.9344 & 0.8747 \\ 0.2810 & -0.2215 & 0.9344 & 5.6801 & -0.4018 \\ -0.0588 & 0.5065 & 0.8747 & -0.4018 & 5.8331 \end{bmatrix},$$

$$Y_1(2) = \begin{bmatrix} 0.4791 & -1.2090 & -1.4337 & 2.6291 & -3.6058 \\ -0.1601 & -0.3069 & -0.6302 & -1.7478 & -1.8294 \end{bmatrix},$$

$$Y_2(2) = \begin{bmatrix} 481.9997 & 0 \\ 33.5383 & 227.2379 \\ -758.6700 & 22.7698 \\ -151.8563 & -107.0235 \\ 120.5349 & -155.7936 \end{bmatrix},$$

$$\varepsilon_1(2) = 0.9593,\ \varepsilon_2(2) = 3.9902,\ \alpha_2 = 3.7487.$$

In the simulations shown below, we have taken $d_1(t) = d_2(t) = d(t)$ where $d(t)$ is a band-limited white noise such that $\sup_{t\geq 0} \|d(t)\| \leq 15$. In Fig. 1 we have represented the noisy outputs of the switched system. In Figs. 2-4, the performance of the system and the observer are reported. We see that the high-gain observer is robust against the system and the output uncertainties and is able not only to reproduce the unmeasured states, but to steer the system to the origin with a good uncertainty attenuation.

6. Conclusion

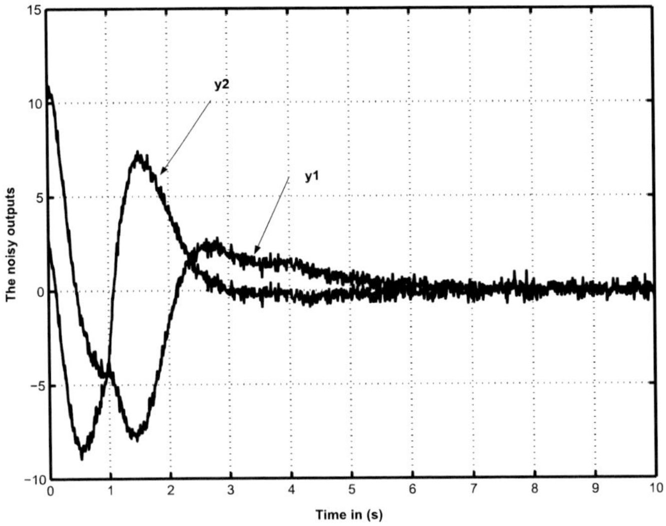

Figure 1. The noisy outputs of the switched system.

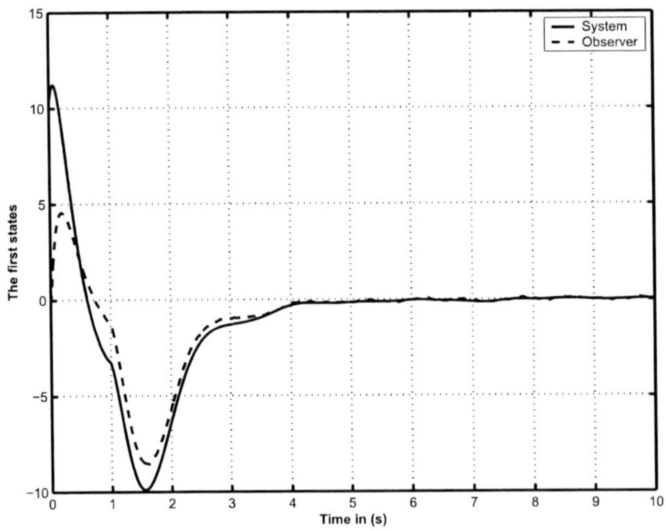

Figure 2. The observer and the system states: $x_1(t)$ and $\hat{x}_1(t)$.

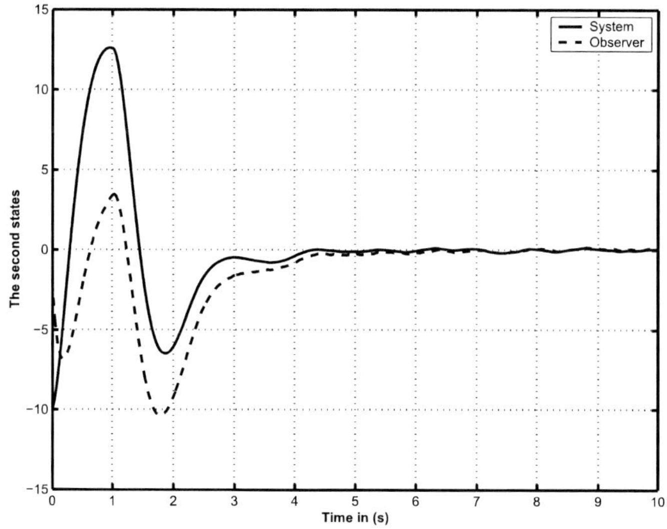

Figure 3. The observer and the system states: $x_2(t)$ and $\hat{x}_2(t)$.

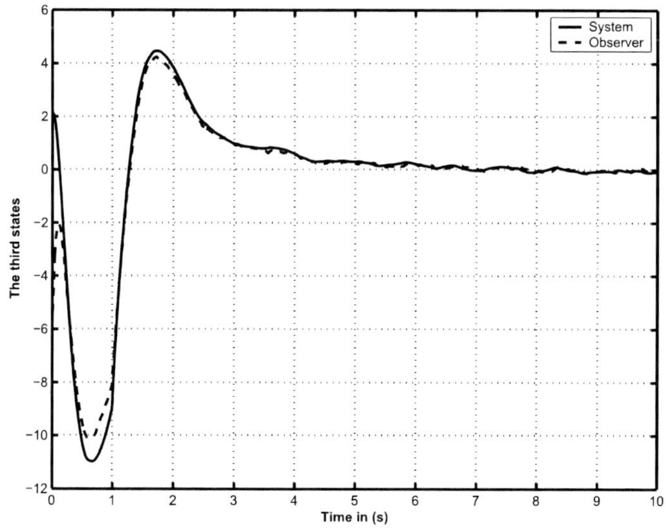

Figure 4. The observer and the system states: $x_3(t)$ and $\hat{x}_3(t)$.

New necessary and sufficient LMIs conditions for the existence of high-gain switching nonlinear observers are provided. The presented results can be seen as an extension of numerous results that have appeared in case of observation of single-mode nonlinear systems. Thanks to these results, the dynamic output feedback of switched nonlinear systems is formulated in convex optimization setting. We showed that the robust observer-based stabilization issue of this class of systems can be decomposed into two main problems: robust stability problem with full measured state feedback, and robust observer design issue with partial state measurements. These main problems were formulated in term of a set of linear matrix inequalities. Accordingly, our future research will include more optimality conditions for stabilization of nonlinear systems by means of nonlinear observers. Extension of the presented results for wide classes of switched nonlinear systems still an open problem for further investigation.

References

Aboky, C., Sallet, G., & Vivalda, L.-C. (2002). Observers for lipschitz nonlinear systems. *International Journal of control*, **75**(3), 204-212.

Balluchiy, A., Benvenuti, L., Benedetto, M. D. D., & Vincentelli, A. L. S. (2002, March). Design of observers for hybrid systems. *In proccedings of Hybrid Systems: Computation and Control Conference (HSCC 2002).*

Boyd, S., Ghaoui, L. E., Feron, E., & Balakrishnan, V. (1994). *Linear matrix inequality in systems and control theory*. SIAM.

Ciccarella, G., Mora, M. D., & Germani, A. (1993). A luenberger-like observer for nonlinear systems. *Int. J. of Control*, **57**(3), 537-556.

Cuzzola, F. A., & Morari, M. (2002). An lmi approach for h_∞ analysis and control of discrete-time piecewise affine systems. *International Journal of Control*, **75**(16-17), 1293-1301.

Daafouz, J., Riedinger, P., & Iung, C. (2002, November). Stability analysis and control synthesis for switched systems: A switched lyapunov function approach. *IEEE Transactions on Automatic Control*, **47**(11), 1883-1887.

Farra, N. H. E., Mhaskar, P., & Christofides, P. D. (2005). Output feedback control of switched nonlinear systems using multiple lyapunov functions. *Systems & Control Letters*, **54**, 1163-1182.

Huijberts, H. J. C., Lilge, T., & Nijmeijer, H. (2001). Nonlinear discrete-time synchronization via extended observers. *International Journal of Bifurcation and Chaos*, **11**(7), 1997-2006.

Ibrir, S. (2005, February). Less conservative linear matrix inequalities for observation of continuous-time and discrete-time nonlinear systems. *In proceedings of the first international conference on modeling, simulation and applied optimization*(paper 220). (American University of Sharjah, United Arab Emirates)

Li, Z. G., Wen, C. Y., & Soh, Y. (2003). Observer-based stabilization of switching linear systems. *Automatica*, **39**, 517-524.

Liu, Y. (1997, December). Switching observer design for uncertain nonlinear systems. *IEEE Transactions on Automatic Control*, **42**(12), 1699-1703.

Miller, C. M., & Rubinovich, E. Y. (2003). Kalman filter for controlled hybrid systems. *Systems & Control Letters*, **50**, 39-50.

Petersen, I. R. (1987). A stabilization algorithm for a class of uncertain linear systems. *Systems & Control Letters*, **8**, 351-357.

Pettersson, S., & Lennartson, B. (2002). Hybrid system stablity and robustness verification using linear matrix inequalities. *International Journal of Control*, **75**(16-17), 1335-1355.

Pogromsky, A. Y., & Nijmeijer, H. (1998). Observer-based robust synchronization of dynamical systems. *International Journal of Bifurcation and Chaos*, **8**(11), 2243-2254.

Raghavan, S., & Hedrick, J. K. (1994). Observer design for a class of nonlinear systems. *Int. J. Control*, **59**(2), 515-528.

Rajamani, R. (1998). Observers for lipschitz nonlinear systems. *IEEE Transactions on Automatic Control*, **43**(3), 397-400.

Rajamani, R., & Cho, Y. M. (1998). Existence and design of observers for nonlinear systems: relation to distance to unobservability. *International Journal of Control*, **69**(5), 717-731.

Savkin, A. V., Petersen, I. R., Skafidas, E., & Evans, R. J. (1996). Hybrid dynamical systems: robust control synthesis problems. *Systems & Control Letters*, **29**, 81-90.

Thau, F. E. (1973). Observing the state of nonlinear dynamic systems. *International Journal of Control*, **17** , 471-479.

In: Nonlinear Analysis Research Trends
Editor: Inès N. Roux, pp. 235-255
ISBN 978-1-60456-358-0

Chapter 7

Multiple Solutions of Nonlinear Elliptic Variational Problems

Siegfried Carl[1]****and Dumitru Motreanu***[2]†
[1]Institut für Mathematik
Martin-Luther-Universität Halle-Wittenberg
06099 Halle, Germany
[2]Départment de Mathématiques, Université de Perpignan
52 Avenue de Villeneuve, 66860 Perpignan, France

Abstract

The existence of multiple nontrivial solutions of semilinear elliptic eigenvalue problems under Dirichlet- or Neumann boundary conditions has received much attention over the last decades. It is well known that the multiplicity of solutions basically depends on the value of the eigenvalue parameter (with respect to the eigenvalues of an associated linear eigenvalue problem), and the structure and growth conditions imposed on the nonlinearity involved. In recent years the study of multiple solutions for quasilinear elliptic problems involving the p-Laplacian has become a subject of growing interest. The main goal of this article is to present new multiple solutions result for quasilinear elliptic variational problems involving the p-Laplacian which, in general, depend on two parameters, and whose nonlinearity may be even multi-valued and given in the form of Clarke's generalized gradient. The dependence of the problem under consideration on two parameters, which are studied in relation to the Fučik spectrum of the p-Laplacian, allows for a more detailed analysis of the multiplicity of solutions. The multi-valued problem considered here requires advanced tools from Nonsmooth Analysis. Our approach will be based on comparison principles for multi-valued elliptic problems and variational methods for nonsmooth functionals such as, e.g., nonsmooth critical point theory and second deformation lemma for nonsmooth and locally Lipschitz functionals.

*E-mail address: siegfried.carl@mathematik.uni-halle.de
†E-mail address: motreanu@univ-perp.fr

1. Introduction

Let $\Omega \subset \mathbb{R}^N$ be a bounded domain with a C^2-boundary $\partial\Omega$, and denote by $V_0 = W_0^{1,p}(\Omega)$, $1 < p < +\infty$, the usual Sobolev space whose elements have generalized homogeneous boundary values, as well as denote by $V_0^* = W^{-1,p'}(\Omega)$ ($\frac{1}{p} + \frac{1}{p'} = 1$) the dual space of V_0. We consider the following nonlinear multivalued elliptic boundary value problem under Dirichlet boundary condition: Find $u \in V_0 \setminus \{0\}$ and constants $a \in \mathbb{R}$, $b \in \mathbb{R}$ such that

$$-\Delta_p u \in a(u^+)^{p-1} - b(u^-)^{p-1} - \partial G(x, u) \quad \text{in } V_0^*, \tag{1.1}$$

where $\Delta_p u = \operatorname{div}(|\nabla u|^{p-2}\nabla u)$ is the p-Laplacian, and $u^+ = \max\{u, 0\}$ and $u^- = \max\{-u, 0\}$ are the positive and negative part of u, respectively, and $G : \Omega \times \mathbb{R} \to \mathbb{R}$ is assumed to be the primitive of a function $g : \Omega \times \mathbb{R} \to \mathbb{R}$ that is merely bounded on bounded sets, i.e., $g \in L^\infty_{\text{loc}}(\Omega \times \mathbb{R})$ and G is given by

$$G(x, s) := \int_0^s g(x, t)\, dt.$$

The multivalued function $s \mapsto \partial G(x, s)$ stands for the generalized Clarke's gradient of the locally Lipschitz function $s \mapsto G(x, s)$. Let $\langle \cdot, \cdot \rangle$ denote the dual pairing between V_0 and V_0^*, then the negative p-Laplacian operator $-\Delta_p$ defined by

$$\langle -\Delta_p u, v \rangle = \int_\Omega |\nabla u(x)|^{p-2} \nabla u(x) \nabla v(x)\, dx \quad \text{for all } u, v \in V_0$$

is known to be a continuous, strictly monotone and coercive operator from V_0 into its dual V_0^*. The norms in V_0 and $L^p(\Omega)$ are denoted by $\|\cdot\|$ and $\|\cdot\|_p$, respectively. Obviously, if $g : \Omega \times \mathbb{R} \to \mathbb{R}$ is a Carathéodory function, i.e., $x \mapsto g(x, s)$ is measurable in Ω for all $s \in \mathbb{R}$ and $s \mapsto g(x, s)$ is continuous in $\mathbb{R}$ for a.a. $x \in \Omega$ then $\partial G(x, s) = \{g(x, s)\}$ is single-valued, and thus problem (1.1) reduces to the following nonlinear elliptic problem depending on parameters a and b: Find $u \in V_0 \setminus \{0\}$ and constants $a \in \mathbb{R}$, $b \in \mathbb{R}$ such that

$$-\Delta_p u = a(u^+)^{p-1} - b(u^-)^{p-1} - g(x, u) \quad \text{in } V_0^*. \tag{1.2}$$

Multiple solution results for (1.2) were obtained by the authors in [9]. Furthermore, note that

$$|u|^{p-2}u = |u|^{p-2}(u^+ - u^-) = (u^+)^{p-1} - (u^-)^{p-1}.$$

Therefore, if one assumes, in addition, $a = b =: \lambda$ then (1.2) reduces to the following nonlinear elliptic eigenvalue problem: Find $u \in V_0 \setminus \{0\}$ and a constant $\lambda \in \mathbb{R}$ such that

$$-\Delta_p u = \lambda |u|^{p-2}u - g(x, u) \quad \text{in } V_0^*. \tag{1.3}$$

In a recent paper (see [8]) the authors considered the eigenvalue problem (1.3) for a Carathéodory function g. Combining the method of sub-supersolution with variational techniques and assuming certain growth conditions of $s \mapsto g(x, s)$ at infinity and at zero the authors were able to prove the existence of at least three nontrivial solutions including one that changes sign. The results in [8] improve among others recent results obtained in [32].

For $a = b =: \lambda$ (1.1) reduces to the corresponding multivalued eigenvalue problem: Find $u \in V_0 \setminus \{0\}$ and a constant $\lambda \in \mathbb{R}$ such that

$$-\Delta_p u \in \lambda |u|^{p-2}u - \partial G(x,u) \quad \text{in } V_0^*. \tag{1.4}$$

The existence of multiple solutions for (1.4) has been shown recently in [4] where techniques for single-valued problems developed in [8] and hemivariational methods applied in [26] have been used.

The existence of multiple solutions for semilinear and quasilinear elliptic problems have been studied by a number of authors, such as, e.g., [1, 2, 5, 15, 19, 20, 21, 24, 23, 33, 35, 36, 38, 39, 40]. All these papers deal with nonlinearities $(x,s) \mapsto g(x,s)$ that are sufficiently smooth.

We impose the following hypotheses on the nonlinearity $g : \Omega \times \mathbb{R} \to \mathbb{R}$ whose primitive is G of problem (1.1):

(g1) $(x,s) \mapsto g(x,s)$ is measurable in each variable separately.

(g2) There exists $c > 0$, and $q \in [p, p^*)$ such that

$$|g(x,s)| \le c(1 + |s|^{q-1})$$

for almost all (a.a.) $x \in \Omega$ and for all $s \in \mathbb{R}$, where p^* denotes the critical Sobolev exponent which is $p^* = Np/(n-p)$ if $p < N$, and $p^* = +\infty$ if $p \ge N$.

(g3)

$$\lim_{s \to 0} \frac{g(x,s)}{|s|^{p-2}s} = 0 \quad \text{uniformly with respect to a.a. } x \in \Omega\,.$$

(g4)

$$\lim_{|s| \to \infty} \frac{g(x,s)}{|s|^{p-2}s} = +\infty \quad \text{uniformly with respect to a.a. } x \in \Omega\,.$$

The main goal of this paper is to provide a detailed multiplicity analysis of the nonsmooth elliptic problem (1.1) in dependence of the two parameters a and b. If λ_1 denotes the first eigenvalue of $(-\Delta_p, V_0)$, then our first result given in Theorem 4.1 roughly states the following:

Assume hypotheses (g1)–(g4). Then for any $(a,b) \in \mathbb{R}^2$ with $a > \lambda_1$ and $b \in \mathbb{R}$ there is a smallest positive solution of (1.1), and for any $(a,b) \in \mathbb{R}^2$ with $b > \lambda_1$ and $a \in \mathbb{R}$ there is a greatest negative solution of (1.1).

Here the notions *smallest* and *greatest* refer to the natural partial ordering of functions. If $a > \lambda_1$ and $b > \lambda_1$, and if, moreover, the pair (a,b) is specified in terms of the Fučik spectrum, an additional sign-changing solution of (1.1) exists. As for the precise formulation of this result we recall the Fučik spectrum, see, e.g., [15].

The set Σ_p of those points $(\mu_1, \mu_2) \in \mathbb{R}^2$ for which the problem

$$u \in V_0 : \quad -\Delta_p\, u = \mu_1\, (u^+)^{p-1} - \mu_2\, (u^-)^{p-1} \quad \text{in } V_0^* \tag{1.5}$$

has a nontrivial solution is called the Fučik spectrum of the negative p-Laplacian on Ω. As is well known the first eigenvalue λ_1 of the p-Laplacian is positive, simple, and admits a positive eigenfunction φ_1 (see [3]), so Σ_p clearly contains the two lines $\lambda_1 \times \mathbb{R}$ and $\mathbb{R} \times \lambda_1$. In addition, the spectrum $\sigma(-\Delta_p)$ of the negative p-Laplacian has an unbounded sequence of variational eigenvalues (λ_l), $l \in \mathbb{N}$, satisfying a standard min-max characterization, and Σ_p contains the corresponding sequence of points (λ_l, λ_l), $l \in \mathbb{N}$. A first nontrivial curve $\mathcal{C}$ in Σ_p through (λ_2, λ_2) asymptotic to $\lambda_1 \times \mathbb{R}$ and $\mathbb{R} \times \lambda_1$ at infinity was constructed and variationally characterized by a mountain-pass procedure by Cuesta, de Figueiredo, and Gossez [14] (see Figure 1), which implies the existence of a continuous path in $\{u \in V_0 : I^{(a,b)}(u) < 0,\ \|u\|_p = 1\}$ joining $-\varphi_1$ and φ_1 provided (a, b) is above the curve $\mathcal{C}$. Here the functional $I^{(a,b)}$ on V_0 is given by

$$I^{(a,b)}(u) = \int_\Omega \Big(|\nabla u|^p - a(u^+)^p - b(u^-)^p \Big)\, dx.$$

The additional hypothesis on the constants a and b that ensures the existence of sign-changing solutions is as follows:

(H) Let $(a, b) \in \mathbb{R}^2_+$ be above the curve $\mathcal{C}$ of the Fučik spectrum constructed in [14], see Figure 1

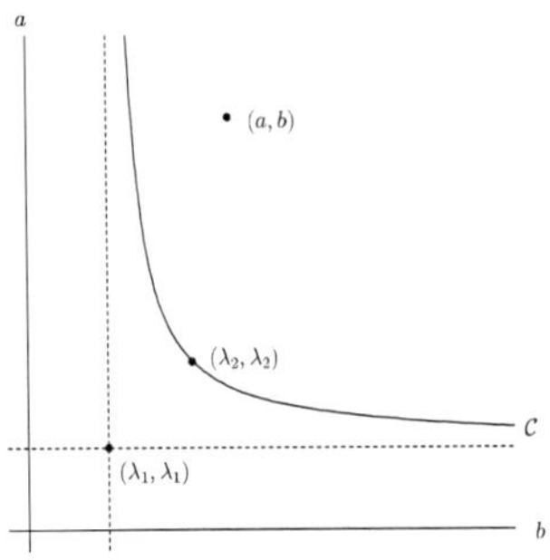

Figure 1. Fučik Spectrum.

Our second main result given in Theorem 6.1 states:
If hypotheses (g1)–(g4) and (H) are satisfied, then problem (1.1) has at least three nontrivial solutions, namely, two extremal opposite constant-sign solutions and a sign-changing solution.

Our approach strongly relies on a combined use of variational and topological arguments on the one hand (e.g. Critical Point Theory and Second Deformation Lemma for nonsmooth functionals), and comparison principles and regularity results for nonlinear multivalued elliptic differential equations on the other hand (sub-supersolution method, truncation techniques, strong maximum principle due to Vazquez [37]).

This paper represents a nonsmooth version of [9, 10] and improves and extends recent results obtained in [4]. The paper is organized as follows: In Section 2 we provide some basic definitions and preliminaries from Nonsmooth Analysis. In Section 3 we prove a new comparison result for the multivalued inclusion (1.1) which is used in the proof of our first main result on the existence of extremal constant-sign solutions given in Section 4. As a preparatory section for the proof of sign-changing solutions we characterize in Section 5 the extremal constant-sign solutions as minimizers of certain related functionals. Finally, in Section 6 our second main result on the existence of sign-changing solutions is proved.

2. Definitions and Preliminary Results

We first recall some basic facts from nonsmooth analysis. Let $(X, \|\cdot\|)$ be a real Banach space, X^* its dual space, and let $\Phi : X \to \mathbb{R}$ be a locally Lipschitz functional. The generalized directional derivative of Φ at u in the direction $v \in X$, denoted $\Phi^o(u; v)$, is defined as

$$\Phi^o(u; v) = \limsup_{y \to u,\ t \downarrow 0} \frac{\Phi(y + t\,v) - \Phi(y)}{t},$$

where t is a positive real, cf., e.g., [12, Chap. 2]. It is known that the function $v \mapsto \Phi^o(u; v)$ is finite, convex, positively homogeneous, and subadditive on X, and satisfies

$$|\Phi^o(u; v)| \leq c(\mathcal{U})\|v\|,$$

where the positive constant $c(\mathcal{U})$ depends only on a neighborhood $\mathcal{U}$ of u. By means of the generalized directional derivative Clarke's generalized gradient of Φ at $u \in X$, denoted $\partial\Phi(u)$, is defined as the subset of X^* given by

$$\partial\Phi(u) = \{\eta \in X^* : \Phi^o(u; v) \geq \langle \eta, v \rangle, \text{ for all } v \in X\},$$

where $\langle \cdot, \cdot \rangle$ denotes the duality pairing between X and X^*. As $v \mapsto \Phi^o(u; v)$ is convex and satisfied $\Phi^o(u; 0) = 0$, the generalized gradient $\partial\Phi(u)$ is nothing but the subdifferential (in the sense of Convex Analysis) of the functional $v \mapsto \Phi^o(u; v)$ at $v = 0$. From [12, Proposition 2.1.2] it follows that $\partial\Phi(u)$ is nonempty, convex, and weak* compact such that $m_\Phi(u)$ given by

$$m_\Phi(u) := \min\{\|u^*\|_{X^*} : u^* \in \Phi(u)\}$$

is well defined. Furthermore it holds

$$\Phi^o(u; v) = \max\{\langle u^*, v \rangle : u^* \in \partial\Phi(u)\}, \quad v \in X.$$

We say that $u \in X$ is a critical point of Φ if $0 \in \partial\Phi(u)$ or equivalently if $\Phi^o(u; v) \geq 0$ for all $v \in X$. One can easily verify that each local minimizer or maximizer of Φ is a critical point. The nonsmooth version of the Palais–Smale condition (PS) given in [11] reads as follows.

Definition 2.1. Let X be real Banach space, and $\Phi : X \to \mathbb{R}$ be a locally Lipschitz functional. We say that Φ satisfies the Palais–Smale condition if any sequence (x_n) with $(\Phi(x_n))$ bounded and $m_\Phi(x_n) \to 0$, has a convergent subsequence.

The following nonsmooth version of Ambrosetti–Rabinowitz's Mountain Pass Theorem is mainly due to Chang (see [11, Theorem 3.4]) and will be used later.

Theorem 2.1. *Let X be a reflexive real Banach space, and let $\Phi : X \to \mathbb{R}$ be a locally Lipschitz functional satisfying (PS). If there exist $x_0,\ x_1 \in X$, $r > 0$ such that $\|x_1 - x_0\| > r$ and $\max\{\Phi(x_0), \Phi(x_1)\} < \inf_{x \in \partial B_r(x_0)} \Phi(x)$, then Φ has a critical point $x_c \in X$ satisfying*

$$\inf_{x \in \partial B_r(x_0)} \Phi(x) \leq \Phi(x_c) = \inf_{\gamma \in \Gamma} \max_{t \in [0,1]} \Phi(\gamma(t)),$$

where $\Gamma = \{\gamma \in C^0([0,1], X) : \gamma(0) = x_0,\ \gamma(1) = x_1\}$, and $\partial B_r(x) = \{u \in X : \|u - x\| = r\}$.

On $V_0 = W_0^{1,p}(\Omega)$ we introduce the norm

$$\|u\| = \Big(\int_\Omega |\nabla u(x)|^p \, dx\Big)^{1/p}.$$

Further we make use of the following notations:

$$L^q(\Omega)_+ := \{u \in L^q(\Omega) : u(x) \geq 0 \ \text{ a.e. in } \Omega\},$$

$$C_0^1(\overline{\Omega})_+ := \{u \in C_0^1(\overline{\Omega}) : u(x) \geq 0 \ \ \forall x \in \Omega\}.$$

It is known (see, e.g., [18, Remark 6.2.10]) that

$$\operatorname{int}(C_0^1(\overline{\Omega})_+) = \left\{u \in C_0^1(\overline{\Omega}) : u(x) > 0, \ \frac{\partial u}{\partial n}(x) < 0 \ \ \forall x \in \Omega\right\},$$

with $n(x)$ being the outward unit normal vector to $\partial\Omega$ at the point $x \in \partial\Omega$.

Let λ_1 be the first eigenvalue of $-\Delta_p : V_0 \to V_0^*$. The following properties of λ_1 and $-\Delta_p$ are well known and can be found, e.g., in [18, Section 6.2].

(e1) $0 < \lambda_1$.

(e2) There exists an eigenfunction φ_1 corresponding to λ_1 such that $\varphi_1 \in \operatorname{int}(C_0^1(\overline{\Omega})_+)$ as well as $\|\varphi_1\|_{L^p(\Omega)} = 1$.

(e3) The operator $-\Delta_p : V_0 \to V_0^*$ is maximal monotone, coercive, and of type (S_+).

In view of assumptions (g1) and (g2) the function $s \mapsto G(x, s)$ is locally Lipschitz and the functional $\mathcal{G} : L^q(\Omega) \to \mathbb{R}$ defined by

$$\mathcal{G}(u) := \int_\Omega G(x, u(x)) \, dx, \quad u \in L^q(\Omega) \tag{2.1}$$

is well defined and locally Lipschitz continuous as well. The generalized gradients $\partial G(x, \cdot)$ and $\partial \mathcal{G}$ can be characterized as follows: Define for every $(x, t) \in \Omega \times \mathbb{R}$,

$$g_1(x, t) := \lim_{\delta \to 0^+} \operatorname*{ess\,inf}_{|\tau - t| < \delta} g(x, \tau), \quad g_2(x, t) := \lim_{\delta \to 0^+} \operatorname*{ess\,sup}_{|\tau - t| < \delta} g(x, \tau).$$

Proposition 1.7 in [29] ensures that

$$\partial G(x, \xi) = [g_1(x, \xi), g_2(x, \xi)], \tag{2.2}$$

while Theorem 4.5.19 of [18] implies

$$\partial \mathcal{G}(u) \subseteq \{w \in L^{q'}(\Omega) : g_1(x, u(x)) \leq w(x) \leq g_2(x, u(x)) \ \text{a.e. in } \Omega\}, \tag{2.3}$$

with $q' := q/(q-1)$. The next result is an immediate consequence of [12, Proposition 2.1.5].

Lemma 2.1. *Suppose $u_n \to u$ in V_0, $w_n \rightharpoonup w$ in $L^{q'}(\Omega)$, and $w_n \in \partial \mathcal{G}(u_n)$ for all $n \in \mathbb{N}$. Then $w \in \partial \mathcal{G}(u)$.*

Definition 2.2. A function $u \in V_0$ is called a solution of (1.1) if there is an $\eta \in L^{q'}(\Omega)$ such that

$$\eta(x) \in \partial G(x, u(x)) \text{ for a.a. } x \in \Omega, \tag{2.4}$$

$$\int_\Omega |\nabla u|^{p-2}\nabla u \nabla\varphi\, dx + \int_\Omega \Big(\eta - a(u^+)^{p-1} + b(u^-)^{p-1}\Big)\varphi\, dx = 0, \quad \forall\varphi \in V_0. \tag{2.5}$$

Remarks 2.2. Due to assumption (g3) we have $g_1(x,0) \le 0 \le g_2(x,0)$ for almost all $x \in \Omega$. Hence, in view of (2.2), problem (1.1) always possesses the trivial solution.

Definition 2.3. A function $\underline{u} \in V$ is called a subsolution of (1.1) if $\underline{u}|_{\partial\Omega} \le 0$, and if there is an $\underline{\eta} \in L^{q'}(\Omega)$ such that

$$\underline{\eta}(x) \in \partial G(x, \underline{u}(x)) \text{ for a.a. } x \in \Omega, \tag{2.6}$$

$$\int_\Omega |\nabla \underline{u}|^{p-2}\nabla \underline{u} \nabla\varphi\, dx + \int_\Omega \Big(\underline{\eta} - a(\underline{u}^+)^{p-1} + b(\underline{u}^-)^{p-1}\Big)\varphi\, dx \le 0, \quad \forall\varphi \in V_0 \cap L^p(\Omega)_+. \tag{2.7}$$

Similarly, we define a supersolution as follows.

Definition 2.4. A function $\overline{u} \in V$ is called a supersolution of (1.1) if there is an $\overline{\eta} \in L^{q'}(\Omega)$ such that

$$\overline{\eta}(x) \in \partial G(x, \overline{u}(x)) \text{ for a.a. } x \in \Omega, \tag{2.8}$$

$$\int_\Omega |\nabla \overline{u}|^{p-2}\nabla \overline{u} \nabla\varphi\, dx + \int_\Omega \Big(\overline{\eta} - a(\overline{u}^+)^{p-1} + b(\overline{u}^-)^{p-1}\Big)\varphi\, dx \ge 0, \quad \forall\varphi \in V_0 \cap L^p(\Omega)_+. \tag{2.9}$$

As already mentioned the operator $-\Delta_p : V_0 \to V_0^*$ is continuous, strictly monotone and coercive. Therefore the following Dirichlet problem:

$$e \in V_0 : \quad -\Delta_p e = 1 \quad \text{in } V_0^* \tag{2.10}$$

has a uniquely defined solution e. The nonlinear regularity theory and the nonlinear strong maximum principle due to [37] imply $e \in \operatorname{int}(C_0^1(\overline{\Omega})_+)$, where $\operatorname{int}(C_0^1(\overline{\Omega})_+)$ denotes the interior of the positive cone $C_0^1(\overline{\Omega})_+$ of the Banach space $C_0^1(\overline{\Omega})$ as given above. With λ_1 denoting the first eigenvalue of $(-\Delta_p, V_0)$ and φ_1 the eigenfunction of $(-\Delta_p, V_0)$ corresponding to λ_1 satisfying $\varphi_1 \in \operatorname{int}(C_0^1(\overline{\Omega})_+)$ and $\|\varphi_1\|_p = 1$ we show the following result.

Lemma 2.3. *Let e be the uniqely defined solution of (2.10). If $a > \lambda_1$, then there exists a constant $\alpha_a > 0$ such that for any $b \in \mathbb{R}$ the function $\alpha_a e$ is a positive supersolution of problem (1.1).*

Proof. Let $a > \lambda_1$. By (g4) there is $s_a > 0$ such that

$$\frac{g(x,s)}{s^{p-1}} > a \text{ for a.a. } x \in \Omega \text{ and all } s > s_a,$$

and by (g2) we get

$$|g(x,s) - as^{p-1}| \le |g(x,s)| + as^{p-1} \le c_a, \text{ for a.a. } x \in \Omega \text{ and all } s \in [0, s_a],$$

which implies

$$g(x,s) \geq as^{p-1} - c_a \text{ for a.a. } x \in \Omega \text{ and all } s \geq 0,$$

and thus in view of the definition of g_2 we obtain

$$g_2(x,s) \geq as^{p-1} - c_a \text{ for a.a. } x \in \Omega \text{ and all } s \geq 0,$$

Let $\bar{u} = \alpha_a e$, where α_a is a positive constant to be specified. Then we get

$$\begin{aligned} &-\Delta_p \bar{u} - a(\bar{u}^+)^{p-1} + b(\bar{u}^-)^{p-1} + g_2(x,\bar{u}) \\ &= \alpha_a^{p-1} - a\alpha_a^{p-1} e^{p-1} + g_2(x, \alpha_a e) \geq \alpha_a^{p-1} - c_a, \end{aligned}$$

which shows that for $\alpha_a := c_a^{\frac{1}{p-1}}$ the function $\alpha_a e$ is in fact a supersolution of (1.1). □

In a similar way the following lemma on the existence of a negative subsolution can be proved.

Lemma 2.4. *Let e be the uniqely defined solution of (2.10). If $b > \lambda_1$, then there exists a constant $\beta_b > 0$ such that for any $a \in \mathbb{R}$ the function $-\beta_b e$ is a negative subsolution of problem (1.1).*

In the next lemma we demonstrate that small constant multiples of φ_1 may be sub- and supersolutions of (1.1). More precisely we have the following result.

Lemma 2.5. *Let φ_1 be the normalized positive eigenfunction corresponding to the first eigenvalue λ_1 of $(-\Delta_p, V_0)$. If $a > \lambda_1$, then for $\varepsilon > 0$ sufficiently small and any $b \in \mathbb{R}$ the function $\varepsilon\varphi_1$ is a positive subsolution of problem (1.1). If $b > \lambda_1$, then for $\varepsilon > 0$ sufficiently small and any $a \in \mathbb{R}$ the function $-\varepsilon\varphi_1$ is a negative supersolution of problem (1.1).*

Proof. By (g3) there is a constant $\delta_a > 0$ such that

$$\frac{|g(x,s)|}{|s|^{p-1}} < a - \lambda_1 \text{ for a.a. } x \in \Omega \text{ and all } 0 < |s| \leq \delta_a,$$

which implies

$$|g(x,s)| \leq (a-\lambda_1)s^{p-1} \text{ for a.a. } x \in \Omega, \text{ and all } s : 0 \leq s \leq \delta_a. \tag{2.11}$$

Define $\underline{u} = \varepsilon\varphi_1$ with $\varepsilon > 0$. Applying (2.11) and the definition of g_1 we get

$$\begin{aligned} &-\Delta_p \underline{u} - a(\underline{u}^+)^{p-1} + b(\underline{u}^-)^{p-1} + g_1(x,\underline{u}) \\ &= \lambda_1(\varepsilon\varphi_1)^{p-1} - a(\varepsilon\varphi_1)^{p-1} + g_1(x,\varepsilon\varphi_1) \\ &\leq (\lambda_1 - a)(\varepsilon\varphi_1)^{p-1} + (a-\lambda_1)(\varepsilon\varphi_1)^{p-1} = 0 \end{aligned}$$

provided $0 \leq \varepsilon\varphi_1 \leq \delta_a$. The latter can be satisfied by choosing ε sufficiently small such that $\varepsilon \in (0, \frac{\delta_a}{\|\varphi_1\|_\infty})$, where $\|\varphi_1\|_\infty$ stands for the supremum-norm of φ_1. This proves that $\varepsilon\varphi_1$ is a subsolution if $\varepsilon \in (0, \frac{\delta_a}{\|\varphi_1\|_\infty})$. In a similar way one can show that for ε sufficiently small the function $-\varepsilon\varphi_1$ is a negative supersolution. □

3. Comparison Principle

In this section we provide an existence and comparison result for the nonsmooth problem (1.1) which will be used as a basic tool in the later analysis of the existence of multiple solutions.

Theorem 3.1. *Let hypotheses (g1)–(g2) be satisfied and assume the existence of a subsolution $\underline{u}$ and supersolution $\overline{u}$ of (1.1) such that $\underline{u} \le \overline{u}$. Then there exists a solution u of (1.1) with $\underline{u} \le u \le \overline{u}$.*

Proof. We introduce the following truncation related with the subsolution $\underline{u}$ and supersolution $\overline{u}$ of (1.1):

$$\tau(x,s) = \begin{cases} \underline{u}(x) & \text{if } s < \underline{u}(x) \\ s & \text{if } \underline{u}(x) \le s \le \overline{u}(x) \\ \overline{u}(x) & \text{if } s > \overline{u}(x). \end{cases}$$

The truncation $\tau : \Omega \times \mathbb{R} \to \mathbb{R}$ is a Carathéodory function, and Lipschitzian with respect to s. According to Definition 2.3 and Definition 2.4 the sub- and supersolution $\underline{u}$ and $\overline{u}$ satisfy the inequalities, respectively,

$$\int_\Omega |\nabla \underline{u}|^{p-2} \nabla \underline{u} \nabla \varphi \, dx + \int_\Omega \Big(\underline{\eta} - a(\underline{u}^+)^{p-1} + b(\underline{u}^-)^{p-1} \Big) \varphi \, dx \le 0, \tag{3.1}$$

and

$$\int_\Omega |\nabla \overline{u}|^{p-2} \nabla \overline{u} \nabla \varphi \, dx + \int_\Omega \Big(\overline{\eta} - a(\overline{u}^+)^{p-1} + b(\overline{u}^-)^{p-1} \Big) \varphi \, dx \ge 0, \tag{3.2}$$

for all $\varphi \in V_0 \cap L^p(\Omega)_+$, where $\underline{\eta} \in L^{q'}(\Omega)$ such that

$$\underline{\eta}(x) \in \partial G(x, \underline{u}(x)) \ \text{ for a.a. } x \in \Omega, \tag{3.3}$$

and $\overline{\eta} \in L^{q'}(\Omega)$ such that

$$\overline{\eta}(x) \in \partial G(x, \overline{u}(x)) \ \text{ for a.a. } x \in \Omega, \tag{3.4}$$

By means of $\overline{\eta}$ and $\underline{\eta}$ we introduce the following truncation of the nonlinearity $g : \Omega \times \mathbb{R} \to \mathbb{R}$:

$$g_\tau(x,s) := \begin{cases} \underline{\eta}(x) & \text{if } s < \underline{u}(x)\,, \\ g(x,s) & \text{for } \underline{u}(x) \le s \le \overline{u}(x)\,, \\ \overline{\eta}(x) & \text{when } s > \overline{u}(x)\,, \end{cases}$$

and define the functional E_τ by

$$E_\tau(u) = \frac{1}{p} \|\nabla u\|_p^p - \int_\Omega \int_0^{u(x)} \Big(a\tau^+(x,s)^{p-1} - b\tau^-(x,s)^{p-1} - g_\tau(x,s) \Big) \, ds \, dx,$$

where as before $\tau^+ \ge 0$ and $\tau^- \ge 0$ denote the positive and negative part of τ, respectively. Due to (g2) the functional $E_\tau : V_0 \to \mathbb{R}$ is locally Lipschitz continuous. Moreover, in view of the truncation involved this functional is bounded below, coercive and weakly lower semicontinuous such that its global minimizer exists which is denoted by u. We are going to show that u is a solution of (1.1) satisfying $\underline{u} \le u \le \overline{u}$ which proves the assertion. As u is a critical point of E_τ we have $0 \in \partial E_\tau(u)$ which results in

$$u \in V_0 : \quad -\Delta_p u = a\tau^+(x,u)^{p-1} - b\tau^-(x,u)^{p-1} - w \quad \text{in } V_0^*, \tag{3.5}$$

for some $w \in L^{q'}(\Omega)$ and such that $w(x) \in \partial G_\tau(x, u(x))$ almost everywhere in Ω, with

$$G_\tau(x,\xi) := \int_0^\xi g_\tau(x,t)dt\,, \quad (x,\xi) \in \Omega \times \mathbb{R}\,.$$

The proof is complete provided the critical point u satisfies $\underline{u} \le u \le \overline{u}$. This is because then $\tau^+(x,u(x)) = u^+(x)$, $\tau^-(x,u(x)) = u^-(x)$, and $\partial G_\tau(x,u(x)) \subset \partial G(x,u(x))$, which implies $w \in \partial G(x,u(x))$, and thus u solves (1.1). Let us show that $u \le \overline{u}$ holds. To this end we subtract (3.2) from (3.5) and apply the special test function $\varphi = (u - \overline{u})^+$ which yields

$$\begin{aligned}
&\int_\Omega (|\nabla u|^{p-2}\nabla u - |\nabla \overline{u}|^{p-2}\nabla \overline{u})\nabla(u-\overline{u})^+\,dx \\
&\le \int_\Omega a(\tau^+(x,u)^{p-1} - (\overline{u}^+)^{p-1})(u-\overline{u})^+\,dx \\
&\quad - \int_\Omega b(\tau^-(x,u)^{p-1} - (\overline{u}^-)^{p-1})(u-\overline{u})^+\,dx \\
&\quad - \int_\Omega (w - \overline{\eta})(u-\overline{u})^+\,dx.
\end{aligned} \tag{3.6}$$

The integrals on the right-hand side of (3.6) are all zero, because the integrands are easily seen to be zero. For example, the third integral on the right-hand side of (3.6) becomes zero, because in $\{x \in \Omega : u(x) > \overline{u}(x)\}$ we have $w(x) \in \partial G_\tau(x,u(x)) = \{\overline{\eta}(x)\}$, i.e., $w(x) = \overline{\eta}(x)$ for a.a. $x \in \{x \in \Omega : u(x) > \overline{u}(x)\}$. Thus from (3.6) we obtain

$$\int_\Omega (|\nabla u|^{p-2}\nabla u - |\nabla \overline{u}|^{p-2}\nabla \overline{u})\nabla(u-\overline{u})^+\,dx \le 0,$$

which results in $\nabla(u-\overline{u})^+ = 0$, and as $(u-\overline{u})^+ \in V_0$ we get $(u-\overline{u})^+ = 0$, i.e., $u \le \overline{u}$. Similarly, one shows $\underline{u} \le u$, which completes the proof. □

Let $\mathcal{S}$ denote the set of all solutions of (1.1) within the ordered interval $[\underline{u}, \overline{u}]$. In view of Theorem 3.1 we see that $\mathcal{S} \ne \emptyset$. A solution $u_* \in \mathcal{S}$ is called the *smallest solution* of $\mathcal{S}$ if for any element $u \in \mathcal{S}$ the inequality $u_* \le u$ holds. Likewise, u^* is called the *greatest solution* of $\mathcal{S}$ if $u \le u^*$ holds for all $u \in \mathcal{S}$. We say $\mathcal{S}$ possesses *extremal solutions* if $\mathcal{S}$ has a smallest and greatest solution. Arguing exactly as in the proofs of [7, Lemma 4.23] and [7, Corollary 4.24] we obtain the following result.

Theorem 3.2. *Let hypotheses (g1)–(g2) be satisfied and assume the existence of a subsolution $\underline{u}$ and supersolution $\overline{u}$ of (1.1) such that $\underline{u} \le \overline{u}$. Then there exist extremal solutions of (1.1) within $[\underline{u}, \overline{u}]$.*

4. Existence of Extremal Constant-Sign Solutions

Combining the results of Lemma 2.3, Lemma 2.4, Lemma 2.5 and Theorem 3.1 we immediately deduce the existence of nontrivial positive solutions of problem (1.1) provided the parameter a satisfies $a > \lambda_1$, and the existence of negative solutions of problem (1.1) provided the parameter b satisfies $b > \lambda_1$. Our main goal of this section is to show that problem (1.1) has a smallest positive solution $u_+ \in \operatorname{int}(C_0^1(\overline{\Omega})_+)$, and a greatest negative solution $u_- \in -\operatorname{int}(C_0^1(\overline{\Omega})_+)$. More precisely the following result will be shown.

Theorem 4.1. *Let hypotheses (g1)–(g4) be fulfilled. For every $a > \lambda_1$ and $b \in \mathbb{R}$ there exists a smallest positive solution $u_+ = u_+(a) \in \operatorname{int}(C_0^1(\overline{\Omega})_+)$ of (1.1) within the order interval $[0, \alpha_a e]$ with the constant $\alpha_a > 0$ as in Lemma 2.3. For every $b > \lambda_1$ and $a \in \mathbb{R}$ there exists a greatest negative solution $u_- = u_-(b) \in -\operatorname{int}(C_0^1(\overline{\Omega})_+)$ of (1.1) within the order interval $[-\beta_b e, 0]$ with the constant $\beta_b > 0$ as in Lemma 2.4.*

Proof. Let $a > \lambda_1$. Lemma 2.3 and Lemma 2.5 ensure that $\overline{u} = \alpha_a e \in \operatorname{int}(C_0^1(\overline{\Omega})_+)$ is a supersolution of problem (1.1) and $\underline{u} = \varepsilon\varphi_1 \in \operatorname{int}(C_0^1(\overline{\Omega})_+)$ is a subsolution of problem (1.1) provided $\varepsilon > 0$ is sufficiently small. We may choose $\varepsilon > 0$ such that, in addition, $\varepsilon\varphi_1 \le \alpha_a e$. Thus by Theorem 3.2 there exists a smallest and a greatest solution of (1.1) within the ordered interval $[\varepsilon\varphi_1, \alpha_a e]$. Let us denote the smallest solution by u_ε. Moreover, the nonlinear regularity theory for the p-Laplacian (cf., e.g., [17, Theorem 1.5.6]) and Vázquez's strong maximum principle [37] ensure that $u_\varepsilon \in \operatorname{int}(C_0^1(\overline{\Omega})_+)$. Thus for every positive integer n sufficiently large there is a smallest solution $u_n \in \operatorname{int}(C_0^1(\overline{\Omega})_+)$ of problem (1.1) within $[\frac{1}{n}\varphi_1, \alpha_a e]$. In this way we inductively construct a sequence (u_n) of smallest solutions which is monotone decreasing, i.e., we have

$$u_n \downarrow u_+ \text{ pointwise,} \tag{4.1}$$

with some function $u_+ : \Omega \to \mathbb{R}$ satisfying $0 \le u_+ \le \alpha_a e$.

Claim 1: u_+ is a solution of problem (1.1).

As $u_n \in \operatorname{int}(C_0^1(\overline{\Omega})_+)$ and u_n are solutions of (1.1) we have

$$\|\nabla u_n\|_p^p = \int_\Omega (a u_n(x)^p - \eta_n) u_n(x))\, dx,$$

where $\eta_n \in L^{q'}(\Omega)$ and $\eta_n(x) \in \partial G(x, u_n(x))$ for almost all $x \in \Omega$. Since $u_n \in [\frac{1}{n}\varphi_1, \alpha_a e]$, the last equation together with (g2) imply that the sequence (u_n) is bounded in V_0. Taking into account (4.1) we obtain that $u_+ \in V_0$ and

$$u_n \rightharpoonup u_+ \text{ in } V_0\,, \quad u_n \to u_+ \text{ in } L^p(\Omega) \text{ and a.e. in } \Omega. \tag{4.2}$$

The solution u_n of (1.1) satisfies

$$\langle -\Delta_p u_n, \varphi\rangle = \int_\Omega (a u_n^{p-1} - \eta_n))\varphi\, dx, \quad \forall \varphi \in V_0, \tag{4.3}$$

which yields with $\varphi = u_n - u_+$ in (4.3) the equation

$$\langle -\Delta_p u_n, u_n - u_+\rangle = \int_\Omega (a u_n^{p-1} - \eta_n))(u_n - u_+)\, dx.$$

Using the convergence properties (4.2) of (u_n) and (g2) as well as the uniform boundedness of the sequence (u_n), we get by applying Lebesgue's dominated convergence

$$\lim_{n\to\infty} \langle -\Delta_p u_n, u_n - u_+ \rangle = 0,$$

which by the S_+-property of $-\Delta_p$ on V_0 implies

$$u_n \to u_+ \ \text{ in } V_0 \text{ as } n \to \infty\,. \tag{4.4}$$

Since u_n are uniformly bounded, from (g2) we see that there exists a constant $c > 0$ such that

$$|\eta_n(x)| \le c \quad \text{a.e. in } \Omega, \forall\, n \in \mathbb{N}\,, \tag{4.5}$$

and thus we get (for some subsequence if necessary) $\eta_n \rightharpoonup \eta_+$ in $L^{q'}(\Omega)$. By the strong convergence (4.4), Lemma 2.1 can be applied to get $\eta_+(x) \in \partial G(x, u_+(x))$ for almost every $x \in \Omega$. Passing to the limit in (4.3) (for some subsequence if necessary) proves Claim 1.

As u_+ belongs, in particular, to $L^\infty(\Omega)$, Claim 1 and assumption (g2) implies $\Delta_p u_+ \in L^\infty(\Omega)$. The nonlinear regularity theory (cf., e.g., Theorem 1.5.6 in [17]) ensures that $u_+ \in C^{1,\gamma}(\overline{\Omega})$ for some $\gamma \in (0,1)$, so $u_+ \in C_0^1(\overline{\Omega})$. In view of (g2), (g3) a constant $\tilde{c}_a > 0$ can be found such that

$$|g(x,s)| \le \tilde{c}_a s^{p-1} \ \text{ for a.a. } x \in \Omega \text{ and all } 0 \le s \le \alpha_a \|e\|_\infty, \tag{4.6}$$

which yields in conjunction with Claim 1 that

$$\Delta_p u_+ \le \tilde{c}_a u_+^{p-1}. \tag{4.7}$$

We now apply Vázquez's strong maximum principle [37] where in its statement the function β is chosen as $\beta(s) = \tilde{c}_a s^{p-1}$ for all $s > 0$, which is possible because $\int_{0^+} \frac{1}{(s\beta(s))^{\frac{1}{p}}} ds = +\infty$. This result guarantees that if $u_+ \neq 0$, then $u_+ > 0$ in Ω and $\frac{\partial u_+}{\partial n} < 0$ on $\partial\Omega$ which means that $u_+ \in \operatorname{int}(C_0^1(\overline{\Omega})_+)$.

Claim 2: $u_+ \in \operatorname{int}(C_0^1(\overline{\Omega})_+)$.

Suppose Claim 2 does not hold. Then by Vázquez's strong maximum principle we must have $u_+ = 0$, and thus the sequence (u_n) satisfies

$$u_n(x) \downarrow 0 \ \text{ for all } x \in \Omega\,. \tag{4.8}$$

Setting

$$\tilde{u}_n = \frac{u_n}{\|\nabla u_n\|_p} \quad \text{for all } n,$$

we may suppose that along a relabelled subsequence one has

$$\tilde{u}_n \rightharpoonup \tilde{u} \ \text{ in } V_0\,, \ \ \tilde{u}_n \to \tilde{u} \ \text{ in } L^p(\Omega) \text{ and a.e in } \Omega, \tag{4.9}$$

with some $\tilde{u} \in V_0$, and there is a function $w \in L^p(\Omega)_+$ such that

$$|\tilde{u}_n(x)| \le w(x) \quad \text{for almost all } x \in \Omega. \tag{4.10}$$

Since u_n are positive solutions of (1.1) we get for $\tilde{u}_n$ the following variational equation:

$$\langle -\Delta_p \tilde{u}_n, \varphi \rangle = a \int_\Omega \tilde{u}_n^{p-1} \varphi \, dx - \int_\Omega \frac{\eta_n}{u_n^{p-1}} \tilde{u}_n^{p-1} \varphi \, dx, \quad \forall \varphi \in V_0. \tag{4.11}$$

With the special test function $\varphi = \tilde{u}_n - \tilde{u}$ in (4.11) we obtain

$$\langle -\Delta_p \tilde{u}_n, \tilde{u}_n - \tilde{u} \rangle = a \int_\Omega \tilde{u}_n^{p-1} (\tilde{u}_n - \tilde{u}) \, dx - \int_\Omega \frac{\eta_n}{u_n^{p-1}} \tilde{u}_n^{p-1} (\tilde{u}_n - \tilde{u}) \, dx. \tag{4.12}$$

From (4.6) and (4.10) we get the estimate

$$\frac{|\eta_n(x))|}{u_n(x)^{p-1}} \tilde{u}_n^{p-1}(x) |\tilde{u}_n(x) - \tilde{u}(x)| \leq \tilde{c}_a w(x)^{p-1} (w(x) + |\tilde{u}(x)|) \text{ for a.a. } x \in \Omega .$$

As the right-hand side of the last inequality is in $L^1(\Omega)$, we may apply Lebesgue's dominated convergence theorem, which in conjunction with (4.9) yields

$$\lim_{n \to \infty} \int_\Omega \frac{\eta_n}{u_n^{p-1}} \tilde{u}_n^{p-1} (\tilde{u}_n - \tilde{u}) \, dx = 0.$$

From (4.9) and (4.12) we conclude

$$\lim_{n \to \infty} \langle -\Delta_p \tilde{u}_n, \tilde{u}_n - \tilde{u} \rangle = 0,$$

which in view of the S_+-property of $-\Delta_p$ on V_0 results in

$$\tilde{u}_n \to \tilde{u} \text{ in } V_0 \text{ as } n \to \infty , \tag{4.13}$$

and therefore, in particular, $\|\nabla \tilde{u}\|_p = 1$. Taking into account (g3), (4.8), and (4.13), we may pass to the limit in (4.11) which results in

$$\langle -\Delta_p \tilde{u}, \varphi \rangle = a \int_\Omega \tilde{u}^{p-1} \varphi \, dx, \quad \forall \varphi \in V_0 \tag{4.14}$$

As $\tilde{u} \neq 0$, relation (4.14) expresses the fact that $\tilde{u} \geq 0$ is an eigenfunction of $(-\Delta_p, V_0)$ corresponding to the eigenvalue a. As $a > \lambda_1$ this is impossible according to Anane [3], because $\tilde{u}$ must change sign. This contradiction proves that the Claim 2 holds true.

Claim 3: $u_+ \in \text{int}\,(C_0^1(\overline{\Omega})_+)$ is the smallest positive solution of (1.1) in $[0, \alpha_a\, e]$.

We already know that $u_+ \in [0, \alpha_a\, e]$. Assume $u \in V_0$ is any positive solution of (1.1) belonging to $[0, \alpha_a\, e]$. Since $u \in L^\infty(\Omega)$, then by (1.1) and (g3) we deduce $\Delta_p u \in L^\infty(\Omega)$. Using [17, Theorem 1.56] we derive $u \in C_0^1(\overline{\Omega})$, and applying Vázquez's strong maximum principle [37] we infer $u \in \text{int}\,(C_0^1(\overline{\Omega})_+)$, which yields $u \in [\frac{1}{n}\varphi_1, \alpha_a\, e]$ for n sufficiently large. This in conjunction with the fact that u_n is the least solution of (1.1) in $[\frac{1}{n}\varphi_1, \alpha_a\, e]$ ensures $u_n \leq u$ if n is large enough. In view of (4.1), we obtain $u_+ \leq u$, which proves Claim 3.

The proof of the existence of the greatest negative solution $u_- = u_-(b) \in -\text{int}\,(C_0^1(\overline{\Omega})_+)$ of (1.1) within the ordered interval $[-\beta_b\, e, 0]$ can be done in a similar way This completes the proof of the theorem. □

5. Variational Characterization of Extremal Solutions

Let hypotheses (g1)–(g4) be satisfied throughout this section. Theorem 4.1 ensures the existence of extremal positive and negative solutions of (1.1) for all $a > \lambda_1$ and $b > \lambda_1$ denoted by $u_+ = u_+(a) \in \operatorname{int}(C_0^1(\overline{\Omega})_+)$ and $u_- = u_-(b) \in -\operatorname{int}(C_0^1(\overline{\Omega})_+)$, respectively. In this section we are going to characterize these extremal solutions as global (local) minimizers of certain locally Lipschitz functionals that are generated by truncation procedures. To this end let us introduce truncation functions related with the extremal solutions u_+ and u_- as follows:

$$\tau_+(x,s) = \begin{cases} 0 & \text{if } s < 0 \\ s & \text{if } 0 \le s \le u_+(x) \\ u_+(x) & \text{if } s > u_+(x) \end{cases}, \quad \tau_-(x,s) = \begin{cases} u_-(x) & \text{if } s < u_-(x) \\ s & \text{if } u_-(x) \le s \le 0 \\ 0 & \text{if } s > 0 \end{cases},$$

$$\tau_0(x,s) = \begin{cases} u_-(x) & \text{if } s < u_-(x) \\ s & \text{if } u_-(x) \le s \le u_+(x) \\ u_+(x) & \text{if } s > u_+(x). \end{cases}$$

The truncations $\tau_+, \tau_-, \tau_0 : \Omega \times \mathbb{R} \to \mathbb{R}$ are continuous, uniformly bounded, and Lipschitzian with respect to s. The extremal positive and negative solutions u_+ and u_- of (1.1), respectively, ensured by Theorem 4.1 satisfy

$$-\Delta_p u_+ = a(u_+)^{p-1} - \eta_+, \quad -\Delta_p u_- = -b(u_-)^{p-1} - \eta_- \ \text{ in } V_0^*, \tag{5.1}$$

where $\eta_+, \eta_- \in L^{q'}(\Omega)$ such that

$$\eta_+(x) \in \partial G(x, u_+(x)), \quad \eta_-(x) \in \partial G(x, u_-(x))$$

for almost every $x \in \Omega$. By means of η_+, η_- we introduce the following truncations of the nonlinearity $g : \Omega \times \mathbb{R} \to \mathbb{R}$:

$$g_+(x,s) := \begin{cases} 0 & \text{if } s < 0, \\ g(x,s) & \text{for } 0 \le s \le u_+(x), \\ \eta_+(x) & \text{when } s > u_+(x), \end{cases}$$

$$g_-(x,s) := \begin{cases} \eta_-(x) & \text{if } s < u_-(x), \\ g(x,s) & \text{for } u_-(x) \le s \le 0, \\ 0 & \text{when } s > 0, \end{cases}$$

$$g_0(x,s) := \begin{cases} \eta_-(x) & \text{if } s < u_-(x), \\ g(x,s) & \text{for } u_-(x) \le s \le u_+(x), \\ \eta_+(x) & \text{when } s > u_+(x), \end{cases}$$

and define functionals E_+ , E_- , E_0 by

$$E_+(u) = \frac{1}{p}\|\nabla u\|_p^p - \int_\Omega \int_0^{u(x)} \Big(a\tau_+(x,s)^{p-1} - g_+(x,s)\Big)\, ds\, dx,$$

$$E_-(u) = \frac{1}{p}\|\nabla u\|_p^p + \int_\Omega \int_0^{u(x)} \Big(b|\tau_-(x,s)|^{p-1} + g_-(x,s)\Big)\, ds\, dx,$$

$$E_0(u) = \frac{1}{p}\,\|\nabla u\|_p^p - \int_\Omega \int_0^{u(x)} \Big(a\tau_+(x,s)^{p-1} - b|\tau_-(x,s)|^{p-1} - g_0(x,s)\Big)\, ds\, dx.$$

Due to (g2) the functionals E_+ , E_- , E_0 : $V_0 \to \mathbb{R}$ are locally Lipschitz continuous. Moreover, in view of the truncations involved these functionals are bounded below, coercive and weakly lower semicontinuous such that their global minimizers exist. The following lemma provides a characterization of the critical points of these functionals.

Lemma 5.1. *Let u_+ and u_- be the extremal constant-sign solutions of (1.1). Then the following holds:*

(i) A critical point $v \in V_0$ of E_+ is a (nonnegative) solution of (1.1) satisfying $0 \le v \le u_+$.

(ii) A critical point $w \in V_0$ of E_- is a (nonpositive) solution of (1.1) satisfying $u_- \le w \le 0$.

(iii) A critical point $z \in V_0$ of E_0 is a solution of (1.1) satisfying $u_- \le z \le u_+$.

Proof. To prove (i) let v be a critical point of E_+, i.e., $0 \in \partial E_+(v)$, which results in

$$v \in V_0 : \quad -\Delta_p v = a\tau_+(x,v)^{p-1} - w \quad \text{in } V_0^*, \tag{5.2}$$

for some $w \in L^{q'}(\Omega)$ and such that $w(x) \in \partial G_+(x, v(x))$ almost everywhere in Ω, with

$$G_+(x,\xi) := \int_0^\xi g_+(x,t)dt\,, \quad (x,\xi) \in \Omega \times \mathbb{R}\,.$$

Let us show that $v \le u_+$ holds. As u_+ is a positive solution of (1.1) it satisfies the first equation in (5.1), and by subtracting that equation from (5.2) and applying the special test function $\varphi = (v - u_+)^+$ we get

$$\begin{aligned}&\int_\Omega (|\nabla v|^{p-2}\nabla v - |\nabla u_+|^{p-2}\nabla u_+)\nabla(v-u_+)^+\, dx\\ &= \int_\Omega [a(\tau_+(x,v)^{p-1} - u_+^{p-1}) - (w - \eta_+)](v-u_+)^+\, dx.\end{aligned} \tag{5.3}$$

By the definition of the truncations introduced above we have $\tau_+(x,v)^{p-1}(x) - u_+^{p-1}(x) = 0$ and $w(x) - \eta_+(x) = 0$ for a.a. $x \in \{v > u_+\}$, and thus the right-hand side of (5.3) is zero which leads to

$$\int_\Omega (|\nabla v|^{p-2}\nabla v - |\nabla u_+|^{p-2}\nabla u_+)\nabla(v-u_+)^+\, dx = 0,$$

and hence it follows $\nabla(v-u_+)^+ = 0$. Because $(v-u_+)^+ \in V_0$ this implies $(v-u_+)^+ = 0$ which proves $v \le u_+$. To prove $0 \le v$ we test (5.2) with $\varphi = v^- = \max\{-v, 0\} \in V_0$ and get

$$\int_\Omega |\nabla v|^{p-2}\nabla v \nabla v^-\, dx = \int_\Omega (a\tau_+(x,v)^{p-1} - w)v^-\, dx = 0, \tag{5.4}$$

which results in $\|\nabla v^-\|_p^p = 0$, and thus $v^- = 0$, i.e., $0 \le v$. Thus the critical point v of E_+ which is a solution of the Dirichlet problem (5.2) satisfies $0 \le v \le u_+$, and therefore $\tau_+(x, v) = v$. Because $\partial G_+(x, v(x)) \subset \partial G(x, v(x))$ it follows $w \in \partial G(x, v(x))$, and therefore v must be a solution of (1.1). This proves (i). In just the same way one can prove also (ii) and (iii) which is omitted. □

The following lemma provides a variational characterization of the extremal constant-sign solutions u_+ and u_-

Lemma 5.2. *Let $a > \lambda_1$ and $b > \lambda_1$. Then the extremal positive solution u_+ of (1.1) is the unique global minimizer of the functional E_+, and the extremal negative solution u_- of (1.1) is the unique global minimizer of the functional E_-. Both u_+ and u_- are local minimizers of E_0.*

Proof. The functional $E_+ : V_0 \to \mathbb{R}$ is bounded below, coercive and weakly lower semi-continuous. Thus there exists a global minimizer $v_+ \in V_0$ of E_+, i.e.,

$$E_+(v_+) = \inf_{u \in V_0} E_+(u)\,, \tag{5.5}$$

As v_+ is a critical point of E_+, so by Lemma 5.1 it is a nonnegative solution of (1.1) satisfying $0 \le v_+ \le u_+$. Since $a - \lambda_1 > 0$ there is a $\nu_a > 0$ such that $a - \lambda_1 - \nu_a > 0$. By (g3) we infer the existence of a $\tilde{\delta}_a > 0$ such that

$$|g(x, s)| \le (a - \lambda_1 - \nu_a)\, s^{p-1}, \quad \forall\, s : 0 < s \le \tilde{\delta}_a,$$

and thus for $\varepsilon > 0$ sufficiently small such that

$$\varepsilon\varphi_1 \le u_+\,, \quad \varepsilon\|\varphi_1\|_{L^\infty(\Omega)} \le \tilde{\delta}_a\,,$$

we obtain (note $\|\varphi_1\|_p = 1$)

$$\begin{aligned} E_+(v_+) \le E_+(\varepsilon\varphi_1) &= \frac{\lambda_1}{p}\varepsilon^p + \int_\Omega \int_0^{\varepsilon\varphi_1(x)} (-a\, s^{p-1} + g(x, s))\, ds\, dx \\ &\le \frac{\lambda_1}{p}\varepsilon^p + \int_\Omega \int_0^{\varepsilon\varphi_1(x)} (-\lambda_1 - \nu_a)\, s^{p-1}\, ds\, dx < 0. \end{aligned}$$

Hence it follows that $E_+(v_+) < 0$, and thus $v_+ \ne 0$. Applying nonlinear regularity theory for the p-Laplacian (cf., e.g., [17, Theorem 1.5.6]) and Vázquez's strong maximum principle we see that $v_+ \in \operatorname{int}(C_0^1(\overline{\Omega})_+)$. As u_+ is the smallest positive solution of (1.1) in $[0, \alpha_a e]$ and $0 \le v_+ \le u_+$, it follows $v_+ = u_+$, which shows that the global minimizer v_+ must be unique and equal to u_+. By similar arguments one can show that the global minimizer v_- of E_- must be unique and $v_- = u_-$. It remains to prove that u_+ and u_- are local minimizers of E_0. Let us show this last assertion for u_+ only. By definition we have

$$E_+(u) = E_0(u) \quad \text{for all } u \in V_0 \text{ with } u \ge 0.$$

Since u_+ is a global minimizer of E_+ and $u_+ \in \operatorname{int}(C_0^1(\overline{\Omega})_+)$, it follows that u_+ is a local minimizer of E_0 with respect to the C^1 topology. Due to a result by Motreanu and Papageorgiou in [25, Proposition 4], we conclude that u_+ is also a local minimizer of E_0 with respect to the V_0 topology. This completes the proof of the lemma. □

Lemma 5.3. *The functional $E_0 : V_0 \to \mathbb{R}$ has a global minimizer v_0 which is a nontrivial solution of (1.1) satisfying $u_- \le v_0 \le u_+$.*

Proof. Apparently $E_0 : V_0 \to \mathbb{R}$ is coercive and weakly lower semicontinuous, and thus a global minimizer v_0 exists which is a critical point of E_0. Apply Lemma 5.1 (iii) and note that, e.g., $E_0(u_+) = E_+(u_+) < 0$, which shows that $v_0 \neq 0$. □

6. Existence of Sign-Changing Solutions

Theorem 4.1 ensures the existence of a smallest positive solution $u_+ \in \text{int}\,(C_0^1(\overline{\Omega})_+)$ in $[0, \alpha_a\, e]$ and a greatest negative solution $u_- \in -\text{int}\,(C_0^1(\overline{\Omega})_+)$ of (1.1) in $[-\beta_b\, e, 0]$. This result in conjunction with variational and topological arguments will be used to prove the existence of sign-changing solutions in the following theorem. The idea is to show the existence of a nontrivial solution u_0 of (1.1) satisfying $u_- \le u_0 \le u_+$ with $u_0 \neq u_-$ and $u_0 \neq u_+$, which then must be sign-changing, because u_+ and u_- are the extremal constant-sign solutions.

Theorem 6.1. *Let hypotheses (g1)–(g4) and (H) be satisfied. Then problem (1.1) has a smallest positive solution $u_+ \in \text{int}\,(C_0^1(\overline{\Omega})_+)$ in $[0, \alpha_a\, e]$, a greatest negative solution $u_- \in -\text{int}\,(C_0^1(\overline{\Omega})_+)$ in $[-\beta_b\, e, 0]$, and a nontrivial sign-changing solution $u_0 \in C_0^1(\overline{\Omega})$ with $u_- \le u_0 \le u_+$.*

Proof. Clearly the existence of the extremal positive and negative solution u_+ and u_- follows from Theorem 4.1, because (H), in particular, implies that $a > \lambda_1$ and $b > \lambda_1$. As for the existence of a sign-changing solution we first note that by Lemma 5.3 it follows that the global minimizer v_0 of E_0 is a nontrivial solution of (1.1) satisfying $u_- \le v_0 \le u_+$. Therefore, if $v_0 \neq u_+$ and $v_0 \neq u_-$ then $v_0 = u_0$ must be a sign-changing solution as asserted, because u_- is the greatest negative and u_+ is the smallest positive solution of (1.1). Thus, we still need to prove the existence of sign-changing solutions in case that either $v_0 = u_-$ or $v_0 = u_+$.

Let us consider the case $v_0 = u_+$ only, since the case $v_0 = u_-$ can be treated quite similarly. By Lemma 5.2 u_- is a local minimizer of E_0. Without loss of generality we may even assume that u_- is a strict local minimizer of E_0, because on the contrary we would find (infinitely many) critical points z of E_0 that are sign-changing solutions thanks to $u_- \le z \le u_+$ and the extremality of the solutions u_-, u_+ obtained in Theorem 4.1 which proves the assertion.

Therefore, it remains to prove the existence of sign-changing solutions under the assumptions that the global minimizer v_0 of E_0 is equal to u_+, and u_- is a strict local minimizer of E_0. This implies the existence of $\rho \in (0, \|u_- - u_+\|)$ such that

$$E_0(u_+) \le E_0(u_-) < \inf\{E_0(u) \,:\, u \in \partial B_\rho(u_-)\}, \tag{6.1}$$

where $\partial B_\rho(u_-) = \{u \in V_0 \,:\, \|u - u_-\| = \rho\}$. The functional E_0 satisfies the Palais-Smale condition(PS), because it is bounded below, locally Lipschitz continuous, and coercive; see e.g. [28, Corollary 2.4]. Thus in view of (6.1) we may apply Theorem 2.1 which ensures the existence of a critical point $u_0 \in V_0$ satisfying $0 \in \partial E_0(u_0)$ and

$$\inf\{E_0(u) \,:\, u \in \partial B_\rho(u_-)\} \le E_0(u_0) = \inf_{\gamma \in \Gamma} \max_{t \in [-1,1]} E_0(\gamma(t)), \tag{6.2}$$

where

$$\Gamma = \{\gamma \in C([-1,1], V_0) : \gamma(-1) = u_-, \gamma(1) = u_+\}.$$

It is clear from (6.1) and (6.2) that $u_0 \neq u_-$ and $u_0 \neq u_+$, and thus u_0 is a sign-changing solution provided $u_0 \neq 0$. To prove the latter we claim

$$E_0(u_0) < 0 \tag{6.3}$$

for which it suffices to construct a path $\hat{\gamma} \in \Gamma$ such that

$$E_0(\hat{\gamma}(t)) < 0 \text{ for all } t \in [-1,1]. \tag{6.4}$$

The construction of such a path $\hat{\gamma}$ can be done by appropriately modifying the approach in [9] and applying the Second Deformation Lemma for locally Lipschitz functionals as it can be found in [13, Theorem 2.10]. This completes the proof. □

Remarks 6.1. The multiplicity results and the existence of sign-changing solutions obtained in the preceding sections generalize very recent results due to the authors obtained in [9, 4]. Moreover, if the function $t \mapsto g(x,t)$ is continuous on $\mathbb{R}$ and $a = b = \lambda$, then $\partial G(x,\xi) = \{g(x,\xi)\}$, and problem (1.1) reduces to

$$u \in V_0 : \quad -\Delta_p u = \lambda |u|^{p-2} u - g(x,u) \text{ in } V_0^*. \tag{6.5}$$

However, even in this setting the results obtained here are more general than [32, Theorem 3.9], because we do not assume that $g(x,t)t \geq 0$ for all $t \in \mathbb{R}$.

Remarks 6.2. Theorem 6.1 improves also Corollary 3.2 of [26]. In fact, let $p = 2$, let $u \in V_0$ be a solution of (1.1) in case $a = b = \lambda$ and $g(x,t) \equiv g(t)$, $(x,t) \in \Omega \times \mathbb{R}$ with $\eta \in L^{q'}(\Omega)$ satisfying $\eta(x) \in \partial G(u(x))$. By definition of Clarke's gradient we have, for any $\varphi \in V_0$,

$$\eta(x)\varphi(x) \leq G^0(u(x); \varphi(x)) \quad \text{a.e. in } \Omega .$$

As u is a solution the following holds: $u \in V_0$ and $(p = 2)$

$$\int_\Omega \nabla u \nabla \varphi \, dx = \lambda \int_\Omega u\varphi, dx - \int_\Omega \eta\varphi, dx,$$

which yields

$$-\int_\Omega \nabla u \nabla \varphi \, dx + \lambda \int_\Omega u\varphi \, dx = \int_\Omega \eta\varphi \, dx \leq \int_\Omega G^0(u; \varphi \, dx, \quad \forall \varphi \in V_0,$$

i.e., u turns out to be a solution of the hemivariational inequality studied in [26]. Since the hypotheses of [26, Corollary 3.2] imply (g1)–(g4), the assertion follows.

References

[1] A. Ambrosetti and D. Lupo, On a class of nonlinear Dirichlet problems with multiple solutions, *Nonlinear Anal.* **8** (1984), 1145–1150.

[2] A. Ambrosetti and G. Mancini, Sharp nonuniqueness results for some nonlinear problems, Nonlinear Anal. **3** (1979), 635–645.

[3] A. Anane, Etude des valeurs propres et de la résonance pour l'opérateur p-Laplacien, *C. R. Acad. Sci. Paris Sér. I Math.,* **305**(16) (1987), 725–728.

[4] D. Averna, S.A. Marano, and D. Motreanu, Multiple solutions for a Dirichlet problem with p-Laplacian and set-valued nonlinearity, *Bull. Australian Math. Soc.* (to appear).

[5] T. Bartsch and Z. Liu, Multiple sign changing solutions of a quasilinear elliptic eigenvalue problem involving the p-Laplacian, *Commun. Contemp. Math.* **6** (2004), 245–258.

[6] H. Brezis and L. Nirenberg, H^1 versus C^1 local minimizers, *C.R. Acad. Sci. Paris Sér. I Math.,* **317** (1993), 465–472

[7] S. Carl, Vy K. Le, and D. Motreanu, Nonsmooth Variational Problems and Their Inequalities.Comparison Principles and Applications, *Springer Monogr. Math.*, Springer, New York, 2007.

[8] S. Carl and D. Motreanu, Constant-sign and sign-changing solutions for nonlinear eigenvalue problems, *Nonlinear Anal.* (2007), doi: 10.1016/j.na.2007.02.013.

[9] S. Carl and D. Motreanu, Sign-changing and extremal constant-sign solutions of nonlinear elliptic problems with supercritical nonlinearities, *Commun. Appl. Nonlinear Anal.* **14** (2007), 85–100.

[10] S. Carl and K. Perera, Sign-changing and multiple solutions for the p-Laplacian, *Abstract Appl. Anal.* **7** (2002), 613–625.

[11] K.-C. Chang, Variational methods for non-differentiable functionals and their applications to partial differential equations, *J. Math. Anal. Appl.* **80** (1981), 102–129.

[12] F. H. Clarke, Optimization and Nonsmooth Analysis, *Classics Appl. Math.* **5**, SIAM, Philadelphia, 1990.

[13] J.-N. Corvellec, Morse theory for continuous functionals, *J. Math. Anal. Appl.* **196** (1995), 1050–1072.

[14] M. Cuesta, D. de Figueiredo, and J.-P. Gossez, The beginning of the Fučik spectrum for the p-Laplacian, *J. Differential Equations* **159** (1999), 212–238.

[15] N. Dancer and K. Perera, Some remarks on the Fučik spectrum of the p-Laplacian and critical groups, *J. Math. Anal. Appl.* **254** (2001), 164–177.

[16] J.P. Garcia Azorero, I. Peral Alonso, and J.J. Manfredi, Sobolev versus Hölder local minimizers and global multiplicity for some quasilinear elliptic equations, *Commun. Contemp. Math.* **6** (2000), 385–404.

[17] L. Gasiński and N. S. Papageorgiou, *Nonsmooth critical point theory and nonlinear boundary value problems*, Chapman & Hall/CRC, Boca Raton, FL, 2005.

[18] L. Gasiński and N. S. Papageorgiou, *Topics in Nonlinear Analysis,* Chapman & Hall/CRC, Boca Raton, FL, 2006.

[19] Q. Jiu and J. Su, Existence and multiplicity results for Dirichlet problems with p-Laplacian, *J. Math. Anal. Appl.* **281** (2003), 587–601.

[20] E. Koizumi and K. Schmitt, Ambrosetti-Prodi-type problems for quasilinear elliptic problems, *Differential Integral Equations,* **18** (2005), 241–262.

[21] C. Li and S. Li, Multiple solutions and sign-changing solutions of a class of nonlinear elliptic equations with Neumann boundary condition, *J. Math. Anal. Appl.* **298** (2004), 14–32.

[22] P. Lindqvist, On the equation div $(|\nabla u|^{p-2}\nabla u) + \lambda|u|^{p-2}u = 0$, *Proc. Amer. Math. Soc.*, **109**(1)(1990), 157–164, *Addendum: Proc. Amer. Math. Soc.* **116**(2) (1992), 583–584.

[23] S. Liu, Multiple solutions for coercive p-Laplacian equations, *J. Math. Anal. Appl.* **316** (2006), 229–236.

[24] Z. Liu, Multiplicity of solutions of semilinear elliptic boundary value problems with jumping nonlinearities at zero, *Nonlinear Anal.* **48** (2002), 1051–1063.

[25] D. Motreanu and N.S. Papageorgiou, Multiple solutions for nonlinear elliptic equations at resonance with a nonsmooth potential, *Nonlinear Anal.* **56** (2004), 1211–1234.

[26] S. A. Marano, G. Molica Bisci, and D. Motreanu, Multiple solutions for a class of elliptic hemivariational inequalities, *J. Math. Anal. Appl.* (2007), doi: 10.1016/j.jmaa.2007.03.077.

[27] S. A. Marano and N. S. Papageorgiou, On some elliptic hemivariational and variational-hemivariational inequalities, *Nonlinear Anal.* **62** (2005), 757–774.

[28] D. Motreanu, V. V. Motreanu, and D. Pasca, A version of Zhong's coercivity result for a general class of nonsmooth functionals, *Abstract Appl. Anal.* **7** (2002), 601–612.

[29] D. Motreanu and P. D. Panagiotopoulos, Minimax Theorems and Qualitative Properties of the Solutions of Hemivariational Inequalities, *Nonconvex Optim. Appl.* **29**, Kluwer, Dordrecht, 1998.

[30] D. Motreanu and N. S. Papageorgiou, Multiple solutions for nonlinear elliptic equations at resonance with a nonsmooth potential, *Nonlinear Anal.* **56** (2004), 1211–1234.

[31] D. Motreanu and V. Radulescu, *Variational and Non-Variational Methods in Nonlinear Analysis and Boundary Value Problems, Nonconvex Optim. Appl.* vol. 67, Kluwer, Dordrecht, 2003.

[32] E.H. Papageorgiou and N.S. Papageorgiou, A multiplicity theorem for problems with the p-Laplacian, *J. Funct. Anal.* **244** (2007), 63–77.

[33] A. Qian and S. Li, Multiple sign-changing solutions of an elliptic eigenvalue problem, *Discrete Contin. Dyn. Syst.* **12** (2005), 737–746.

[34] P. H. Rabinowitz, Minimax Methods in Critical Point Theory with Applications to Differential Equations, *CBMS Regional Conference Series in Mathematics,* vol. **65**, American Mathematical Society, Providence, RI, 1986.

[35] M. Struwe, A note on a result of Ambrosetti and Mancini, *Ann. Mat. Pura Appl.* **131** (1982), 107–115.

[36] M. Struwe, Variational Methods. *Applications to Nonlinear Partial Differential Equations and Hamiltonian Systems*, 2nd edn, Springer-Verlag, Berlin, 1996.

[37] J. Vázquez, A strong maximum principle for some quasilinear elliptic equations, *Appl. Math. Optim.,* **12**(3) (1984), 191–202.

[38] Z. Zhang and S. Li, On sign-changing and multiple solutions of the p-Laplacian, *J. Functional Analysis* **197** (2003), 447–468.

[39] Z. Zhang, J Chen, and S. Li, Construction of pseudo-gradient vector field and sign-changing multiple solutions involving p-Laplacian, *J. Differential Equations* **201** (2004), 287–303.

[40] Z. Zhang and X. Li, Sign-changing solutions and multiple solutions theorems for semilinear elliptic boundary value problems with a reaction term nonzero at zero, *J. Differential Equations* **178** (2002), 298–313.

In: Nonlinear Analysis Research Trends
Editor: Inès N. Roux, pp. 257-279
ISBN 978-1-60456-358-0

Chapter 8

WAVE EQUATIONS AND REACTION–DIFFUSION EQUATIONS WITH TWO NONLINEAR SOURCE TERMS OF DIFFERENT SIGN

Xu Runzhang[1,2]***, Zhang Wenying***[1,2] ***and Wu Weining***[3]
[1]College of Science, Harbin Engineering University,
Harbin 150001, People's Republic of China
[2]College of Automation, Harbin Engineering University,
Harbin 150001, People's Republic of China
[3]College of Computer Science and Technology, Harbin Engineering University,
Harbin 150001, People's Republic of China

Abstract

In this paper we study the initial boundary value problem of wave equations and reaction-diffusion equations with two nonlinear source terms of different sign. By introducing a family of potential wells W_δ and corresponding outside sets V_δ of W_δ we first obtain the invariant sets and vacuum isolating of solutions. Then we get the threshold result of global existence and nonexistence of solutions. Finally we prove the global existence of solutions for the problem with critical initial conditions .

Key Words: semilinear wave equations, reaction-diffusion equations, potential wells, global existence, nonexistence, vacuum isolating

1. Introduction

As is well known the most important models of semilinear wave equations are

$$u_{tt} - \Delta u = -|u|^{p-1}u$$

and

$$u_{tt} - \Delta u = |u|^{p-1}u.$$

In each physical system described by one of above equations, there is only one source, where in first equation the sign of source term and u are different and in second equation the sign of source term and u are the same. There are a lot of results [1-14] on the global existence and nonexistence of solutions for above equations and their generalizations. Recently in [10] by introducing a family of potential wells, the initial boundary value problem for above second equation was studied. And some new results on global existence of solutions and the invariant sets were obtained. In particular,the vacuum isolating behaviors of solutions were discovered in [10].

If in a physical system there are two nonlinear source of different sign, then the following wave equations with two nonlinear source terms of different sign should be considered

$$u_{tt} - \Delta u = f(u) \equiv a|u|^{p-1}u - b|u|^{q-1}u, \quad x \in \Omega,\ t > 0, \tag{1.1}$$

where $p \neq q$,$p > 1$,$q > 1$, a and b are two positive constants. To our knowledge, equation (1.1) can not be included by any known results. For example, in [14] Tsutsumi studied the global existence and finite time blow up of solutions for the following semilinear hyperbolic equations

$$u_{tt} + Au = f(u).$$

He required $f(u)$ to be homogeneous of degree $p > 2$, i.e.

$$f(\lambda u) = \lambda^p f(u), \quad \forall \lambda > 0.$$

Clearly $f(u)$ in equation (1.1) does not satisfy this assumption. In [4] Glassey studied the finite time blow up of solutions for semilinear hyperbolic equations

$$u_{tt} - \Delta u = f(u),$$

where $f(u)$ must satisfy the condition: there exists a convex function $g(u)$ such that $f(u) \geq g(u)$ for $-\infty < u < \infty$. Hence this condition is not applicable for $f(u)$ in equation (1.1). In [11] Payne and Sattinger also studied the finite time blow up of solutions for above semilinear hyperbolic equations by using potential well method. Where $f(u)$ must satisfy the condition: a) $f(u)$ is monotone and convex for $u > 0$, concave for $u < 0$; or b) $f(u)$ is convex for $-\infty < u < \infty$. Clearly $f(u)$ in equation (1.1) does not satisfy any one of above two conditions, even if for $u > 0$.

In this paper we study the initial boundary value problem of equation (1.1) with initial boundary conditions

$$u(x,0) = u_0(x),\ u_t(x,0) = u_1(x), \quad x \in \Omega, \tag{1.2}$$

$$u(x,t) = 0, \quad x \in \partial\Omega,\ t \geq 0 \tag{1.3}$$

and the initial boundary value problem of reaction-diffusion equation with two nonlinear source terms of different sign

$$u_t - \Delta u = f(u) \equiv a|u|^{p-1}u - b|u|^{q-1}u, \quad x \in \Omega,\ t > 0 \tag{1.4}$$

with

$$u(x,0) = u_0(x), \quad x \in \Omega \tag{1.5}$$

and (1.3), where $\Omega \subset \mathbf{R}^n$ is a bounded domain, $a > 0$, $b > 0$, p and q satisfy

(H) $1 < q < p < \infty$ if $n = 1, 2$; $1 < q < p \le \dfrac{n+2}{n-2}$ if $n \ge 3$.

First we introduce a family of potential wells W_δ and corresponding outside sets V_δ of W_δ and give a series of the properties of W_δ and V_δ. Then by using W_δ and V_δ we prove the invariance of some sets under the flow of (1.1)—(1.3) and vacuum isolating of solutions. After that we obtain a threshold result of global existence and nonexistence of solutions. Finally we discuss the global existence of solutions for the problem with critical initial conditions $I(u_0) \ge 0$, $E(0) = d$ (or $J(u_0) = d$) where d is the depth of potential well $W = W_1$ (see Definition 2.1).

In this paper we also denote $\|\cdot\|_{L^p(\Omega)}$ by $\|\cdot\|_p$, $\|\cdot\| = \|\cdot\|_2$ and $(u, v) = \int_\Omega uv\mathrm{d}x$.

2. Introduction of the Family of Potential Wells

In this paper we always assume that p and q satisfy (H). For problem (1.1)—(1.3) and problem (1.4), (1.5), (1.3) we define

$$J(u) = \frac{1}{2}\|\nabla u\|^2 - \frac{a}{p+1}\|u\|_{p+1}^{p+1} + \frac{b}{q+1}\|u\|_{q+1}^{q+1},$$
$$I(u) = \|\nabla u\|^2 - a\|u\|_{p+1}^{p+1} + b\|u\|_{q+1}^{q+1}$$

and

$$I_\delta(u) = \delta\|\nabla u\|^2 - a\|u\|_{p+1}^{p+1} + b\|u\|_{q+1}^{q+1}, \quad \delta \ge 0.$$

Lemma 2.1 Let $\|\nabla u\| \ne 0$. Then

(i) $\lim\limits_{\lambda\to 0} J(\lambda u) = 0$, $\lim\limits_{\lambda\to+\infty} J(\lambda u) = -\infty$.

(ii) In the interval $0 < \lambda < \infty$ there exists a unique $\bar\lambda = \bar\lambda(u)$ such that

$$\frac{\mathrm{d}}{\mathrm{d}\lambda}J(\lambda u)\Big|_{\lambda=\bar\lambda} = 0.$$

(iii) $J(\lambda u)$ is increasing on $0 \le \lambda \le \bar\lambda$, decreasing on $\bar\lambda \le \lambda < \infty$ and takes the maximum at $\lambda = \bar\lambda$.

(iv) $I(\lambda u) > 0$ for $0 < \lambda < \bar\lambda$, $I(\lambda u) < 0$ for $\bar\lambda < \lambda < \infty$ and $I(\bar\lambda u) = 0$.

Proof First $\|\nabla u\| \ne 0$ gives $\|u\|_{p+1} \ne 0$, $\|u\|_{q+1} \ne 0$.

(i) The conclusion follows from

$$J(\lambda u) = \frac{\lambda^2}{2}\|\nabla u\|^2 - \frac{a\lambda^{p+1}}{p+1}\|u\|_{p+1}^{p+1} + \frac{b\lambda^{q+1}}{q+1}\|u\|_{q+1}^{q+1}.$$

(ii) First note that

$$\frac{\mathrm{d}}{\mathrm{d}\lambda}J(\lambda u) = \lambda\|\nabla u\|^2 - a\lambda^p\|u\|_{p+1}^{p+1} + b\lambda^q\|u\|_{q+1}^{q+1} = 0, \quad \lambda > 0$$

is equivalent to

$$a\lambda^{p-1}\|u\|_{p+1}^{p+1} - b\lambda^{q-1}\|u\|_{q+1}^{q+1} = \|\nabla u\|^2. \tag{2.1}$$

Let

$$h(\lambda) = a\lambda^{p-1}\|u\|_{p+1}^{p+1} - b\lambda^{q-1}\|u\|_{q+1}^{q+1} = \lambda^{q-1}\left(a\lambda^{p-q}\|u\|_{p+1}^{p+1} - b\|u\|_{q+1}^{q+1}\right),$$

$$\lambda_0 = \left(\frac{b\|u\|_{q+1}^{q+1}}{a\|u\|_{p+1}^{p+1}}\right)^{\frac{1}{p-q}}.$$

Then $h(\lambda) < 0$ for $0 < \lambda < \lambda_0$, $h(\lambda) > 0$ for $\lambda_0 < \lambda < \infty$ and $h(\lambda)$ is increasing for $\lambda > \lambda_0$. Hence for any $\|\nabla u\| > 0$ there exists a unique $\bar{\lambda} > \lambda_0$ such that (2.1) holds.

(iii) Note that

$$\frac{\mathrm{d}}{\mathrm{d}\lambda}J(\lambda u) = \lambda\left(\|\nabla u\|^2 - h(\lambda)\right).$$

By the proof of (ii) it follows that if $0 < \lambda \le \lambda_0$, then $h(\lambda) \le 0$; if $\lambda_0 < \lambda < \bar{\lambda}$, then $0 < h(\lambda) < \|\nabla u\|^2$; if $\bar{\lambda} < \lambda < \infty$, then $h(\lambda) > \|\nabla u\|^2$. Hence we have $\frac{\mathrm{d}}{\mathrm{d}\lambda}J(\lambda u) > 0$ for $0 < \lambda < \bar{\lambda}$, $\frac{\mathrm{d}}{\mathrm{d}\lambda}J(\lambda u) < 0$ for $\bar{\lambda} < \lambda < \infty$. From this, the conclusion of (ii) follows.

(iv) The conclusion follows from the proof of (iii) and

$$I(\lambda u) = \lambda^2\|\nabla u\|^2 - a\lambda^{p+1}\|u\|_{p+1}^{p+1} + b\lambda^{q+1}\|u\|_{q+1}^{q+1} = \lambda\frac{\mathrm{d}}{\mathrm{d}\lambda}J(\lambda u).$$

Lemma 2.2 Let $\delta > 0$. If $0 < \|\nabla u\| \le r(\delta)$, then $I_\delta(u) > 0$. In particular, if $0 < \|\nabla u\| \le r(1)$, then $I(u) > 0$, where

$$r(\delta) = \left(\frac{\delta}{aC_p^{p+1}}\right)^{\frac{1}{p-1}}, \quad C_p = \sup_{u \in H_0^1(\Omega)} \frac{\|u\|_{p+1}}{\|\nabla u\|}.$$

Proof From $0 < \|\nabla u\| < r(\delta)$ we have $\|u\|_{q+1} > 0$ and by

$$\begin{aligned}&a\|u\|_{p+1}^{p+1} - b\|u\|_{q+1}^{q+1}\\ <&a\|u\|_{p+1}^{p+1} \le aC_p^{p+1}\|\nabla u\|^{p+1}\\ =&aC_p^{p+1}\|\nabla u\|^{p-1}\|\nabla u\|^2 \le \delta\|\nabla u\|^2\end{aligned}$$

we get $I_\delta(u) > 0$. ∎

Lemma 2.3. Let $\delta > 0$. If $I_\delta(u) < 0$, then $\|\nabla u\| > r(\delta)$. In particular, if $I(u) < 0$, then $\|\nabla u\| > r(1)$.

Proof $I_\delta(u) < 0$ gives

$$\delta\|\nabla u\|^2 < a\|u\|_{p+1}^{p+1} - b\|u\|_{q+1}^{q+1} < a\|u\|_{p+1}^{p+1} \le aC_p^{p+1}\|\nabla u\|^{p-1}\|\nabla u\|^2. \tag{2.2}$$

Eq. (2.2) implies $\|\nabla u\| > r(\delta)$. ∎

Lemma 2.4 Let $\delta > 0$. If $I_\delta(u) = 0$, then $\|\nabla u\| > r(\delta)$ or $\|\nabla u\| = 0$. In particular, if $I(u) = 0$, then $\|\nabla u\| > r(1)$ or $\|\nabla u\| = 0$.

Proof If $\|\nabla u\| = 0$, then $I_\delta(u) = 0$. If $I_\delta(u) = 0$ and $\|\nabla u\| \ne 0$, then by

$$\delta\|\nabla u\|^2 = a\|u\|_{p+1}^{p+1} - b\|u\|_{q+1}^{q+1} < a\|u\|_{p+1}^{p+1} \le aC_p^{p+1}\|\nabla u\|^{p-1}\|\nabla u\|^2$$

we get $\|\nabla u\| > r(\delta)$. ∎

Definition 2.1 For $\delta \geq 0$ we define

$$d(\delta) = \inf J(u)$$

subject to the conditions $u \in H_0^1(\Omega),\ I_\delta(u) = 0,\ \ \|\nabla u\| \neq 0$ and $d = d(1)$.

Lemma 2.5 As a function of δ $d(\delta)$ possesses the following properties

(i) $d(\delta) > a(\delta)r^2(\delta)$ for $0 < \delta < \dfrac{p+1}{2}$, $a(\delta) = \dfrac{1}{2} - \dfrac{\delta}{p+1}$;

(ii) $\lim\limits_{\delta\to 0} d(\delta) = d(0) > 0$ and there exists a unique $\delta_0 > \dfrac{p+1}{2}$ such that $d(\delta_0) = 0$ and $d(\delta) > 0$ for $0 < \delta < \delta_0$.

(iii) $d(\delta)$ is increasing on $0 \leq \delta \leq 1$, decreasing on $1 \leq \delta \leq \delta_0$ and takes the maximum $d = d(1)$ at $\delta = 1$.

(iv) $d(\delta)$ is continuous for $0 < \delta < \delta_0$.

Proof (i) If $I_\delta(u) = 0$ and $\|\nabla u\| \neq 0$, then by Lemma 2.4 we have $\|\nabla u\| > r(\delta)$, and

$$\begin{aligned}
J(u) &= \frac{1}{2}\|\nabla u\|^2 - \frac{a}{p+1}\|u\|_{p+1}^{p+1} + \frac{b}{q+1}\|u\|_{q+1}^{q+1} \\
&> \frac{1}{2}\|\nabla u\|^2 - \frac{1}{p+1}\left(a\|u\|_{p+1}^{p+1} - b\|u\|_{q+1}^{q+1}\right) \\
&= \left(\frac{1}{2} - \frac{\delta}{p+1}\right)\|\nabla u\|^2 + \frac{1}{p+1}I_\delta(u) = a(\delta)\|\nabla u\|^2 > a(\delta)r^2(\delta).
\end{aligned}$$

(ii) First $I_0(u) = 0$ implies $a\|u\|_{p+1}^{p+1} = b\|u\|_{q+1}^{q+1}$ and

$$\begin{aligned}
J(u) &= \frac{1}{2}\|\nabla u\|^2 - \frac{a}{p+1}\|u\|_{p+1}^{p+1} + \frac{b}{q+1}\|u\|_{q+1}^{q+1} \\
&= \frac{1}{2}\|\nabla u\|^2 + a\frac{p-q}{(p+1)(q+1)}\|u\|_{p+1}^{p+1} \\
&= \frac{1}{2}\|\nabla u\|^2 + b\frac{p-q}{(p+1)(q+1)}\|u\|_{q+1}^{q+1}.
\end{aligned}$$

Hence we have $d(0) \geq 0$. If $d(0) = 0$, then there exist $\{u_m\} \in H_0^1(\Omega)$ such that $\|\nabla u_m\| \neq 0$, $m = 1, 2, \cdots$ and $\lim\limits_{m\to\infty}\|\nabla u_m\| = \lim\limits_{m\to\infty}\|u_m\|_{p+1} = \lim\limits_{m\to\infty}\|u_m\|_{q+1} = 0$ and $a\|u_m\|_{p+1}^{p+1} = b\|u_m\|_{q+1}^{q+1}$. This is impossible from $\|u_m\|_{q+1} = O\left(\|u_m\|_{p+1}\right)$. For example, for any $u \in H_0^1(\Omega)$, $\|\nabla u\| \neq 0$ and $\lambda_m \to 0$, $m \to \infty$ we have $\lim\limits_{m\to\infty}\|\nabla(\lambda_m u)\| = \lim\limits_{m\to\infty}\|\lambda_m u\|_{p+1} = \lim\limits_{m\to\infty}\|\lambda_m u\|_{q+1} = 0$, but $a\|\lambda_m u\|_{p+1}^{p+1} = b\|\lambda_m u\|_{q+1}^{q+1}$ can not hold for sufficiently large m.

Next for any $u \in H_0^1(\Omega)$, $\|\nabla u\| \neq 0$ and $\delta > 0$, we define $\lambda = \lambda(\delta)$ such that

$$\delta\|\nabla(\lambda u)\|^2 = a\|\lambda u\|_{p+1}^{p+1} - b\|\lambda u\|_{q+1}^{q+1}. \tag{2.3}$$

Then $I_\delta(\lambda u) = 0$ and

$$\delta\|\nabla u\|^2 = a\lambda^{p-1}\|u\|_{p+1}^{p+1} - b\lambda^{q-1}\|u\|_{q+1}^{q+1}. \tag{2.4}$$

From the proof of (ii) of Lemma 2.1 it follows that for any $\delta > 0$ there exists a unique $\lambda = \lambda(\delta)$ such that (2.3) and (2.4) hold. Again by the proof of Lemma 2.1 we have $\lim_{\delta\to+\infty} \lambda(\delta) = +\infty$. Hence by (i) of Lemma 2.1 we get $\lim_{\delta\to+\infty} J(\lambda u) = \lim_{\lambda\to+\infty} J(\lambda u) = -\infty$. From this and (i) of this Lemma it follows that there exists a unique $\delta_0 > \frac{p+1}{2}$ such that $d(\delta_0) = 0$ and $d(\delta) > 0$ for $0 \le \delta < \delta_0$.

(iii) We prove that $d(\delta') < d(\delta'')$ for any $0 < \delta' < \delta'' < 1$ or $1 < \delta'' < \delta' < \delta_0$. Clearly it is enough to prove that for any $0 < \delta' < \delta'' < 1$ or $1 < \delta'' < \delta' < b$ and any $u \in H_0^1(\Omega)$, $I_{\delta''}(u) = 0$ and $\|\nabla u\| \neq 0$ there exists a $v \in H_0^1(\Omega)$, $I_{\delta'}(v) = 0$ and $\|\nabla v\| \neq 0$ and a constant $\varepsilon(\delta', \delta'') > 0$ such that $J(v) < J(u) - \varepsilon(\delta', \delta'')$. In fact, for above u we also define $\lambda(\delta)$ by (2.3), then $I_\delta(\lambda(\delta)u) = 0$, $\lambda(\delta'') = 1$ and (2.4) holds. Let $g(\lambda) = J(\lambda u)$. Then

$$\begin{aligned}\frac{\mathrm{d}}{\mathrm{d}\lambda} g(\lambda) &= \frac{1}{\lambda}\Big(\|\nabla(\lambda u)\|^2 - a\|\lambda u\|_{p+1}^{p+1} + b\|\lambda u\|_{q+1}^{q+1}\\ &= \frac{1}{\lambda}\Big((1-\delta)\|\nabla(\lambda u)\|^2 + I_\delta(\lambda u)\Big) = (1-\delta)\lambda\|\nabla u\|^2.\end{aligned}$$

Take $v = \lambda(\delta')u$, then $I_{\delta'}(v) = 0$ and $\|\nabla v\| \neq 0$.

If $0 < \delta' < \delta'' < 1$, then

$$J(u) - J(v) = g(1) - g(\lambda(\delta')) > (1-\delta'')r^2(\delta'')\lambda(\delta')(1-\lambda(\delta')) \equiv \varepsilon(\delta', \delta'').$$

If $1 < \delta'' < \delta' < b$, then

$$J(u) - J(v) = g(1) - g(\lambda(\delta')) > (\delta''-1)r^2(\delta'')\lambda(\delta'')(\lambda(\delta')-1) \equiv \varepsilon(\delta', \delta''). \quad \blacksquare$$

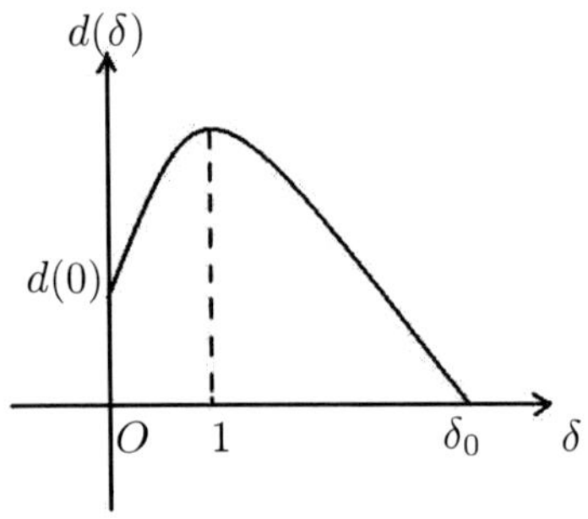

Figure 1.

(iv) Clearly it is enough to prove that for any $\delta_1 \in (0,1)$ and $\delta_2 \in (1, \delta_0)$,$d(\delta)$ is uniformly continuous on $[\delta_1, 1]$ and $[1, \delta_2]$ respectively.

First we prove that for any $\delta_1 \in (0,1)$, $d(\delta)$ is uniformly continuous on $[\delta_1, 1]$. For any given $0 < \varepsilon < d$ and $\delta' \in [\delta_1, 1)$ take a $u_{\delta'}(x) \in H_0^1(\Omega)$ such that $I_{\delta'}(u_{\delta'}) = 0$, $\|\nabla u_{\delta'}\| \neq 0$ and

$$0 \le J(u_{\delta'}) - d(\delta') < \frac{\varepsilon}{2}. \tag{2.5}$$

From the proof of Lemma 2.1 it follows that for each $\delta \in [\delta', 1]$ there exists a unique $\lambda = \lambda_{\delta'}(\delta)$ satisfying

$$\delta\|\nabla(\lambda u_{\delta'})\|^2 = a\|\lambda u_{\delta'}\|_{p+1}^{p+1} - b\|\lambda u_{\delta'}\|_{q+1}^{q+1}, \tag{2.6}$$

which implies that

$$I_\delta\left(\lambda_{\delta'}(\delta)u_{\delta'}\right) = 0, \;\; \lambda_{\delta'}(\delta') = 1$$

and

$$\delta = \frac{a\lambda^{p-1}\|u_{\delta'}\|_{p+1}^{p+1} - b\lambda^{q-1}\|u_{\delta'}\|_{q+1}^{q+1}}{\|\nabla u_{\delta'}\|^2} \stackrel{\text{def}}{\equiv} \varphi_{\delta'}(\lambda), \;\; 1 \leq \lambda \leq \lambda_{\delta'}(1). \tag{2.7}$$

Again by the proof of Lemma 2.1 we know that $\varphi_{\delta'}(\lambda)$ is increasing and continuous on $[1, \lambda_{\delta'}(1)]$, hence its inverse function $\lambda = \lambda_{\delta'}(\delta)$ is increasing and continuous on $[\delta', 1]$. Let $g_{\delta'}(\lambda) = J(\lambda u_{\delta'}) = J(\lambda_{\delta'}(\delta)u_{\delta'})$ for $\delta' \leq \delta \leq 1$. Then as in (iii) of this lemma we can get

$$\frac{\mathrm{d}}{\mathrm{d}\lambda}g_{\delta'}(\lambda) = (1-\delta)\lambda\|\nabla u_{\delta'}\|^2 \geq 0, \;\; \delta' \leq \delta \leq 1.$$

Hence for $\delta' \leq \delta \leq 1$ and $\lambda = \lambda_{\delta'}(\delta)$ we get

$$\begin{aligned}0 &\leq J(\lambda u_{\delta'}) - J(u_{\delta'}) = g_{\delta'}(\lambda) - g_{\delta'}(1) = (1-\tilde{\delta})\tilde{\lambda}\|\nabla u_{\delta'}\|^2(\lambda - 1)\\ &\leq (1-\delta_1)\|\nabla u_{\delta'}\|^2\lambda(\lambda-1), \;\; \delta' \leq \tilde{\delta} \leq \delta, \; 1 \leq \tilde{\lambda} \leq \lambda.\end{aligned}$$

From

$$\begin{aligned}\left(\frac{1}{2} - \frac{1}{p+1}\right)\|\nabla u_{\delta'}\|^2 &\leq \left(\frac{1}{2} - \frac{\delta'}{p+1}\right)\|\nabla u_{\delta'}\|^2 + \frac{1}{p+1}I_{\delta'}(u_{\delta'})\\ &\leq J(u_{\delta'}) < d(\delta') + \frac{\varepsilon}{2} < 2d\end{aligned}$$

we get

$$\|\nabla u_{\delta'}\|^2 \leq 4\frac{p+1}{p-1}d, \;\; \delta_1 \leq \delta' \leq 1.$$

Hence for $\delta_1 \leq \delta' \leq \delta \leq 1$ and $\lambda = \lambda_{\delta'}(\delta)$ we have

$$0 \leq J\left(\lambda_{\delta'}(\delta)u_{\delta'}\right) - J(u_{\delta'}) \leq 4(1-\delta_1)\frac{p+1}{p-1}d\lambda(\lambda-1).$$

Take $\lambda_1 > 1$ such that

$$4(1-\delta_1)\frac{p+1}{p-1}d\lambda_1(\lambda_1 - 1) = \frac{\varepsilon}{2}.$$

Then from

$$\begin{aligned}\varphi'_{\delta'}(\lambda) &= \frac{1}{\|\nabla u_{\delta'}\|^2}\left((p-1)\lambda^{p-2}a\|u_{\delta'}\|_{p+1}^{p+1} - (q-1)\lambda^{q-2}b\|u_{\delta'}\|_{q+1}^{q+1}\right)\\ &\geq \frac{p-1}{\|\nabla(\lambda u_{\delta'})\|^2}\left(a\|\lambda u_{\delta'}\|_{p+1}^{p+1} - b\|\lambda u_{\delta'}\|_{q+1}^{q+1}\right) = (p-1)\delta' \geq (p-1)\delta_1,\end{aligned}$$

we get

$$0 \leq \varphi_{\delta'}(\lambda_1) - \varphi_{\delta'}(1) \geq (p-1)\delta_1(\lambda_1 - 1) \stackrel{\text{def}}{=} \Delta_1 \text{ for all } \delta' \in [\delta_1, 1].$$

Hence when $\delta'' \in [\delta', 1]$ and $0 \le \delta'' - \delta' < \Delta_1$ we have $0 \le \lambda_{\delta'}(\delta'') - 1 < \lambda_1 - 1$ and

$$0 \le J\left(\lambda_{\delta'}(\delta'')u_{\delta'}\right) - J(u_{\delta'}) < \frac{\varepsilon}{2}.$$

Therefore we get

$$\begin{aligned} 0 &\le d(\delta'') - d(\delta') \le J\left(\lambda_{\delta'}(\delta'')u_{\delta'}\right) - d(\delta') \\ &= J\left(\lambda_{\delta'}(\delta'')u_{\delta'}\right) - J(u_{\delta'}) + J(u_{\delta'}) - d(\delta') < \frac{\varepsilon}{2} + \frac{\varepsilon}{2} = \varepsilon. \end{aligned}$$

Next we prove that for any $\delta_2 \in (1, \delta_0)$, $d(\delta)$ is uniformly continuous on $[1, \delta_2]$. For any given $\varepsilon > 0$ and $\delta' \in [1, \delta_2]$ take a $u_{\delta'}(x) \in H_0^1(\Omega)$ such that $I_{\delta'}(u_{\delta'}) = 0$, $\|\nabla u_{\delta'}\| \ne 0$ and satisfies (2.5). Again by the proof of Lemma 2.1 it follows that for each $\delta \in [1, \delta']$ there exists a unique $\lambda = \lambda_{\delta'}(\delta)$ satisfying (2.6), $I_\delta\left(\lambda_{\delta'}(\delta)u_{\delta'}\right) = 0$ and $\lambda_{\delta'}(\delta') = 1$. Again denote the function in (2.7) by $\varphi_{\delta'}(\lambda)$, where $\lambda_{\delta'}(1) \le \lambda \le 1$, then $\varphi_{\delta'}(\lambda)$ on $[\lambda_{\delta'}(1), 1]$ and $\lambda_{\delta'}(\delta)$ on $[1, \delta']$ are increasing and continuous respectively. Again for $\delta \in [1, \delta']$ and $\lambda = \lambda_{\delta'}(\delta)$ define $g_{\delta'}(\lambda) = J(\lambda u_{\delta'})$,then we have

$$\frac{\mathrm{d}}{\mathrm{d}\lambda} g_{\delta'}(\lambda) = (1 - \delta)\lambda\|\nabla u_{\delta'}\|^2 \le 0, \quad 1 \le \delta \le \delta'$$

and

$$\begin{aligned} 0 &\le J(\lambda u_{\delta'}) - J(u_{\delta'}) = g_{\delta'}(\lambda) - g_{\delta'}(1) = \left(\tilde{\tilde{\delta}} - 1\right)\tilde{\tilde{\lambda}}\|\nabla u_{\delta'}\|^2(1 - \lambda) \\ &\le (\delta_2 - 1)\|\nabla u_{\delta'}\|^2(1 - \lambda), \quad \delta \le \tilde{\tilde{\delta}} \le \delta', \ \lambda \le \tilde{\tilde{\lambda}} \le 1. \end{aligned}$$

Now we proof that $\|\nabla u_{\delta'}\|$ are bounded for $1 \le \delta' \le \delta_2$. First in (i) of Lemma 2.1 let $v = \lambda u$ we get

$$\lim_{\|\nabla v\| \to +\infty} J(v) = -\infty.$$

From this and

$$J(u_{\delta'}) \ge d(\delta') \ge d(\delta_2) > 0 \text{ for all } \delta' \in [1, \delta_2]$$

we can obtain $\|\nabla u_{\delta'}\|$ are bounded for $1 \le \delta' \le \delta_2$. Let $\|\nabla u_{\delta'}\|^2 \le M$. then we have

$$0 \le J(\lambda u_{\delta'}) - J(u_{\delta'}) \le (\delta_2 - 1)M(1 - \lambda), \quad \lambda_{\delta'}(1) \le \lambda \le 1.$$

Take λ_2 such that $0 < \lambda_2 < 1$ and

$$(\delta_2 - 1)M(1 - \lambda_2) = \frac{\varepsilon}{2}.$$

As above we can get

$$\varphi'_{\delta'}(\lambda) \ge (p - 1)\delta' \ge p - 1$$

and

$$0 \le \varphi_{\delta'}(1) - \varphi_{\delta'}(\lambda_2) \ge (p - 1)(1 - \lambda_2) \stackrel{\text{def}}{=} \Delta_2 \text{ for all } \delta' \in [1, \delta_2].$$

Hence when $\delta'' \in [1, \delta']$ and $0 \le \delta' - \delta'' < \Delta_2$ we have $0 \le 1 - \lambda_{\delta'}(\delta'') < 1 - \lambda_2$ and

$$0 \le J\left(\lambda_{\delta'}(\delta'')u_{\delta'}\right) - J(u_{\delta'}) < \frac{\varepsilon}{2}.$$

Finally we obtain

$$
\begin{aligned}
0 &\le d(\delta'') - d(\delta') \le J\left(\lambda_{\delta'}(\delta'')u_{\delta'}\right) - d(\delta') \\
&= J\left(\lambda_{\delta'}(\delta'')u_{\delta'}\right) - J(u_{\delta'}) + J(u_{\delta'}) - d(\delta') < \frac{\varepsilon}{2} + \frac{\varepsilon}{2} = \varepsilon.
\end{aligned}
$$

∎

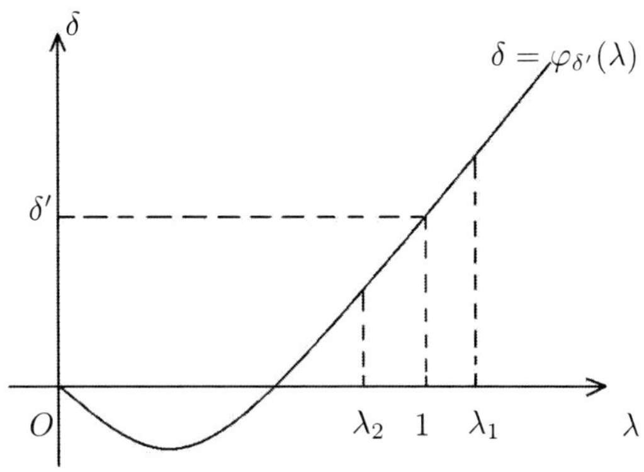

Figure 2.

Lemma 2.6 Let $0 < \delta < \frac{p+1}{2}$. Assume that $J(u) \le d(\delta)$ and $I_\delta(u) > 0$. Then

$$0 < \|\nabla u\|^2 < \frac{d(\delta)}{a(\delta)}.$$

In particular, if $J(u) \le d$ and $I(u) > 0$, then

$$0 < \|\nabla u\|^2 < \frac{2(p+1)}{p-1}d.$$

Proof This lemma follows from

$$a(\delta)\|\nabla u\|^2 + \frac{1}{p+1}I_\delta(u) < J(u) \le d(\delta). \tag{2.8}$$

∎

Again by (2.8) we can obtain the following Lemma 2.7 and Lemma 2.8.

Lemma 2.7 Let $0 < \delta < \frac{p+1}{2}$. Assume that $J(u) \le d(\delta)$ and $\|\nabla u\|^2 \ge \frac{d(\delta)}{a(\delta)}$. Then $I_\delta(u) < 0$. In particular, if $J(u) \le d$ and $\|\nabla u\|^2 \ge \frac{2(p+1)}{p-1}d$, then $I(u) < 0$. ∎

Lemma 2.8 Let $0 < \delta < \frac{p+1}{2}$. Assume that $J(u) \le d(\delta)$ and $I_\delta(u) = 0$. Then $\|\nabla u\|^2 < \frac{d(\delta)}{a(\delta)}$. In particular, if $J(u) \le d$ and $I(u) = 0$, then $\|\nabla u\|^2 < \frac{2(p+1)}{p-1}d$. ∎

Remark 2.1 The results of Lemma 2.2 — Lemma 2.4 and Lemma 2.6 — Lemma 2.8 show that the space $H_0^1(\Omega)$ is divided into two parts $I_\delta(u) > 0$ and $I_\delta(u) < 0$ by the surface

$I_\delta(u) = 0$ ($\|\nabla u\| \neq 0$). The inside part of $I_\delta(u) = 0$ is $I_\delta(u) > 0$ and the outside part of $I_\delta(u) = 0$ is $I_\delta(u) < 0$. Sphere $\|\nabla u\| = r(\delta)$ lies inside of $I_\delta(u) > 0$ and sphere $\|\nabla u\|^2 = \dfrac{d(\delta)}{a(\delta)}$ lies inside of $I_\delta(u) < 0$.(see Fig.3)

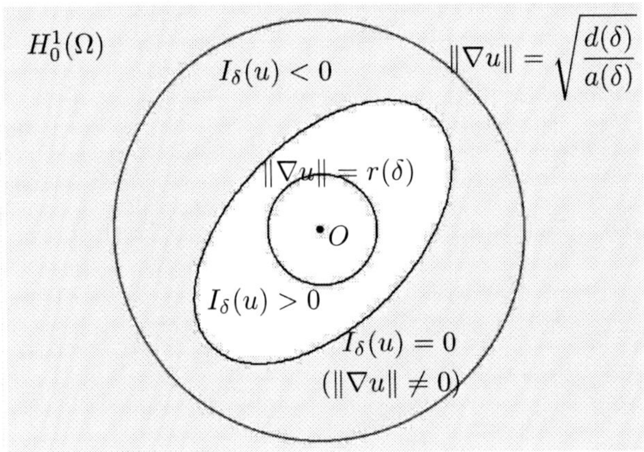

Figure 3.

Now we can define a family of potential wells as follows

$$
\begin{aligned}
&W = \{u \in H_0^1(\Omega) \mid I(u) > 0,\ J(u) < d\} \cup \{0\};\\
&W_\delta = \{u \in H_0^1(\Omega) \mid I_\delta(u) > 0,\ J(u) < d(\delta)\} \cup \{0\}, \quad 0 < \delta < \delta_0;\\
&\overline{W}_\delta = W_\delta \cup \partial W_\delta = \{u \in H_0^1(\Omega) \mid I_\delta(u) \geq 0,\ J(u) \leq d(\delta)\}, \quad 0 < \delta < \delta_0.
\end{aligned}
$$

In addition we define

$$
\begin{aligned}
&V = \{u \in H_0^1(\Omega) \mid I(u) < 0,\ J(u) < d\};\\
&V_\delta = \{u \in H_0^1(\Omega) \mid I_\delta(u) < 0,\ J(u) < d(\delta)\} \quad 0 < \delta < \delta_0;\\
&B_\delta = \{u \in H_0^1(\Omega) \mid \|\nabla u\| < r(\delta)\} \quad 0 < \delta < \delta_0;\\
&\overline{B}_\delta = B_\delta \cup \partial B_\delta = \{u \in H_0^1(\Omega) \mid \|\nabla u\| \leq r(\delta)\} \quad 0 < \delta < \delta_0;\\
&B_\delta^c = \{u \in H_0^1(\Omega) \mid \|\nabla u\| > r(\delta)\} \quad 0 < \delta < \delta_0.
\end{aligned}
$$

Lemma 2.9 Let $0 < \delta < \dfrac{p+1}{2}$. Then

$$
B_{r_1(\delta)} \subset W_\delta \subset B_{r_2(\delta)}, \quad V_\delta \subset B_\delta^c
$$

where

$$
\begin{aligned}
&B_{r_1(\delta)} = \{u \in H_0^1(\Omega) \mid \|\nabla u\|^2 < \min\{r^2(\delta),\ r_0^2(\delta)\},\\
&B_{r_2(\delta)} = \left\{u \in H_0^1(\Omega) \;\middle|\; \|\nabla u\|^2 < \frac{d(\delta)}{a(\delta)}\right\},
\end{aligned}
$$

and $r_0(\delta)$ is the unique real root of equation

$$\frac{1}{2}r^2 + \frac{b}{q+1}C_q^{q+1}r^{q+1} = d(\delta), \quad C_q = \sup_{u \in H_0^1(\Omega)} \frac{\|u\|_{q+1}}{\|\nabla u\|}.$$

Proof First $\|\nabla u\| < r(\delta)$ gives $\|\nabla u\| = 0$ or $I_\delta(u) > 0$. On the other hand,

$$J(u) \le \frac{1}{2}\|\nabla u\|^2 + \frac{b}{q+1}\|u\|_{q+1}^{q+1} \le \frac{1}{2}\|\nabla u\|^2 + \frac{b}{q+1}C_q^{q+1}\|\nabla u\|^{q+1}$$

and $\|\nabla u\|^2 < r_0^2(\delta)$ yield $J(u) < d(\delta)$. Hence we have $B_{r_1(\delta)} \subset W_\delta$. The remainder of this lemma follows from Lemma 2.3 and Lemma 2.6. ∎

From the definition of W_δ, V_δ and Lemma 2.5 we can obtain the following

Lemma 2.10

(i) If $0 < \delta' < \delta'' \le 1$, then $W_{\delta'} \subset W_{\delta''}$.

(ii) If $1 \le \delta'' < \delta' < \delta_0$, then $V_{\delta'} \subset V_{\delta''}$.

Lemma 2.11 Let $d(0) \le J(u) < d$ for some $u \in H_0^1(\Omega)$, $\delta_1 < \delta_2$ be the two roots of equation $d(\delta) = J(u)$. Then the sign of $I_\delta(u)$ is unchangeable for $\delta_1 < \delta < \delta_2$.

Proof First $J(u) > 0$ implies $\|\nabla u\| \neq 0$. If the sign of $I_\delta(u)$ is changed for $\delta_1 < \delta < \delta_2$, then there exists a $\bar{\delta} \in (\delta_1, \delta_2)$ such that $I_{\bar{\delta}}(u) = 0$. Thus by the definition of $d(\delta)$ we have $J(u) \ge d(\bar{\delta})$ which contradicts

$$J(u) = d(\delta_1) = d(\delta_2) < d(\bar{\delta}).$$ ∎

Lemma 2.12 Let $0 < J(u) < d(0)$ for some $u \in H_0^1(\Omega)$, δ_2 be the unique real root of equation $d(\delta) = J(u)$. Then the sign of $I_\delta(u)$ is unchangeable for $0 < \delta < \delta_2$.

Proof First $J(u) > 0$ implies $\|\nabla u\| \neq 0$. If the sign of $I_\delta(u)$ is changed for $0 < \delta < \delta_2$, then there exists a $\bar{\delta} \in (0, \delta_2)$ such that $I_{\bar{\delta}}(u) = 0$. Hence by Lemma 2.5 we have

$$J(u) \ge d(\bar{\delta}) > d(0) > d(\delta_2) \text{ if } 0 < \bar{\delta} \le 1$$

or

$$J(u) \ge d(\bar{\delta}) > d(\delta_2) \text{ if } 1 \le \bar{\delta} < \delta_2$$

which contradicts $J(u) = d(\delta_2)$. ∎

3. Invariant Sets and Vacuum Isolating of Solutions

In this section we discuss the invariance of some sets under the flow of (1.1)—(1.3) and vacuum isolating behavior of solutions for problem (1.1)—(1.3). First for problem (1.1)—(1.3) we define the energy

$$E(t) = \frac{1}{2}\|u_t\|^2 + \frac{1}{2}\|\nabla u\|^2 - \frac{a}{p+1}\|u\|_{p+1}^{p+1} + \frac{b}{q+1}\|u\|_{q+1}^{q+1} \equiv \frac{1}{2}\|u_t\|^2 + J(u).$$

Theorem 3.1 Let $u_0(x) \in H_0^1(\Omega)$, $u_1(x) \in L^2(\Omega)$. Assume that $d(0) \le e < d$, $\delta_1 < \delta_2$ are the two roots of equation $d(\delta) = e$. Then

(i) All solutions of problem (1.1)—(1.3) with $E(0) = e$ belong to W_δ for $\delta_1 < \delta < \delta_2$, provided $I(u_0) > 0$ or $\|\nabla u_0\| = 0$.

(ii) All solutions of problem (1.1)—(1.3) with $E(0) = e$ belong to V_δ for $\delta_1 < \delta < \delta_2$, provided $I(u_0) < 0$.

Proof (i) Let $u(t)$ be any solution of problem (1.1)—(1.3) with $E(0) = e$ and $I(u_0) > 0$ or $\|\nabla u_0\| = 0$, T be the existence time of $u(t)$. If $\|\nabla u_0\| = 0$, then $u_0(x) \in W_\delta$ for $0 < \delta < \delta_0$. If $I(u_0) > 0$, then by the definition of $d(\delta)$ and

$$\frac{1}{2}\|u_1\|^2 + J(u_0) = E(0) = d(\delta_1) = d(\delta_2) < d(\delta), \quad \delta_1 < \delta < \delta_2, \tag{3.1}$$

it follows that $I_\delta(u_0) > 0$ and $J(u_0) < d(\delta)$ *i.e.* $u_0(x) \in W_\delta$ for $\delta_1 < \delta < \delta_2$. Next we prove $u(t) \in W_\delta$ for $\delta_1 < \delta < \delta_2$ and $0 < t < T$. If it is false, then must have a $t_0 \in (0, T)$ such that $u(t_0) \in \partial W_\delta$ for some $\delta \in (\delta_1, \delta_2)$, *i.e.* $I_\delta(u(t_0)) = 0$, $\|\nabla u(t_0)\| \neq 0$ or $J(u(t_0)) = d(\delta)$. From the energy equality

$$\frac{1}{2}\|u_t\|^2 + J(u) = E(0) < d(\delta), \quad \delta_1 < \delta < \delta_2,\ 0 < t < T, \tag{3.2}$$

we see that $J(u(t_0)) = d(\delta)$ is impossible. On the other hand, if $I_\delta(u(t_0)) = 0$ and $\|\nabla u(t_0)\| \neq 0$, then by the definition of $d(\delta)$ we have $J(u(t_0)) \geq d(\delta)$ which contradicts (3.2).

(ii) Let $u(t)$ be any solution of problem (1.1)—(1.3) with $E(0) = e$ and $I(u_0) < 0$, T be the existence time of $u(t)$. First by (3.1) we can obtain $I_\delta(u_0) < 0$ and $J(u_0) < d(\delta)$, *i.e.* $u_0(x) \in V_\delta$ for $\delta_1 < \delta < \delta_2$. Next we prove $u(t) \in V_\delta$ for $\delta_1 < \delta < \delta_2$ and $0 < t < T$. If it is false, let $t_0 \in (0, T)$ be the first time such that $u(t) \in V_\delta$ for $0 \leq t < t_0$ and $u(t_0) \in \partial V_\delta$, *i.e.* $I_\delta(u(t_0)) = 0$ or $J(u(t_0)) = d(\delta)$ for some $\delta \in (\delta_1, \delta_2)$. Again (3.2) shows that $J(u(t_0)) = d(\delta)$ is impossible. If $I_\delta(u(t_0)) = 0$, then $I_\delta(u(t)) < 0$ for $0 < t < t_0$ and Lemma 2.3 yield $\|\nabla u(t)\| > r(\delta)$ and $\|\nabla u(t_0)\| \geq r(\delta)$. Hence by the definition of $d(\delta)$ we get $J(u(t_0)) \geq d(\delta)$ which contradicts (3.2). ∎

From Theorem 3.1 and Lemma 2.5 we can obtain the following

Corollary 3.2 If in Theorem 3.1 the assumption $E(0) = e$ is replaced by $d(0) \leq E(0) \leq e$, then the conclusion of Theorem 3.1 also holds.

Theorem 3.3 Let $u_0(x) \in H_0^1(\Omega)$, $u_1(x) \in L^2(\Omega)$. Assume that $0 < e < d(0)$, δ_2 is the unique real root of equation, $d(\delta) = e$. Then

(i) All solutions of problem (1.1)—(1.3) with $E(0) = e$ belong to W_δ for $0 < \delta < \delta_2$, provided $I(u_0) > 0$ or $\|\nabla u_0\| = 0$.

(ii) All solutions of problem (1.1)—(1.3) with $E(0) = e$ belong to V_δ for $0 < \delta < \delta_2$, provided $I(u_0) < 0$.

Proof The proof of this theorem is similar to the proof of Theorem 3.1 where δ_1 is replaced by 0 and (3.1) and (3.2) are replaced by

$$\frac{1}{2}\|u_1\|^2 + J(u_0) = E(0) < d(\delta), \quad 0 < \delta < \delta_2 \tag{3.1$'$}$$

and

$$\frac{1}{2}\|u_t\|^2 + J(u) = E(0) < d(\delta), \quad 0 < \delta < \delta_2,\ 0 < t < T. \tag{3.2$'$}$$

∎

Corollary 3.4 If in Theorem 3.3 the assumption $E(0) = e$ is replaced by $0 < E(0) \leq e$, then the conclusion of Theorem 3.3 also holds.

Corollary 3.5 If in Corollary 3.2 the assumption $d(0) \leq E(0) < e$ is replaced by $0 < E(0) \leq e$, then the conclusion of Corollary 3.2 also holds.

Theorem 3.6 Let $u_i(x)$ $(i = 0, 1)$, e and δ_i $(i = 1, 2)$ be the same as those in Theorem 3.1. Then for any $\delta \in (\delta_1, \delta_2)$ both sets W_δ and V_δ are invariant, thereby both sets

$$W_{\delta_1\delta_2} = \bigcup_{\delta_1<\delta<\delta_2} W_\delta, \text{ and } V_{\delta_1\delta_2} = \bigcup_{\delta_1<\delta<\delta_2} V_\delta$$

are invariant respectively under the flow of (1.1)—(1.3), provided $0 < E(0) \leq e$.

Theorem 3.7 Let $u_i(x)$ $(i = 0, 1)$, e and δ_2 be the same as those in Theorem 3.3. Then for any $\delta \in (0, \delta_2)$ both sets W_δ and V_δ are invariant, thereby both sets

$$W_{0,\delta_2} = \bigcup_{0<\delta<\delta_2} W_\delta, \text{ and } V_{0,\delta_2} = \bigcup_{0<\delta<\delta_2} V_\delta$$

are invariant respectively under the flow of (1.1)—(1.3), provided $0 < E(0) \leq e$.

From the proof of Theorem 3.1 and Theorem 3.3 we see that if $0 < E(0) < d$, then $I(u_0) = 0$ and $\|\nabla u_0\| \neq 0$ is impossible. So the results of Corollary 3.2 and Corollary 3.4 show that for the set of all solutions of problem (1.1)—(1.3) with $0 < E(0) \leq e$ there is a vacuum region

$$U_e = \{u \in H_0^1(\Omega) \mid \|\nabla u\| \neq 0,\ I_\delta(u) = 0,\ \delta_1 < \delta < \delta_2\} \text{ if } d(0) \leq e < d$$

or

$$U_e = \{u \in H_0^1(\Omega) \mid \|\nabla u\| \neq 0,\ I_\delta(u) = 0,\ 0 < \delta < \delta_2\} \text{ if } 0 < e < d(0)$$

such that there is no any solution of problem (1.1)—(1.3)(see Fig.4).

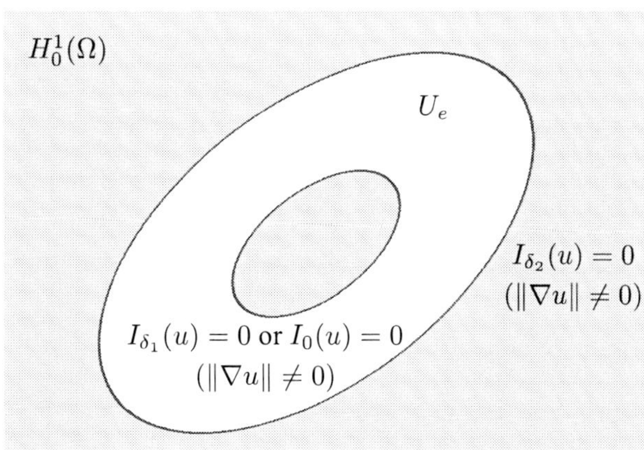

Figure 4.

The vacuum region U_e become bigger and bigger with decreasing of e. As the limit case we obtain

$$U_0 = \{u \in H_0^1(\Omega) \mid \|\nabla u\| \neq 0,\ I_\delta(u) = 0,\ 0 < \delta < \delta_0\}.$$

In fact, we have the following

Theorem 3.8 All nontrivial solutions $u(t)$ of problem (1.1)—(1.3) with $E(0) = 0$ belong to

$$B_{r_0}^c = \left\{ u \in H_0^1(\Omega) \mid \|\nabla u\| > r_0 = \left(\frac{p+1}{2aC_*^{p+1}}\right)^{\frac{1}{p-1}} \right\}$$

in the existence interval of $u(t)$.

Proof Let $u(t)$ be any solution of problem (1.1)—(1.3) with $E(0) = 0$, T be the existence time of $u(t)$. From the energy equality

$$\frac{1}{2}\|u_t\|^2 + J(u) \equiv E(0) = 0$$

we get $J(u) \leq 0$ for $0 \leq t < T$. Hence by

$$\begin{aligned} \frac{1}{2}\|\nabla u\|^2 &\leq \frac{a}{p+1}\|u\|_{p+1}^{p+1} - \frac{b}{q+1}\|u\|_{q+1}^{q+1} \leq \frac{a}{p+1}\|u\|_{p+1}^{p+1} \\ &\leq \frac{a}{p+1}C_*^{p+1}\|\nabla u\|^{p-1}\|\nabla u\|^2, \quad 0 \leq t < T \end{aligned}$$

we see that one must have either $\|\nabla u\| = 0$ or $\|\nabla u\| \geq r_0$. If $\|\nabla u_0\| = 0$, then $\|\nabla u\| \equiv 0$ for $0 \leq t < T$. Otherwise there exists a $t_0 \in (0, T)$ such that $0 < \|\nabla u(t_0)\| < r_0$. By a similar argument we can prove that if $\|\nabla u_0\| \geq r_0$, then $\|\nabla u\| \geq r_0$ for $0 < t < T$. ∎

Theorem 3.9 Let $u_0(x) \in H_0^1(\Omega)$, $u_1(x) \in L^2(\Omega)$. Assume that $E(0) < 0$ or $E(0) = 0$, $\|\nabla u_0\| \neq 0$. Then all solutions of problem (1.1)—(1.3) belong to V_δ for $0 < \delta < \dfrac{p+1}{2}$.

Proof Let $u(t)$ be any solution of problem (1.1)—(1.3) with $E(0) < 0$ or $E(0) = 0$, $\nabla u_0 \neq 0$, T be the existence time of $u(t)$. The energy equality gives

$$\frac{1}{2}\|u_t\|^2 + a(\delta)\|\nabla u\|^2 + \frac{1}{p+1}I_\delta(u) \leq \frac{1}{2}\|u_t\|^2 + J(u) = E(0) \quad 0 < \delta < \frac{p+1}{2} \tag{3.3}$$

From (3.3) it follows that if $E(0) < 0$, then $I_\delta(u) < 0$, $J(u) < 0 < d(\delta)$ for $0 < \delta < \dfrac{p+1}{2}$; if $E(0) = 0$, $\|\nabla u_0\| \neq 0$, then by Theorem 3.8 we have $\|\nabla u\| > r_0$ for $0 \leq t < T$. Again by (3.3) we get $I_\delta(u) < 0$, $J(u) < 0 < d(\delta)$ for $0 < \delta < \dfrac{p+1}{2}$. Hence for above two cases we always have $u(t) \in V_\delta$ for $0 < \delta < \dfrac{p+1}{2}$, $0 \leq t < T$. ∎

Corollary 3.10 Under the conditions of Theorem 3.9 all solutions of problem (1.1)—(1.3) belong to $\bar{B}^c_{\frac{p+1}{2}}$.

Proof Let $u(t)$ be any solution of problem (1.1)—(1.3) with $E(0) < 0$ or $E(0) = 0$, $\|\nabla u_0\| \neq 0$, T be the existence time of $u(t)$. Then Theorem 3.9 gives

$$u(t) \in V_\delta \ \text{ for } \ 0 < \delta < \frac{p+1}{2}, 0 \leq t < T.$$

From this and Lemma 2.3 we get $\|\nabla u\| > r(\delta)$ for $0 < \delta < \frac{p+1}{2}$, $0 \le t < T$. Letting $\delta \to \frac{p+1}{2}$ we obtain $\|\nabla u\| \ge r\left(\frac{p+1}{2}\right)$ for $0 \le t < T$. ∎

4. Global Existence and Blow up of Solutions for Problem (1.1)—(1.3)

In this section we shall prove the global existence and finite time blow up of solutions for problem (1.1)—(1.3) by using the family of potential wells introduced above.

Theorem 4.1 Let $u_0(x) \in H_0^1(\Omega)$, $u_1(x) \in L^2(\Omega)$. Assume that $0 < E(0) < d$ and $I(u_0) > 0$ or $\|\nabla u_0\| = 0$. Then problem (1.1)—(1.3) admits a global weak solution $u(t) \in L^\infty(0,\infty; H_0^1(\Omega))$ with $u_t(t) \in L^\infty(0,\infty; L^2(\Omega))$ and $u(t) \in W$ for $0 \le t < \infty$.

Proof Construct the approximate solutions $u_m(x,t)$ of problem (1.1)—(1.3) as we did in [10]. Then by the same arguments used in the proof of Theorem 3.2 in [10] we can obtain

$$\frac{1}{2}\|u_{mt}\|^2 + J(u_m) = E_m(0) < d, \quad 0 \le t < \infty \tag{4.1}$$

and $u_m(t) \in W$ for sufficiently large m and $0 \le t < \infty$. From (4.1) and

$$\begin{aligned} J(u_m) &\ge \frac{1}{2}\|\nabla u_m\|^2 - \frac{1}{p+1}\left(a\|u_m\|_{p+1}^{p+1} - b\|u_m\|_{q+1}^{q+1}\right) \\ &= \left(\frac{1}{2} - \frac{1}{p+1}\right)\|\nabla u_m\|^2 + \frac{1}{p+1}I(u_m) \ge \frac{p-1}{2(p+1)}\|\nabla u_m\|^2 \end{aligned}$$

it follows that

$$\frac{1}{2}\|u_{mt}\|^2 + \frac{p-1}{2(p+1)}\|\nabla u_m\|^2 < d, \quad 0 \le t < \infty \tag{4.2}$$

for sufficiently large m. Eq. (4.2) gives

$$\|\nabla u_m\|^2 < \frac{2(p+1)}{p-1}d, \quad 0 \le t < \infty; \tag{4.3}$$

$$\|u_m\|_{r_k+1}^2 \le C_r^2\|\nabla u_m\|^2 < C_r^2\frac{2(p+1)}{p-1}d, \quad r = p, q,\ 0 \le t < \infty; \tag{4.4}$$

$$\|u_{mt}\|^2 < 2d, \quad 0 \le t < \infty. \tag{4.5}$$

From (4.3)—(4.5) and compactness method it follows that problem (1.10—(1.3) admits a global weak solution $u(t) \in L^\infty\left(0,\infty; H_0^1(\Omega)\right)$ with $u_t(t) \in L^\infty\left(0,\infty; L^2(\Omega)\right)$. Finally by Theorem 3.1 and Theorem 3.3 we have $u(t) \in W$ for $0 \le t < \infty$. ∎

Corollary 4.2 Under the conditions of Theorem 4.1 we further have $u(t) \in W_\delta$ for $\delta_1 < \delta < \delta_2$, $0 \le t < \infty$, provided $d(0) \le E(0) < d$, where $\delta_1 < \delta_2$ are the two roots of equation $d(\delta) = E(0)$.

Corollary 4.3 Under the conditions of Theorem 4.1 we further have $u(t) \in W_\delta$ for $0 < \delta < \delta_2$, $0 \le t < \infty$, provided $0 < E(0) < d(0)$, where δ_2 is the unique root of equation $d(\delta) = E(0)$.

Corollary 4.4 If in Corollary 4.2 and Corollary 4.3 the assumption $I(u_0) > 0$ is replaced by $I_{\delta_2}(u_0) > 0$, then the conclusions of Corollary 4.2 and Corollary 4.3 also hold.

Theorem 4.5 Let $u_0(x) \in H_0^1(\Omega)$, $u_1(x) \in L^2(\Omega)$. Assume that $0 < E(0) < d$, $\|\nabla u\| \leq r(\delta_2)$, where δ_2 is the larger real root (if $d(0) \leq E(0) < d$) or the unique real root (if $0 < E(0) < d(0)$) of equation $d(\delta) = E(0)$. Then problem (1.1)—(1.3) admits a global weak solution $u(t) \in L^\infty(0,\infty; H_0^1(\Omega))$ with $u_t(t) \in L^\infty(0,\infty; L^2(\Omega))$ and $\|\nabla u(t)\|^2 < \dfrac{d(\delta)}{a(\delta)}$ for $\delta_1 < \delta < \min\left\{\delta_2, \dfrac{p+1}{2}\right\}$ (if $d(0) \leq E(0) < d$) or $\|\nabla u(t)\|^2 < \dfrac{d(\delta)}{a(\delta)}$ for $0 < \delta < \min\left\{\delta_2, \dfrac{p+1}{2}\right\}$ (if $0 < E(0) < d(0)$) and $\|u_t(t)\|^2 \leq 2E(0)$, $0 \leq t < \infty$.

Proof First from $\|\nabla u_0\| \leq r(\delta_2)$ and Lemma 2.2 we have $I_{\delta_2}(u_0) > 0$ or $\|\nabla u_0\| = 0$. Hence from Corollary 4.4 it follows that problem (1.1)—(1.3) admits a global weak solution $u(t) \in L^\infty(0,\infty; H_0^1(\Omega))$ with $u_t(t) \in L^\infty(0,\infty; L^2(\Omega))$ and $u(t) \in W_\delta$ for $\delta_1 < \delta < \delta_2$ (if $d(0) \leq E(0) < d$) or $0 < \delta < \delta_2$ (if $0 < E(0) < d(0)$). Finally Theorem 4.5 follows from Lemma 2.9 and (3.2).

Corollary 4.6 Under the conditions of Theorem 4.5 we further have $\|\nabla u(t)\|^2 \leq \dfrac{d(\delta_1)}{a(\delta_1)}$ for $d(0) \leq E(0) < d$ or $\|\nabla u(t)\|^2 \leq \dfrac{d(0)}{a(0)} = 2d(0)$ for $0 < E(0) < d(0)$.

Theorem 4.7 Let $u_0(x) \in H_0^1(\Omega)$, $u_1(x) \in L^2(\Omega)$. Assume that $E(0) < d$ and $I(u_0) < 0$. Then the solution of problem (1.1)—(1.3) must blow up in finite time, *i.e.* there exists a $T > 0$ such that

$$\lim_{t\to T} \|u(t)\| = +\infty. \tag{4.6}$$

Proof Let $u(t)$ be any solution of problem (1.1)—(1.3) with $E(0) < d$ and $I(u_0) < 0$, $F(t) = \|u(t)\|^2$. Then $\dot{F}(t) = 2(u_t, u)$,

$$\ddot{F}(t) = 2\|u_t\|^2 + 2(u_{tt}, u) = 2\|u_t\|^2 - 2I(u). \tag{4.7}$$

Hence

$$\begin{aligned}\ddot{F}(t) &\geq 2\|u_t\|^2 - 2\|\nabla u\|^2 + 2(p+1)\left(\frac{a}{p+1}\|u\|_{p+1}^{p+1} - \frac{b}{q+1}\|u\|_{q+1}^{q+1}\right)\\ &= (p+3)\|u_t\|^2 + (p-1)\|\nabla u\|^2 - 2(p+1)E(0)\\ &\geq (p+3)\|u_t\|^2 + (p-1)\lambda_1 F(t) - 2(p+1)E(0),\end{aligned} \tag{4.8}$$

where $\lambda_1 > 0$ is the first eigenvalue of problem $\Delta\varphi + \lambda\varphi = 0$, $\varphi|_{\partial\Omega} = 0$.

(i) If $E(0) \leq 0$, then (4.8) gives

$$\ddot{F}(t) \geq (p+3)\|u_t\|^2. \tag{4.9}$$

(ii) If $0 < E(0) < d$, then from Theorem 3.1 and Theorem 3.3 it follows that $u(t) \in V_\delta$ for $1 < \delta < \delta_2$ and $t > 0$, where δ_2 is the same as those in Theorem 3.1 or Theorem 3.3. Hence $I_\delta(u) < 0$ and (by Lemma 2.3) $\|\nabla u\| > r(\delta)$ for $1 < \delta < \delta_2$ and $t > 0$. So we get $I_{\delta_2}(u) \leq 0$ and $\|\nabla u\| \geq r(\delta_2)$ for $t > 0$. And by (4.7) we obtain

$$\begin{aligned}&\ddot{F}(t) \geq 2(\delta_2 - 1)\|\nabla u\|^2 - 2I_{\delta_2}(u) \geq 2(\delta_2-1)r^2(\delta_2) > 0,\\ &\dot{F}(t) \geq 2(\delta_2-1)r^2(\delta_2)t + \dot{F}(0),\end{aligned} \tag{4.10}$$

which shows that there exists a $t_0 \geq 0$ such that $\dot{F}(t_0) > 0$ and $F(t) \geq \dot{F}(t_0)(t - t_0) + F(t_0)$, $t \geq t_0$. So for sufficiently large t we have $(p-1)\lambda_1 F(t) > 2(p+1)E(0)$ and (4.9) holds. Finally (4.9) gives (see [5], P293)

$$F(t)\ddot{F}(t) - \frac{p+3}{4}(\dot{F}(t))^2 \geq (p+3)(\|u\|^2\|u_t\|^2 - (u, u_t)^2) \geq 0,$$

$$(F^{-\alpha}(t))'' = -\frac{\alpha}{F^{\alpha+2}(t)}(F(t)\ddot{F}(t) - (\alpha+1)(\dot{F}(t))^2) \leq 0, \quad \alpha = \frac{p-1}{4}$$

and (4.6) for some $T > 0$. ∎

Corollary 4.8 Under the conditions of Theorem 4.7 we further have

$$\lim_{t\to T} \|\nabla u(t)\| = +\infty \tag{4.11}$$

and

$$\lim_{t\to T} \|u(t)\|_r = +\infty, \tag{4.12}$$

where $2 < r \leq \infty$ if $n = 1$; $2 < r < \infty$ if $n = 2$; $2 < r \leq \frac{2n}{n-2}$ if $n \geq 3$.

Proof (4.11) follows from (4.6) and $\|\nabla u\|^2 \geq \lambda_1\|u\|^2$. From Sobolev embedding theorem and $\|\nabla u(t)\| < C(t)$ for $0 \leq t < T$ we get $\|u\|_r < C(t)$ for $0 \leq t < T$, where r is defined above. And by Hölder inequality we get

$$\|u\| \leq |\Omega|^{\frac{r-2}{2r}}\|u\|_r. \tag{4.13}$$

Hence (4.12) follows from (4.13) and (4.6). ∎

Note that from

$$\frac{1}{2}\|u_1\|^2 + \frac{p-1}{2(p+1)}\|\nabla u_0\|^2 + \frac{1}{p+1}I(u_0) \leq \frac{1}{2}\|u_1\|^2 + J(u_0) = E(0)$$

we see that if $0 < E(0) < d$, then $I(u_0) = 0$, $\|\nabla u_0\| \neq 0$ is impossible; if $E(0) = 0$, then $I(u_0) > 0$ or $I(u_0) = 0$, $\|\nabla u_0\| \neq 0$ is impossible; if $E(0) < 0$, then $I(u_0) \geq 0$ is impossible. Hence all possible cases already have been discussed in Theorem 4.1 and Theorem 4.7 and a threshold result already has been obtained as follows

Corollary 4.9 Let $u_0(x) \in H_0^1(\Omega)$, $u_1(x) \in L^2(\Omega)$. Assume that $E(0) < d$. Then when $I(u_0) \geq 0$ problem (1.1)—(1.3) admits a global weak solution and when $I(u_0) < 0$ the problem does not admit any global weak solution.

Remark 4.1 If in the proof of Theorem 4.7 we do not use the family of sets V_δ and only use the set V as did in [11], then for the case $0 < E(0) < d$ we can only obtain $\ddot{F}(t) > 0$ and $\dot{F}(t)$ is increasing, but can not prove $\dot{F}(t) > 0$ and $F(t)$ is increasing, thereby can not obtain $(p-1)\lambda_1 F(t) > 2(p+1)E(0)$ and (4.9) for sufficiently large t, unless we further assume $\dot{F}(0) = 2(u_0, u_1) > 0$. Therefor in the case $0 < E(0) < d$, $I(u_0) < 0$ in order to prove the finite time blow up of solutions for problem $(1.1) - (1,3)$ without the assumption $(u_0, u_1) > 0$ the introducing of the family of set V_δ is necessary.

5. Problem (1.1)—(1.3) with Critical Initial Conditions $I(u_0) \geq 0$, $E(0) = d$

In this section we prove the global existence of solutions for problem (1.1)—(1.3) with critical initial conditions $I(u_0) \geq 0$, $E(0) = d$.

Theorem 5.1 Let $u_0(x) \in H_0^1(\Omega)$, $u_1(x) \in L^2(\Omega)$. Assume that $E(0) = d$ and $I(u_0) \geq 0$. Then problem (1.1)—(1.3) admits a global weak solution $u(t) \in L^\infty(0,\infty; H_0^1(\Omega))$ with $u_t(t) \in L^\infty(0,\infty; L^2(\Omega))$ and $u(t) \in \overline{W}$ for $0 \leq t < \infty$.

Proof We prove this theorem for the following two cases (i) and (ii).

(i) $\|\nabla u_0\| \neq 0$.

Let $\lambda_m = 1 - \dfrac{1}{m}$ and $u_{0m} = \lambda_m u_0$, $m = 2, 3, \cdots$. Consider the initial conditions

$$u(x,0) = u_{0m}(x), \quad u_t(x,0) = u_1(x) \tag{5.1}$$

and corresponding problem (1.1), (5.1), (1.3). From $I(u_0) \geq 0$ and Lemma 2.1 we have $\bar{\lambda} = \bar{\lambda}(u_0) \geq 1$. Hence $I(u_{0m}) > 0$,

$$\begin{aligned} J(u_{0m}) &\geq \frac{1}{2}\|\nabla u_{0m}\|^2 - \frac{1}{p+1}\left(a\|u_{0m}\|_{p+1}^{p+1} - b\|u_{0m}\|_{q+1}^{q+1}\right) \\ &= \left(\frac{1}{2} - \frac{1}{p+1}\right)\|\nabla u_{0m}\|^2 + \frac{1}{p+1} I(u_{0m}) > 0 \end{aligned}$$

and $J(u_{0m}) = J(\lambda_m u_0) < J(u_0)$. And

$$0 < E_m(0) \equiv \frac{1}{2}\|u_1\|^2 + J(u_{0m}) < \frac{1}{2}\|u_1\|^2 + J(u_0) = E(0) = d.$$

So it follows from Theorem 4.1 that for each m problem (1.1), (5.1), (1.3) admits a global weak solution $u_m(t) \in L^\infty(0,\infty; H_0^1(\Omega))$ with $u_{mt}(t) \in L^\infty(0,\infty; L^2(\Omega))$ and $u_m(t) \in W$ for $0 \leq t < \infty$ satisfying

$$\begin{aligned} (u_{mt}, v) + \int_0^t (\nabla u_m, \nabla v)\mathrm{d}\tau &= \int_0^t (f(u_m), v)\mathrm{d}\tau + (u_1, v), \\ &\forall\, v \in H_0^1(\Omega),\ 0 \leq t < \infty \end{aligned} \tag{5.2}$$

and

$$\frac{1}{2}\|u_{mt}\|^2 + J(u_m) = E_m(0) < d. \tag{5.3}$$

From (5.3) and

$$J(u_m) \geq \left(\frac{1}{2} - \frac{1}{p+1}\right)\|\nabla u_m\|^2 + \frac{1}{p+1} I(u_m) \geq \frac{p-1}{2(p+1)}\|\nabla u_m\|^2$$

we can obtain

$$\|\nabla u_m\|^2 \leq \frac{2(p+1)}{p-1} d, \quad 0 \leq t < \infty,$$

$$\|u_m\|_{r+1}^2 \leq C_r^2 \|\nabla u_m\|^2 \leq C_r^2 \frac{2(p+1)}{p-1} d, \quad r = p, q,\ 0 \leq t < \infty$$

and

$$\|u_{mt}\|^2 \leq 2d, \quad 0 \leq t < \infty.$$

Hence there exist a $u(x,t) \in \overline{W}$ and a subsequence $\{u_\nu\}$ of $\{u_m\}$ such that as $\nu \to \infty$

$$u_\nu \longrightarrow u \text{ in } L^\infty(0,\infty; H_0^1(\Omega)) \text{ weak star and } a.e. \text{ in } Q = \Omega \times [0,\infty),$$
$$u_{\nu t} \longrightarrow u_t \text{ in } L^\infty(0,\infty; L^2(\Omega)) \text{ weak star},$$
$$|u_\nu|^{r-1}u_\nu \longrightarrow |u|^{r-1}u \text{ in } L^\infty(0,\infty; L^{r'}(\Omega)) \text{ weak star}, \; r' = \frac{r+1}{r}, \; r = p, q.$$

Letting $m - \nu \to \infty$ in (5.2) we get

$$(u_t, v) + \int_0^t (\nabla u, \nabla v)\mathrm{d}\tau = \int_0^t (f(u), v)\mathrm{d}\tau + (u_1, v),$$
$$\forall\, v \in H_0^1(\Omega), \; 0 \leq t < \infty. \tag{5.4}$$

Eq. (5.4) implies that $u(x,t)$ satisfies equation (1.1) and $u_t(x,0) = u_1(x)$. On the other hand, clearly we have $u(x,0) = u_0(x)$. So $u(x,t)$ is a global weak solution of problem (1.1)—(1.3).

(ii) $\|\nabla u_0\| = 0$.

Note that $\|\nabla u_0\| = 0$ implies $J(u_0) = 0$ and $\frac{1}{2}\|u_1\|^2 = E(0) = d$. Let $\lambda_m = 1 - \frac{1}{m}$, $u_{1m}(x) = \lambda_m u_1(x)$, $m = 2, 3, \cdots$. Consider the initial conditions

$$u(x,0) = u_0(x), \quad u_t(x,0) = u_{1m}(x) \tag{5.5}$$

and corresponding problem (1.1), (5.5), (1.3). From $\|\nabla u_0\| = 0$,

$$0 < E_m(0) = \frac{1}{2}\|u_{1m}\|^2 + J(u_0) = \frac{1}{2}\|\lambda_m u_1\|^2 < E(0) = d$$

and Theorem 4.1 it follows that for each m problem (1.1), (5.5), (1.3) admits a global weak solution $u_m(t) \in L^\infty(0,\infty; H_0^1(\Omega))$ with $u_{mt}(t) \in L^\infty(0,\infty; L^2(\Omega))$ and $u_m(t) \in W$ for $0 \leq t < \infty$ satisfying (5.2) and (5.3). The remainder of proof is the same as that in the part (i) of proof of this theorem. ∎

6. Reaction–Diffusion Equations

In this section we discuss the invariant sets, vacuum isolating of solutions, global existence and blow up of solutions and the problem with critical initial conditions for problem (1.4), (1.5), (1.6). First as did in [11] we assume the weak solution $u(x,t)$ of problem (1.4), (1.5), (1.3) satisfies the following condition

$$\int_0^t \|u_t\|^2\mathrm{d}\tau + J(u) \leq J(u_0), \quad \forall\, t \in [0,T), \tag{6.1}$$

where T is the existence time of solution $u(x,t)$. Exam carefully the proof of Theorem 3.1 — Corollary 3.10, Theorem 4.1 — Corollary 4.6 and Theorem 5.1 we see that if in these theorems the assumptions "$u_1(x) \in L^2(\Omega)$" and "or $\|\nabla u_0\| = 0$" are taken out, $E(0)$ is replaced by $J(u_0)$ and the energy equality $E(t) = E(0)$ is replaced by (6.1), and the conclusion $u_t(t) \in L^\infty\left(0,\infty; L^2(\Omega)\right)$ is replaced by $u_t(t) \in L^2\left(0,\infty; L^2(\Omega)\right)$, then the conclusions of these theorems also hold. Next we prove the blow up of solutions for problem (1.4), (1.5), (1.3).

Theorem 6.1 Let $u_0(x) \in H_0^1(\Omega)$. Assume that $J(u_0) < d$ and $I(u_0) < 0$. Then the solution of problem (1.4), (1.5), (1.3) must blow up in finite time *i.e.* there exists a $T > 0$ such that

$$\lim_{t\to T}\int_0^t \|u\|^2 \mathrm{d}\tau = +\infty. \tag{6.2}$$

Proof Let $u(t)$ be any solution of problem (1.4), (1.5), (1.3), with $J(u_0) < d$, $I(u_0) < 0$,

$$F_1(t) = \int_0^t \|u\|^2 \mathrm{d}\tau.$$

Then $\dot{F}_1(t) = \|u\|^2$,

$$\ddot{F}_1(t) = 2(u_t, u) = 2((f(u), u) - \|\nabla u\|^2) = -2I(u). \tag{6.3}$$

From (6.3), (6.1) and

$$(f(u), u) \geq (p+1)\left(\frac{a}{p+1}\|u\|_{p+1}^{p+1} - \frac{b}{q+1}\|u\|_{q+1}^{q+1}\right)$$

we can obtain

$$\begin{aligned}\ddot{F}_1(t) &\geq 2(p+1)\int_0^t \|u_t\|^2 \mathrm{d}\tau + (p-1)\|\nabla u\|^2 - 2(p+1)J(u_0)\\ &\geq 2(p+1)\int_0^t \|u_t\|^2 \mathrm{d}\tau + (p-1)\lambda_1 \dot{F}_1(t) - 2(p+1)J(u_0)\end{aligned}$$

and

$$\begin{aligned}F_1\ddot{F}_1 - \frac{p+1}{2}(\dot{F}_1)^2 \geq & 2(p+1)\left[\int_0^t \|u\|^2 \mathrm{d}\tau \int_0^t \|u_t\|^2 \mathrm{d}\tau - \left(\int_0^t (u, u_t)\mathrm{d}\tau\right)^2\right]\\ &+ (p-1)\lambda_1 F_1\dot{F}_1 - (p+1)\|u_0\|^2\dot{F}_1\\ &- 2(p+1)J(u_0)F_1 + \frac{p+1}{2}\|u_0\|^2,\end{aligned}$$

Hölder inequality gives

$$\begin{aligned}&F_1\ddot{F}_1 - \frac{p+1}{2}(\dot{F}_1)^2\\ \geq & (p-1)\lambda_1 F_1\dot{F}_1 - (p+1)\|u_0\|^2\dot{F}_1 - 2(p+1)J(u_0)F_1 + \frac{p+1}{2}\|u_0\|^2.\end{aligned} \tag{6.4}$$

(i) If $J(u_0) \leq 0$, then

$$F_1\ddot{F}_1 - \frac{p+1}{2}(\dot{F}_1)^2 \geq (p-1)\lambda_1 F_1\dot{F}_1 - (p+1)\|u_0\|^2\dot{F}_1.$$

First we prove $I(u) < 0$ for $t > 0$. If it is false, then must have $t_0 > 0$ such that $I(u(t_0)) = 0$. Let $t_0 > 0$ be the first time such that $I(u(t_0)) = 0$, then $I(u(t)) < 0$ for $0 \le t < t_0$. By Lemma 2.3 we have $\|\nabla u\| > r(1)$ for $0 < t < t_0$. Hence we have $\|\nabla u(t_0)\| \ge r(1)$ and $J(u(t_0)) \ge d$ which contradicts (6.1). So by (6.3) we get $\ddot{F}_1(t) > 0$ for $t > 0$. From this and $\dot{F}_1(0) = \|u_0\|^2 \ge 0$ it follows that there exists a $t_0 \ge 0$ such that $\dot{F}_1(t_0) > 0$ and

$$F_1(t) \ge \dot{F}_1(t_0)(t - t_0) + F_1(t_0) \ge \dot{F}_1(t_0)(t - t_0), t \ge t_0.$$

Hence for sufficiently large t we have $(p - 1)\lambda_1 F_1 > (p + 1)\|u_0\|^2$ and

$$F_1\ddot{F}_1 - \frac{p+1}{2}(\dot{F}_1)^2 > 0. \tag{6.5}$$

(ii) If $0 < J(u_0) < d$, then by an argument similar to that in the proof of Theorem 4.7 we can obtain $I_{\delta_2}(u) \le 0$ and $\|\nabla u\| \ge r(\delta_2)$ for $t \ge 0$, where $\delta_2 > 1$ is the larger real root (if $d(0) \le J(u_0) < d$) or the unique real root (if $0 < J(u_0) < d(0)$) of equation $d(\delta) = J(u_0)$. Thus (6.3) gives

$$\begin{aligned}
&\ddot{F}_1(t) = -2I(u) = 2(\delta_2 - 1)\|\nabla u\|^2 - 2I_{\delta_2}(u) \ge 2(\delta_2 - 1)r^2(\delta_2) > 0, \quad t \ge 0,\\
&\dot{F}_1(t) \ge 2(\delta_2 - 1)r^2(\delta_2)t + \dot{F}_1(0) \ge 2(\delta_2 - 1)r^2(\delta_2)t, \quad t \ge 0,\\
&F_1(t) \ge (\delta_2 - 1)r^2(\delta_2)t^2 + F_1(0) = (\delta_2 - 1)r^2(\delta_2)t^2, \quad t \ge 0.
\end{aligned}$$

Hence for sufficiently large t we have

$$\begin{aligned}
&\frac{1}{2}(p - 1)\lambda_1 F_1(t) > (p + 1)\|u_0\|^2,\\
&\frac{1}{2}(p - 1)\lambda_1 \dot{F}_1(t) > 2(p + 1)J(u_0).
\end{aligned}$$

Therefor by (6.4) we again obtain (6.5) for sufficiently large t. The remainder of proof is the same as that in [5]. ∎

Corollary 6.2 Under the conditions of Theorem 6.1 we further have

$$\overline{\lim_{t\to T}} \|u\|_r = +\infty, \tag{6.6}$$

and

$$\overline{\lim_{t\to T}} \|\nabla u\| = +\infty. \tag{6.7}$$

where $2 \le r \le \infty$ if $n = 1$; $2 \le r < \infty$ if $n = 2$; $2 \le r \le \dfrac{2n}{n-2}$ if $n \ge 3$.

Proof From (6.2) we obtain

$$\overline{\lim_{t\to T}} \|u\| = +\infty.$$

The remainder of proof is similar to that in the proof of Corollary 4.8. ∎

Corollary 6.3 Let $u_0(x) \in H_0^1(\Omega)$. Assume that $J(u_0) < d$. Then when $I(u_0) \ge 0$ problem (1.4),(1.5),(1.3) admits a global weak solution, when $I(u_0) < 0$ the problem does not admit any global weak solution.

Remark 6.1 From (6.4) we see that in the case $0 < J(u_0) < d$ in order to prove (6.5) we must prove

$$(p-1)\lambda_1 F_1 \dot{F}_1 \geq (p+1)\|u_0\|^2 \dot{F}_1 + 2(p+1)J(u_0)F_1 \tag{6.8}$$

at least for sufficiently large t. Since $\dot{F}_1(t) > 0$ for $t > 0$, (6.8) yields

$$(p-1)\lambda_1 F_1 \dot{F}_1 > 2(p+1)J(u_0)F_1$$

and

$$\dot{F}_1(t) > \frac{2(p+1)}{(p-1)\lambda_1} J(u_0) \tag{6.9}$$

for sufficiently large t. However if we do not use the family of sets V_δ and only use the set V as did in [11], then by (6.4) we can only obtain $\ddot{F}_1(t) > 0$ and $\dot{F}_1(t)$ is increasing, but can not obtain (6.9) even if for sufficiently large t. Therefore in the case $0 < J(u_0) < d$, $I(u_0) < 0$ in order to prove the finite time blow up of solutions for problem (1.4), (1.5), (1.3) the introducing of the family of sets V_δ is necessary.

References

[1] J.M.Ball, Remarks on blow–up and nonexistence theorems for nonlinear evolution equations, *Quart. J.Math.* **28** (1977), 473–486.

[2] V.A. Galaktionov, S.I. Pohozaev, Blow–up and critical exponents for nonlinear hyperbolic equations , *Nonlinear Analysis*, **53** (2003), 453–466.

[3] V.Geogev, G.Todorova. Existence of a solution of the wave equation with nonlinear damping and source terms, *Journal of Differential Equations.* **109**(1994), 295–308.

[4] R.T.Glassay, Blow–up theorems for nonlinear wave equations, *Math.Z.* **32** (1973), 183–203.

[5] M.Grillakis,Regularity for the wave equation with critical nonlinearity, *Ann.Math.* **132** (1990), 485–509.

[6] Hiroshi Uesaka, Oscillation or nonosillation property for semiliear wave equations, *Journal of Computational and Applied Mathematics.* **164-165** (2004), 723–730.

[7] H.A.Levine, Instablity and nonexistence of global solutions to nonlinear wave equations of the form $Pu = Au + F(u)$, *Trans. of Math, AMS,* **92** (1974), 1–21.

[8] H.A.Levine, Some additional remarks on the nonexistence of global solutions to nonlinear wave equations, *SIAM J. Math. Anal.* **5** (1974), 138–146.

[9] J.L. Lions, Quelques méthodes de résolution des problèmes aux limites non linéaires, Dunod Gauthier-Villars. Paris (1969).

[10] LIU Yacheng, On potential wells and vacuum isolating of solutions for semilinear wave equations, *Journal of Differential Equations,* **192** (2003),155–169.

[11] L.E. Payne and D.H. Sattinger, Sadle points and instability of nonlinear hyperbolic equations, *Israel.J. Math.* **22** (1975), 273–303.

[12] D.H. Sattinger, On global solution of nonlinear hyperbolic equations, *Arch. Rat. Mech. Anal.* **30** (1968), 148–172.

[13] J. Shatah and M. Struwe, Regularity results for nonlinear wave equations, *Ann. Math.* **138** (1993), 503–518.

[14] M. Tsutsumi, On solutions of semilinear differential equations in a Hilbert space, *Math. Japan,* **17** (1972), 173–193.

In: Nonlinear Analysis Research Trends
Editor: Inès N. Roux, pp. 281-297
ISBN 978-1-60456-358-0

Chapter 9

STATIONARY SOLUTIONS OF PHASE TRANSITIONS IN A COUPLED VISCOELASTIC SYSTEM

***Ming Mei*[1,2]*, *Yau Shu Wong*[3] *and Liping Liu*[4]**
[1]Department of Mathematics, Champlain College St.-Lambert
St.-Lambert, Quebec, J4P 3P2, Canada
[2]Department of Mathematics and Statistics, McGill University
Montreal, Quebec, H3A 2K6, Canada
[3]Department of Mathematical and Statistical Sciences, University of Alberta
Edmonton, Alberta, T6G 2G1, Canada
[4]Department of Mathematics, North Carolina Agricultural
and Technical State University
Greensboro, NC 27411, USA

Abstract

This study focuses on a coupled 2×2 system of mixed type for viscosity-capillarity with periodic initial-boundary condition in viscoelastic material. The main concern in the present study is the stationary solutions. It is shown that the stationary system has one or multiple stationary solutions depending on the value of the viscosities ε_1 and ε_2. In particular, the non-trivial solutions always present phase transitions. The criteria for the type of system responses (one or multiple solutions) are specified. Furthermore, the calculation formula for the number of multiple solutions is provided. Finally, numerical simulations on the prototypical system are reported to verify the theoretical results.

Keywords: Phase transitions, system of viscosity-capillarity, mixed type, stationary solutions, periodic boundary conditions.

AMS: 35R10, 35B40, 34K30, 58D25

1. Introduction and Main Results

The viscous-capillarity system in the viscoelastic material dynamics (resp. the compressible van der Waals fluids) can be written as a system of 2×2 viscous conservation laws of mixed

*Corresponding author.

type:

$$\begin{cases} v_t - u_x = \varepsilon_1 v_{xx}, \\ u_t - \sigma(v)_x = \varepsilon_2 u_{xx}, \end{cases} \quad (x,t) \in R \times R_+. \tag{1.1}$$

In this paper, we study the coupled system with the initial condition

$$(v,u)|_{t=0} = (v_0, u_0)(x), \quad x \in (-\infty, \infty) \tag{1.2}$$

and the $2L$-periodic boundary condition

$$(v,u)(x,t) = (v,u)(x+2L,t), \quad (x,t) \in (-\infty,\infty) \times (0,\infty) \tag{1.3}$$

where $L > 0$ is a given constant. Note that from the compatibility condition, we have

$$v_0(x) = v_0(x+2L), \quad u_0(x) = u_0(x+2L). \tag{1.4}$$

Here $v(x,t)$ is the strain (resp. specific volume), $u(x,t)$ the velocity, $\varepsilon_1 > 0$ and $\varepsilon_2 > 0$ the viscous constants, $\sigma(v)$ the stress function (resp. pressure function), which is assumed to be sufficiently smooth and non-monotonic.

As a prototype, c.f. [15, 16], the simplest function:

$$\sigma(v) = v^3 - v \tag{1.5}$$

is considered in this study. This function captures the basic features for the phase transition models. For such a stress function $\sigma(v)$, it has only two critical points $\pm\frac{1}{\sqrt{3}}$ such that $\sigma'(\pm\frac{1}{\sqrt{3}}) = 0$, and $\sigma'(v) > 0$ for $v \in (-\infty, -\frac{1}{\sqrt{3}}) \cup (\frac{1}{\sqrt{3}}, \infty)$, $\sigma'(v) < 0$ for $v \in (-\frac{1}{\sqrt{3}}, \frac{1}{\sqrt{3}})$. Physically, this determines three phases, for example, water, vapor, and water-vapor mixture phases in van der Waals fluids. Mathematically, Eq.(1.1) with $\varepsilon_1 = \varepsilon_2 = 0$ is hyperbolic in $(-\infty, -\frac{1}{\sqrt{3}}) \cup (\frac{1}{\sqrt{3}}, \infty)$ and elliptic in $(-\frac{1}{\sqrt{3}}, \frac{1}{\sqrt{3}})$. Here, $v = \pm\frac{1}{\sqrt{3}}$ are the two phase boundaries. In the case of the van der Waals fluids, the pressure is exactly given by $-\sigma(v) = \frac{R\theta}{v-b} - \frac{a}{v^2}$ with positive constants R, θ, a and b satisfying $R\theta b/a < (2/3)^3$ and $v > b > 0$. It is known that there are also two critical points of $\sigma(v)$, says v_1 and v_2, such that $\sigma'(v_1) = \sigma'(v_2) = 0$, $\sigma'(v) > 0$ for $v \in (b, v_1) \cup (v_2, \infty)$ and $\sigma'(v) < 0$ for $v \in (v_1, v_2)$. The region (b, v_1) is the water-region, (v_2, ∞) is the vapor-region, and (v_1, v_2) is the water-vapor mixture region, c.f. [1, 13].

Since the periodic solutions $(v,u)(x,t)$ of (1.1)-(1.3) in the entire space $(-\infty,\infty)$ can be regarded as $2L$-periodic extensions of that on $[0, 2L]$, let us focus the system (1.1) on the bounded interval $[0, 2L]$. Integrating (1.1) over $[0, 2L] \times [0, t]$ and using the periodic boundary condition (1.3), we obtain

$$\int_0^{2L} v(x,t)dx = \int_0^{2L} v_0(x)dx, \quad \int_0^{2L} u(x,t)dx = \int_0^{2L} u_0(x)dx. \tag{1.6}$$

Let

$$m_0 := \frac{1}{2L}\int_0^{2L} v_0(x)dx, \quad m_1 := \frac{1}{2L}\int_0^{2L} u_0(x)dx, \tag{1.7}$$

then

$$\int_0^{2L} [v(x,t) - m_0]dx = 0, \qquad \int_0^{2L} [u(x,t) - m_1]dx = 0. \tag{1.8}$$

Therefore, (m_0, m_1) is the average of the initial value $(v_0, u_0)(x)$ over $[0, 2L]$, and so the average of the solution $(v, u)(x, t)$ over $[0, 2L]$.

Phase transitions are very common and interesting phenomena arising thermodynamics, and have been the hottest spots in mathematical and physical communities. A lot of great progress in, for example, the construction of the solutions and their stationary solutions, the behaviors of the solutions, as well as the asymptotic stabilities of the solutions, has been made by many mathematicians and physicians, see [1]-[25] and the references therein. In this paper, we are interested in the steady-state solutions of the system (1.1)-(1.3). The corresponding stationary problem of (1.1)-(1.3) is

$$\begin{cases} -U_x = \varepsilon_1 V_{xx}, \\ -\sigma(V)_x = \varepsilon_2 U_{xx}, \\ (V, U)(x) = (V, U)(x + 2L), \\ \frac{1}{2L}\int_0^{2L} V(x)dx = m_0, \\ \frac{1}{2L}\int_0^{2L} U(x)dx = m_1, \end{cases} \tag{1.9}$$

where $(V, U) = (V, U)(x)$. It is easy to see that the average initial data (m_0, m_1) is a trivial stationary solution of (1.9). In the present paper, we prove that, if m_0 is in the hyperbolic region $(-\infty, -\frac{1}{\sqrt{3}})$ or $(\frac{1}{\sqrt{3}}, \infty)$ but not so close to the phase boundaries $v = \pm\frac{1}{\sqrt{3}}$, or when the viscosities ε_1 and ε_2 are suitably large (regardless of the value of m_0), then the stationary problem (1.9) has and only has the trivial solution (m_0, m_1). On the other hand, if m_0 is in the elliptic region $(-\frac{1}{\sqrt{3}}, \frac{1}{\sqrt{3}})$ and the viscosities ε_1 and ε_2 are sufficiently small, then the stationary problem (1.9) has multiple non-trivial solutions, and these solutions always admit phase transitions. The number of those multiple solutions will be counted shortly. Here we provide a necessary preparation for our further study in [15, 16] on the asymptotic behaviors of the solution of (1.1), specifically, the convergence of the solution $(v, u)(x, t)$ to the stationary solution $(V, U)(x)$. We further present some numerical simulations in different cases which confirm also our theoretical results.

Our main results are stated as follows.

Theorem 1.1 (Stationary Solutions)

1. *If the viscosity is large, such that*

$$\varepsilon_1\varepsilon_2 > \frac{L^2}{\pi^2}\Big(1 - \frac{3}{4}m_0^2\Big), \tag{1.10}$$

then the stationary problem (1.9) *has a unique solution* $(V, U)(x) = (m_0, m_1)$. *No phase transition exhibits.*

2. *If the viscosity is sufficiently small such that*

$$\varepsilon_1\varepsilon_2 < \frac{L^2}{\pi^2}(1 - 3m_0^2), \tag{1.11}$$

then besides the trivial solution (m_0, m_1), the stationary problem (1.9) has at least one non-trivial solution, which is oscillating periodically, and must occur phase transitions.

Theorem 1.2 (Number of Non-Trivial Stationary Solutions)

1. *When $m_0 = 0$ and $\varepsilon_1\varepsilon_2 < \frac{L^2}{\pi^2}(1-3m_0^2)$, then there exist $(2N_0+1)$ solutions for (1.9), where N_0 is the integer of*

$$N_0 = \left[\frac{2L}{\pi\sqrt{\varepsilon_1\varepsilon_2}}\right]. \tag{1.12}$$

Here, $[x]$ denotes the greatest integer $\leq x$.

2. *When $0 < m_0 < \frac{1}{\sqrt{5}}$ and $\varepsilon_1\varepsilon_2 < \frac{L^2}{\pi^2}(1-3m_0^2)$, there is a unique non-trivial stationary solution for (1.9).*

3. *When $\frac{1}{\sqrt{5}} < m_0 < \frac{1}{\sqrt{3}}$, there exists a number $\varepsilon^* > \frac{L^2}{\pi^2}(1-3m_0^2)$ such that, when $\frac{L^2}{\pi^2}(1-3m_0^2) < \varepsilon_1\varepsilon_2 < \varepsilon^*$, there are two non-trivial solutions. However, when $\varepsilon_1\varepsilon_2 \leq \frac{L^2}{\pi^2}(1-3m_0^2)$, there is only one non-trivial solution for (1.9).*

Remark 1.3

1. *In the first part of Theorem 1.1, the condition (1.10) implies large viscosities if m_0 is in the elliptic region. However, when $m_0 > \frac{2}{\sqrt{3}}$ or $m_0 < -\frac{2}{\sqrt{3}}$, which implies that m_0 is in the hyperbolic regions, even if the viscosity disappears, i.e., $\varepsilon_1 = \varepsilon_2 = 0$, the condition (1.10) still holds, and thus the stationary solution of (1.9) is unique and $(V, U)(x) \equiv (m_0, m_1)$. Note that, in the second part of Theorem 1.1, the condition (1.11) implies that the initial average m_0 is in the elliptic region. The theorem indicates that for the initial average m_0 in the elliptic region, when the viscosity $\varepsilon_1\varepsilon_2$ is small, multiple non-trivial stationary solutions $(V, U)(x)$ exist, and they exhibit phase transitions.*

2. *When $m_0 = 0$, the conditions (1.10) and (1.11) become*

$$\varepsilon_1\varepsilon_2 \gtrless \frac{L^2}{\pi^2},$$

which is the critical condition, and it will lead to one or multiple solutions.

3. *Theorem 1.2 gives the exact numbers of the non-trivial stationary solutions for different range of the initial average m_0. These results are the same to [23, 9] for the Cahn-Hilliard equation in Neumann boundary condition.*

2. Proof of Main Theorem

In this section, we are going to prove Theorem 1.1 and Theorem 1.2. Substituting the first equation of (1.9) into the second equation and integrating the resultant equation over $[0, x]$, we obtain the stationary equation for $V(x)$ as follows

$$\begin{cases} \varepsilon_1\varepsilon_2 V_{xx} = \sigma(V) - a \\ V(x) = V(x+2L) \\ \frac{1}{2L}\int_0^{2L} V(x)dx = m_0, \end{cases} \tag{2.1}$$

where $a = \sigma(V(0)) - \varepsilon_1\varepsilon_2 V_{xx}(0)$ is a constant determined automatically by Eq. (2.1). In fact, integrating Eq. (2.1) over $[0, 2L]$ and noticing the periodic boundary condition $V(x) = V(x+2L)$, one verifies

$$a = \frac{1}{2L}\int_0^{2L} \sigma(V(x))dx. \tag{2.2}$$

Eq. (2.1) is similar to the so-called Cahn-Hilliard equation which has been studied widely by researchers in the mathematical physics community, for example, see [8, 9, 18, 19, 20, 23, 24], and the references therein. With the Neumann boundary condition, Zheng [23] studied the existence of the trivial and non-trivial solutions (although he just gave a roughly sufficient conditions on the size of viscosity), in particular, the number of non-trivial solutions for the initial mean $m_0 = 0$ in the case of small viscosity was counted. While in [8, 9, 18, 19, 20] Grinfeld, Novick-Cohen, Peletier, Segel *et al*, specified the values of m_0 and the size of the viscosity for the existence of trivial or nontrivial solutions by the transversality arguments. All of these works will be a great help in dealing with (2.1) for the periodic boundary condition.

Inspiring by [23] and [9], we can testify the existence of the non-trivial solutions by specifying the size of viscosity, i.e., the criteria (1.10) and (1.11), and we further count the number of the non-trivial solutions by specifying the location of the initial average m_0.

Letting

$$\bar{V}(x) := V(x) - m_0, \tag{2.3}$$

and applying (1.8), then we reduce (2.1) into

$$\begin{cases} \varepsilon_1\varepsilon_2 \bar{V}_{xx} = \bar{\sigma}(\bar{V}) - \bar{a} \\ \bar{V}(x) = \bar{V}(x+2L) \\ \int_0^{2L} \bar{V}(x)dx = 0, \end{cases} \tag{2.4}$$

where

$$\bar{\sigma}(\bar{V}) := \sigma(\bar{V} + m_0) - \sigma(m_0), \quad \bar{a} := \frac{1}{2L}\int_0^{2L} \bar{\sigma}(\bar{V}(x))dx. \tag{2.5}$$

Define a periodic Sobolev space by

$$H^1_{per,0} = \left\{ v(y) | v(y) \in H^1_{per}(R), \int_0^{2L} v(y)dy = 0 \right\} \tag{2.6}$$

and a functional on $H^1_{per,0}$ by

$$G(v) = \int_0^{2L} \Big[\frac{\varepsilon_1\varepsilon_2}{2} v_x^2 + H(v)\Big] dx, \tag{2.7}$$

where

$$H(v) = \int_0^v \bar{\sigma}(s) ds = \frac{1}{4} v^4 + m_0 v^3 + \frac{3m_0^2 - 1}{2} v^2,$$

we now have the first lemma as follows.

Lemma 2.1 *The solutions of* (2.4) *are equivalent to the critical points of the functional* $G(v)$ *defined in* (2.7) *over* $H^1_{per,0}$.

Proof. If $\bar{V}(x)$ is a solution of (2.4), then for any $w(x) \in H^1_{per,0}$, multiplying (2.4) by $w(x)$ and integrating it over $[0, 2L]$, and noting $\int_0^{2L} \bar{a} w(x) dx = \bar{a} \int_0^{2L} w(x) dx = 0$, one gets

$$\int_0^{2L} \Big(\varepsilon_1\varepsilon_2 \bar{V}_x w_x + \bar{\sigma}(\bar{V}) w\Big) dx = 0, \quad \text{for any } w \in H^1_{per,0}. \tag{2.8}$$

This implies $\bar{V}(x)$ is a critical point of $G(v)$ over $H^1_{per,0}$.

On the other hand, let $\bar{V}(x)$ be a critical point of $G(v)$ over $H^1_{per,0}$, namely, (2.8) holds, we prove that $\bar{V}(x)$ is a solution of (2.4). For any $\bar{w}(x) \in H^1(0, 2L)$ satisfying $\bar{w}(x) = \bar{w}(x + 2L)$, it can be verified that

$$w(x) := \bar{w}(x) - \frac{1}{2L} \int_0^{2L} \bar{w}(x) dx \in H^1_{per,0} \tag{2.9}$$

and

$$\int_0^{2L} \bar{\sigma}(\bar{V}(x)) \cdot \Big(\frac{1}{2L} \int_0^{2L} \bar{w}(x) dx\Big) dx = \int_0^{2L} \bar{w}(x) \cdot \Big(\frac{1}{2L} \int_0^{2L} \bar{\sigma}(\bar{V}(x)) dx\Big) dx. \tag{2.10}$$

Thus, (2.8)–(2.10) gives

$$\begin{aligned}
0 &= \int_0^{2L} \Big(\varepsilon_1\varepsilon_2 \bar{V}_x w_x + \bar{\sigma}(\bar{V}) w\Big) dx \\
&= \int_0^{2L} \Big[\varepsilon_1\varepsilon_2 \bar{V}_x \Big(\bar{w}(x) - \frac{1}{2L}\int_0^{2L} \bar{w}(x) dx\Big)_x \\
&\quad + \bar{\sigma}(\bar{V}(x)) \Big(\bar{w}(x) - \frac{1}{2L}\int_0^{2L} \bar{w}(x) dx\Big)\Big] dx \\
&= \int_0^{2L} \Big[\varepsilon_1\varepsilon_2 \bar{V}_x \bar{w}_x + \bar{\sigma}(\bar{V}(x)) \bar{w}(x)\Big] dx \\
&\quad - \int_0^{2L} \bar{w}(x) \Big(\frac{1}{2L}\int_0^{2L} \bar{\sigma}(\bar{V}(x)) dx\Big) dx \\
&= \int_0^{2L} \Big[\varepsilon_1\varepsilon_2 \bar{V}_x \bar{w}_x + \Big(\bar{\sigma}(\bar{V}(x)) - \frac{1}{2L}\int_0^{2L} \bar{\sigma}(\bar{V}(x)) dx\Big) \bar{w}(x)\Big] dx \\
&= \int_0^{2L} \Big[\varepsilon_1\varepsilon_2 \bar{V}_x \bar{w}_x + (\bar{\sigma}(\bar{V}(x)) - \bar{a}) \bar{w}(x)\Big] dx,
\end{aligned} \tag{2.11}$$

where we used (2.5). Thus, we prove in (2.11) that $\bar{V}(x)$ is a weak solution of the stationary problem (2.4). Furthermore, by the usual bootstrap argument, we conclude that $\bar{V}(x)$ is the classical solution of (2.4). The detail is omitted. □

We now prove a useful inequality which is somewhat similar to the so-called Poincaré inequality.

Lemma 2.2 *Let $\bar{V}(x) \in H^1_{per,0}$ be the solution of Eq. (2.4). Then there exists at least one point x_* such that $\bar{V}(x_*) = 0$, and*

$$\|\bar{V}\|_{L^2} \le \frac{L}{\pi}\|\bar{V}_x\|_{L^2}. \tag{2.12}$$

Proof. It is easy to verify that 0 is a solution of (2.4). If $\bar{V}(x) \equiv 0$, then (2.12) automatically holds. If $\bar{V}(x) \not\equiv 0$ on $[0, 2L]$, then there must exist at least one point, says $x_* \in [0, 2L]$, such that $\bar{V}(x_*) = 0$. In fact, notice that $\int_0^{2L} \bar{V}(s)ds = 0$ and $\bar{V}(x) \in C^0(0, 2L)$ (because $\bar{V}(x) \in H^1(0, 2L)$), then $\bar{V}(x)$ must change signs on $[0, 2L]$, which implies that $\bar{V}(x_*) = 0$ for some $x_* \in [0, 2L]$. Otherwise, either $\bar{V}(x) > 0$ or $\bar{V}(x) < 0$ on $[0, 2L]$ leads to $\int_0^{2L} \bar{V}(s)ds > 0$ or $\int_0^{2L} \bar{V}(s)ds < 0$, but this is a contradiction with the condition $\int_0^{2L} \bar{V}(s)ds = 0$.

Consider the following eigenvalue problem

$$\begin{cases} -\tilde{v}_{xx} = \beta^2 \tilde{v}, \\ \tilde{v}(x) = \tilde{v}(x + 2L), \\ \int_0^{2L} \tilde{v}(x)dx = 0. \end{cases} \tag{2.13}$$

The eigenvalues are given by

$$\beta_k = \frac{kL}{\pi}, \quad k = 1, 2, 3, \cdots, \tag{2.14}$$

and the corresponding eigenfunctions are

$$\tilde{v}_{1,k}(x) = \frac{1}{\sqrt{L}} \sin \beta_k x, \quad \tilde{v}_{2,k}(x) = \frac{1}{\sqrt{L}} \cos \beta_k x, \quad k = 1, 2, 3, \cdots \tag{2.15}$$

satisfying

$$< \tilde{v}_{i,k}, \tilde{v}_{j,l} >= \begin{cases} 1, & i = j, k = l \\ 0, & \text{otherwise}, \end{cases} \tag{2.16}$$

where

$$< \tilde{v}_{i,k}, \tilde{v}_{j,l} >= \int_0^{2L} \tilde{v}_{i,k}(x)\tilde{v}_{j,l}(x)dx$$

is the inner product of L^2_{per}. It is known that the sequence $\{\tilde{v}_{i,k}(x)\}$ ($i = 1, 2$ and $k = 1, 2, 3, \cdots$) forms an orthonormal basis for the space

$$L^2_{per,0} = \Big\{\tilde{v}(x) | \tilde{v}(x) \in L^2_{per} \text{ and } \int_0^{2L} \tilde{v}(x)dx = 0\Big\}.$$

Therefore, as a periodic function satisfying $\bar{V}(x) = \bar{V}(x+2L)$ and $\int_0^{2L} \bar{V}(x)dx = 0$, $\bar{V}(x)$ is in the space $L^2_{per,0}$, and can be expressed in the Fourier form

$$\bar{V}(x) = \sum_{k=1}^{\infty} (A_k \tilde{v}_{1,k}(x) + B_k \tilde{v}_{2,k}(x)),$$

where the coefficients A_k and B_k are determined by

$$A_k =< \bar{V}, \tilde{v}_{1,k} >, \quad B_k =< \bar{V}, \tilde{v}_{2,k} > .$$

Its derivative is given by

$$\bar{V}_x(x) = \sum_{k=1}^{\infty} \beta_k (A_k \tilde{v}_{2,k}(x) - B_k \tilde{v}_{1,k}(x)).$$

Making the inner products, we have

$$\begin{aligned} \|\bar{V}\|^2 &= \int_0^{2L} \bar{V}^2(x)dx \\ &= \int_0^{2L} \sum_{k=1}^{\infty}\sum_{l=1}^{\infty} (A_k \tilde{v}_{1,k}(x) + B_k \tilde{v}_{2,k}(x))(A_l \tilde{v}_{1,l}(x) + B_l \tilde{v}_{2,l}(x))\, dx \\ &= \sum_{k=1}^{\infty}\sum_{l=1}^{\infty} \Big(A_k A_l \int_0^{2L} \tilde{v}_{1,k}(x)\tilde{v}_{1,l}(x)dx + A_k B_l \int_0^{2L} \tilde{v}_{1,k}(x)\tilde{v}_{2,l}(x)dx \\ &\quad + B_k A_l \int_0^{2L} \tilde{v}_{2,k}(x)\tilde{v}_{1,l}(x)dx + B_k B_l \int_0^{2L} \tilde{v}_{2,k}(x)\tilde{v}_{2,l}(x)dx \Big) \\ &= \sum_{k=1}^{\infty} (A_k^2 + B_k^2) \end{aligned} \tag{2.17}$$

and

$$\begin{aligned} \|\bar{V}_x\|^2 &= \int_0^{2L} \bar{V}_x^2(x)dx \\ &= \int_0^{2L} \sum_{k=1}^{\infty}\sum_{l=1}^{\infty} \beta_k\beta_l (A_k \tilde{v}_{2,k}(x) - B_k \tilde{v}_{1,k}(x))(A_l \tilde{v}_{2,l}(x) - B_l \tilde{v}_{1,l}(x))\, dx \\ &= \sum_{k=1}^{\infty} \beta_k^2 (A_k^2 + B_k^2). \end{aligned} \tag{2.18}$$

Since $\beta_k \geq \beta_1 = \frac{\pi}{L}$ for $k = 1, 2, \cdots$, then (2.17) and (2.18) give

$$\|\bar{V}_x\|^2 = \sum_{k=1}^{\infty} \beta_k^2 (A_k^2 + B_k^2) \geq \beta_1^2 \sum_{k=1}^{\infty} (A_k^2 + B_k^2) = \frac{\pi^2}{L^2} \|\bar{V}\|^2,$$

which implies (2.12). □

Now we are ready to discuss the existence of the trivial and non-trivial stationary solutions, as well as their phase transitions.

Lemma 2.3 *If* (1.10) *holds, then the stationary problem* (2.4) *has and only has the trivial solution* $\bar{V}(x) \equiv 0$.

Proof. Multiplying (2.4) by $\bar{V}(x)$ and integrating by parts with respect to x over $[0, 2L]$, as well as using the periodic boundary condition, we obtain

$$0 = \int_0^{2L} (-\varepsilon_1\varepsilon_2\bar{V}_{xx} + \bar{\sigma}(\bar{V}) - \bar{a})\bar{V}(x)dx = \int_0^{2L} (\varepsilon_1\varepsilon_2\bar{V}_x^2 + \bar{\sigma}(\bar{V})\bar{V})dx. \tag{2.19}$$

Since

$$\begin{aligned}\bar{\sigma}(\bar{V})\bar{V} &= \bar{V}^4 + 3m_0\bar{V}^3 + (3m_0^2 - 1)\bar{V}^2 \\ &= \bar{V}^2\Big(\bar{V} + \frac{3}{2}m_0\Big)^2 + \Big(\frac{3}{4}m_0^2 - 1\Big)\bar{V}^2 \\ &\geq \Big(\frac{3}{4}m_0^2 - 1\Big)\bar{V}^2,\end{aligned} \tag{2.20}$$

applying (2.12) and (2.20) into (2.19), as well as noting (1.10), we have

$$\begin{aligned}0 &= \int_0^{2L} (\varepsilon_1\varepsilon_2\bar{V}_x^2 + \bar{\sigma}(\bar{V})\bar{V})dx \\ &\geq \int_0^{2L} \Big[\frac{\varepsilon_1\varepsilon_2\pi^2}{L^2}\bar{V}^2 + \Big(\frac{3}{4}m_0^2 - 1\Big)\bar{V}^2\Big]dx \\ &= \Big[\frac{\varepsilon_1\varepsilon_2\pi^2}{L^2} - \Big(1 - \frac{3}{4}m_0^2\Big)\Big]\int_0^{2L} \bar{V}^2(x)dx \geq 0,\end{aligned} \tag{2.21}$$

which implies $\int_0^{2L} \bar{V}^2(x)dx \equiv 0$, i.e., $\bar{V}(x) \equiv 0$. Here, in order to obtain the non-negativity of the last step of (2.21), we need

$$\frac{\varepsilon_1\varepsilon_2\pi^2}{L^2} - \Big(1 - \frac{3}{4}m_0^2\Big) > 0,$$

which leads to the sufficient condition (1.10). □

Lemma 2.4 *If* (1.11) *holds, then, except the trivial solution* 0*, the stationary problem* (2.4) *has at least one non-trivial solution.*

Proof. Let $\bar{V}_n(x)$ be a minimizing sequence:

$$G(\bar{V}_n) \to g = \inf_{v\in H^1_{per,0}} G(v), \tag{2.22}$$

where $G(V)$ is defined in (2.7). Notice that $\bar{V}_n$ is bounded in $H^1_{per,0}$, hence there exists a subsequence, still denoting by $\bar{V}_n$, such that

$$\bar{V}_n \to \bar{V} \quad \text{weakly in} \quad H^1_{per,0}, \tag{2.23}$$

and

$$\bar{V}_n \to \bar{V} \quad \text{strongly in} \quad L^p(0, 2L), 1 < p < 6. \tag{2.24}$$

This implies

$$\int_0^{2L} \bar{V}_x^2 dx \leq \underline{\lim}_{n\to\infty} \int_0^{2L} |(\bar{V}_n)_x|^2 dx, \tag{2.25}$$

and

$$\int_0^{2L} H(\bar{V}_n)dx \to \int_0^{2L} H(\bar{V})dx. \tag{2.26}$$

From (2.22), (2.25) and (2.26), we see that $\bar{V}(x)$ is a minimum element of functional G, and by Lemma 2.1, it is also a solution of stationary problem (2.4).

It is known that 0 satisfying $G(0) = 0$ is the trivial solution of (2.4). We are now going to prove that the minimum element $\bar{V}(x)$ is a non-trivial solution. In fact, we can prove $g = G(\bar{V}) < 0$, which implies that $\bar{V}(x) \not\equiv 0$ is non-trivial.

Choosing a test function as $\eta\tilde{v}_{1,1}(x)$, where $\tilde{v}_{1,1}(x) = \frac{1}{\sqrt{L}} \sin \frac{\pi x}{L}$, and $\eta > 0$ is a constant such that

$$\eta^2 < \frac{8L}{3}\Big(1 - 3m_0^2 - \frac{\varepsilon_1\varepsilon_2\pi^2}{L^2}\Big). \tag{2.27}$$

It is known that $\eta\tilde{v}_{1,1}(x) \in H^1_{per,0}$. A straightforward calculation gives

$$\int_0^{2L} |(\tilde{v}_{1,1})_x|^2 dx = \frac{\pi^2}{L^2}\int_0^{2L} \tilde{v}_{1,1}^2 dx.$$

and

$$\int_0^{2L} \tilde{v}_{1,1}^2(x)dx = 1, \quad \int_0^{2L} \tilde{v}_{1,1}^3(x)dx = 0, \quad \int_0^{2L} \tilde{v}_{1,1}^4(x)dx = \frac{3}{4L}.$$

Hence,

$$\begin{aligned} G(\eta\tilde{v}_{1,1}) &= \int_0^{2L} \Big[\frac{\varepsilon_1\varepsilon_2\eta^2}{2}|(\tilde{v}_{1,1})_x|^2 + H(\eta\tilde{v}_{1,1})\Big]dx \\ &= \int_0^{2L} \Big[\frac{\varepsilon_1\varepsilon_2\eta^2\pi^2}{2L^2}\tilde{v}_{1,1}^2 + \frac{\eta^4}{4}\tilde{v}_{1,1}^4 + m_0\eta^3\tilde{v}_{1,1}^3 + \frac{(3m_0^2-1)\eta^2}{2}\tilde{v}_{1,1}^2\Big]dx \\ &= \Big(\frac{\varepsilon_1\varepsilon_2\pi^2}{2L^2} - \frac{1-3m_0^2}{2}\Big)\eta^2 \int_0^{2L}\tilde{v}_{1,1}^2 dx + \frac{\eta^4}{4}\int_0^{2L}\tilde{v}_{1,1}^4 dx + m_0\eta^3\int_0^{2L}\tilde{v}_{1,1}^3 dx \\ &= -\Big(\frac{1-3m_0^2}{2} - \frac{\varepsilon_1\varepsilon_2\pi^2}{2L^2} - \frac{3}{16L}\eta^2\Big)\eta^2 \\ &< 0, \end{aligned} \tag{2.28}$$

because of (1.11) and (2.27). Since $\bar{V}(x)$ is the minimum element of $G(v)$ on H^1_{per}, (2.28) implies $G(\bar{V}) = g < 0$. □

Lemma 2.5 *The non-trivial solutions are oscillatory, and always exhibit phase transitions.*

Proof. According to the periodicity, the non-trivial solution is always oscillatory.

Now we are going to prove that the non-trivial solutions always exhibit phase transitions. In fact, differentiating (2.1) with respect to x gives

$$\sigma'(V)V_x = \varepsilon_1\varepsilon_2 V_{xxx}. \tag{2.29}$$

Notice that $V_{xxx}(x)$ is non-monotone, namely, $V_{xxx}(x)$ changes signs on $[0, 2L]$. Let $V_{xxx}(x) \neq 0$ at some points x, we have four possible cases:

Case 1. $V_{xxx}(x) > 0$ and $V_x(x) > 0$;
Case 2. $V_{xxx}(x) > 0$ and $V_x(x) < 0$;
Case 3. $V_{xxx}(x) < 0$ and $V_x(x) > 0$;
Case 4. $V_{xxx}(x) < 0$ and $V_x(x) < 0$.

For Case 1, when $V_{xxx}(x) > 0$ and $V_x(x) > 0$, then from (2.29) we have $\sigma'(V(x)) > 0$, which is equivalent to $|V(x)| > \frac{1}{\sqrt{3}}$, i.e., $V(x) > \frac{1}{\sqrt{3}}$ or $V(x) < -\frac{1}{\sqrt{3}}$. This indicates that the phase transitions exhibit. Similarly, we can verify that, for Case 2, the solution V at x doesn't exhibit a phase transition, i.e., we have $-\frac{1}{\sqrt{3}} < V(x) < \frac{1}{\sqrt{3}}$, while Case 3 like Case 1 exhibits a phase transition, and Case 4 like Case 2 doesn't exhibit a phase transition.

If there are some points such that Case 1 or Case 3 occurs, we immediately prove that the non-trivial solutions exhibit phase transitions. This proves our lemma. If there are only Case 2 and Case 4, but no Case 1 and Case 3, we see that, whatever $V_{xxx} > 0$ or < 0, we always have $V_x < 0$. This is a contradiction to the non-monotonicity of $V(x)$. Therefore, the non-trivial stationary solution $V(x)$ always occurs phase transitions. □

Proof of Theorem 1.1. Thanks to Lemma 2.3, and notice that $V(x) = \bar{V}(x) + m_0$, we proved Part 1 of Theorem 1.1. Furthermore, Lemma 2.4 implies the existence of multiple non-trivial stationary solutions of (2.1) in Part 2 of Theorem 1.1, and Lemma 2.5 proves that all non-trivial stationary solutions are oscillatory and exhibit the phase transitions. Thus, Theorem 1.1 is proved. □

Now we are going to prove Theorem 1.2, i.e., we count the number of the non-trivial solutions for $m_0 \in [0, \frac{1}{\sqrt{3}})$.

Firstly, when $m_0 = 0$, for the stationary equation (2.1), if we change the periodic interval from $[0, 2L]$ to $[-L, L]$, then it is equivalent to

$$\begin{cases} \varepsilon_1\varepsilon_2 V_{xx} = \sigma(V) - a \\ V(x-L) = V(x+L) \\ \frac{1}{2L}\int_{-L}^{L} V(x)dx = m_0. \end{cases} \tag{2.30}$$

Let $y = \frac{x}{L}$, (2.30) can be rewritten as

$$\begin{cases} \frac{\varepsilon_1\varepsilon_2}{L^2} V_{yy} = \sigma(V) - a \\ V(y-1) = V(y+1) \\ \frac{1}{2}\int_{-1}^{1} V(y)dy = m_0. \end{cases} \tag{2.31}$$

By a very similar method as shown in [2, 23] (see [23]: Eqs. (3.16) and (3.17) on page 178 and Theorem 3.4 on page 179), we can prove that the modified stationary equation (2.31) has $2N_0 + 1$ solutions for N_0 satisfying (1.12). Here the detail of the proof is omitted.

Lemma 2.6 *For $m_0 = 0$ and $\varepsilon_1\varepsilon_2 < \frac{L^2}{\pi^2}(1 - 3m_0^2)$, there are $(2N_0 + 1)$ solutions for the stationary equation (2.31), where N_0 is the integer satisfying (1.12), i.e.,*

$$\frac{2L}{\pi\sqrt{\varepsilon_1\varepsilon_2}} - 1 < N_0 \leq \frac{2L}{\pi\sqrt{\varepsilon_1\varepsilon_2}}. \tag{2.32}$$

Furthermore, when $m_0 \neq 0$, by the change of variable $y = \frac{x}{2L} + \frac{1}{2}$, then we deduce (2.30) to

$$\begin{cases} \frac{\varepsilon_1\varepsilon_2}{4L^2} V_{yy} = \sigma(V) - a \\ V(y) = V(y+1) \\ \int_0^1 V(y)dy = m_0. \end{cases} \tag{2.33}$$

By the transversality arguments as presented in [9], based on the size of viscosity $\varepsilon_1\varepsilon_2$ and the location of the initial average m_0, we can very similarly count the number of stationary solutions as follows. Here, the proof is also omitted.

Lemma 2.7

1. *When $m_0 \in (0, \frac{1}{\sqrt{5}})$, if $\varepsilon_1\varepsilon_2 \geq \frac{L^2}{\pi^2}(1 - 3m_0^2)$, then there is no non-trivial stationary solution to Eq.* (2.33)*. However, if $\varepsilon_1\varepsilon_2 < \frac{L^2}{\pi^2}(1 - 3m_0^2)$, then there is only one non-trivial solution to Eq.* (2.33)*.*

2. *When $m_0 \in (\frac{1}{\sqrt{5}}, \frac{1}{\sqrt{3}})$, there exists a positive number $\varepsilon^* > \frac{L^2}{\pi^2}(1 - 3m_0^2)$ such that, if $\varepsilon_1\varepsilon_2 > \varepsilon^*$, then there is no non-trivial solution to Eq.* (2.33)*. However, if $\frac{L^2}{\pi^2}(1 - 3m_0^2) < \varepsilon_1\varepsilon_2 < \varepsilon^*$, then there are only two non-trivial solutions to Eq.* (2.33)*. While, if $\varepsilon_1\varepsilon_2 \leq \frac{L^2}{\pi^2}(1 - 3m_0^2)$, then there is only one non-trivial solution to Eq.* (2.33)*.*

Proof of Theorem 1.2. Combining Lemma 2.6 and Lemma 2.7, we immediately prove Theorem 1.2. □

3. Numerical Simulations

With the central finite difference for the second order derivative and the Simpson's rule for the integral, system (2.1) is discretized as follows:

$$\begin{cases} \varepsilon_1\varepsilon_2(v_{i-1} - 2v_i + v_{i+1}) = h^2(v_i^3 - v_i) - a, \quad i = 1, 2, \cdots, n-1 \\ v_n = v_0 \\ h\sum_{i=1}^{n/2}(2v_{2i-2} + 4v_{2i-1}) = 6Lm_0 \\ a = h^2(v_0^3 - v_0) - \varepsilon_1\varepsilon_2(v_1 - 2v_0 + v_{n-1}) \end{cases} \tag{3.1}$$

where n is even, and $h = \frac{2L}{n}$.

The above scheme is a system of $n + 1$ nonlinear algebraic equations with $n + 1$ variables. The Newton-Raphson's nonlinear solver is used with various initial guessing points. The computer program uses a random generator for the initial guess. For each case, a reasonable number of sets of guessing values has been implemented. The convergent solution could be unique or multiple depending on the system configuration, i.e. the values of the initial average m_0 and the viscosity ε_1 and ε_2. The smaller the space step h is, i.e. the larger the number of grids n is, the more accurate the numerical solutions are. However, when n is large, the computational cost for the nonlinear solver increases dramatically. For the chosen $L = \pi$ for the cases reported in the following, n near 60 (i.e. h near 0.1) provides convergent solutions in a short time (within couple of minutes) and with good accuracy. Further increasing the value n does not change the solutions significantly.

To verify Theorem 1.1, first of all, by plugging $m_0 = 0$ and $\varepsilon_1\varepsilon_2 = 4$ in the numerical program, we obtain only the trivial zero solution $V \equiv 0$. For small viscosities, the case with $m_0 = 0$ and $\varepsilon_1\varepsilon_2 = 0.49$ is examined, and one non-trivial solution is found as shown in Figure 3.1, which also verifies the first item of Theorem 1.2 with $N_0 = 2$.

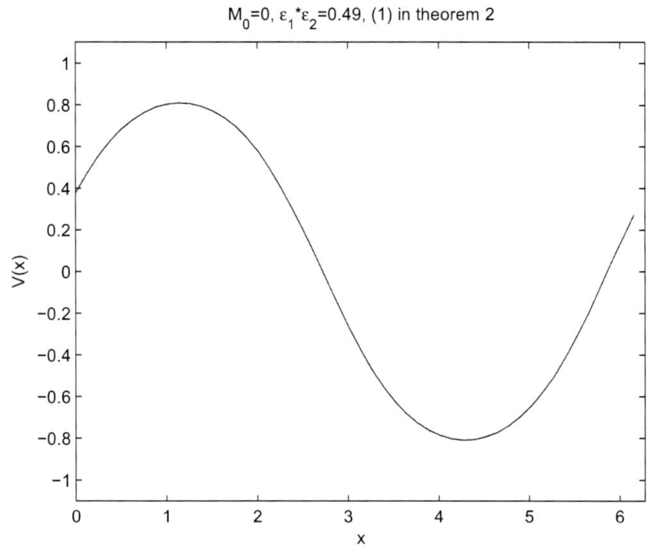

Figure 3.1. Case 1: $m_0 = 0$, $\varepsilon_1\varepsilon_2 = 0.49$, Item 1 in Theorem 1.2.

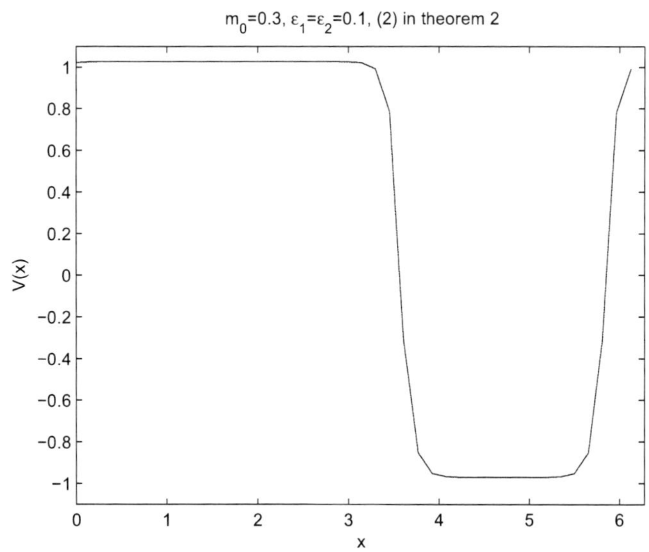

Figure 3.2. Case 2: $m_0 = 0.3$, $\varepsilon_1 = \varepsilon_2 = 0.1$, Item 2 in Theorem 1.2 with a unique non-trivial solution.

For Item 2 in Theorem 1.2, the chosen parameters are $m_0 = 0.3$ and $\varepsilon_1 = \varepsilon_2 = 0.1$, and the unique non-trivial solution is found. See Figure 3.2.

Select $m_0 = 0.5$ and $\varepsilon_1\varepsilon_2 = 0.28$ that satisfies $\frac{L^2}{\pi^2}(1 - 3m_0^2) < \varepsilon_1\varepsilon_2 < \varepsilon^*$, we find two non-trivial solutions as shown in Figure 3.3.

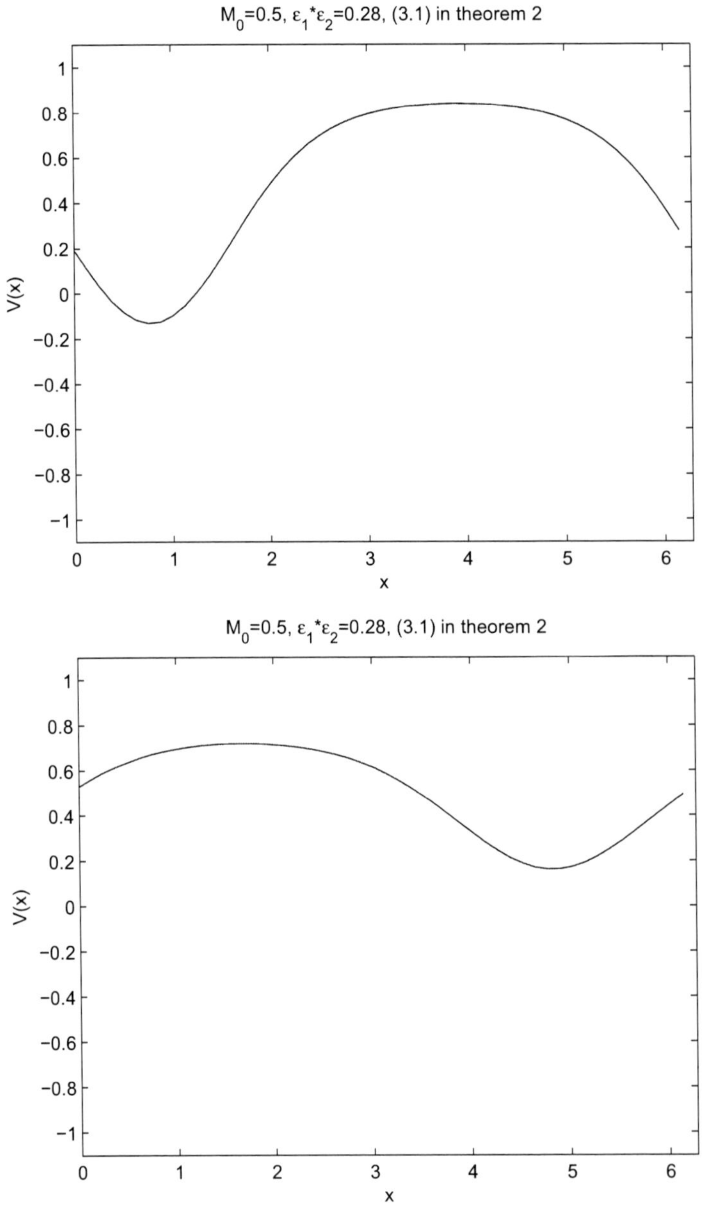

Figure 3.3. Case 3: $m_0 = 0.5$, $\varepsilon_1\varepsilon_2 = 0.28$, the first part of Item 3 in Theorem 1.2 with two non-trivial solutions.

For the same value of the initial average $m_0 = 0.5$, when the viscosity is small as $\varepsilon_1\varepsilon_2 = 0.02$ such that $\varepsilon_1\varepsilon_2 < \frac{L^2}{\pi^2}(1 - 3m_0^2)$, the program only finds one non-trivial solution as shown Figure 3.4.

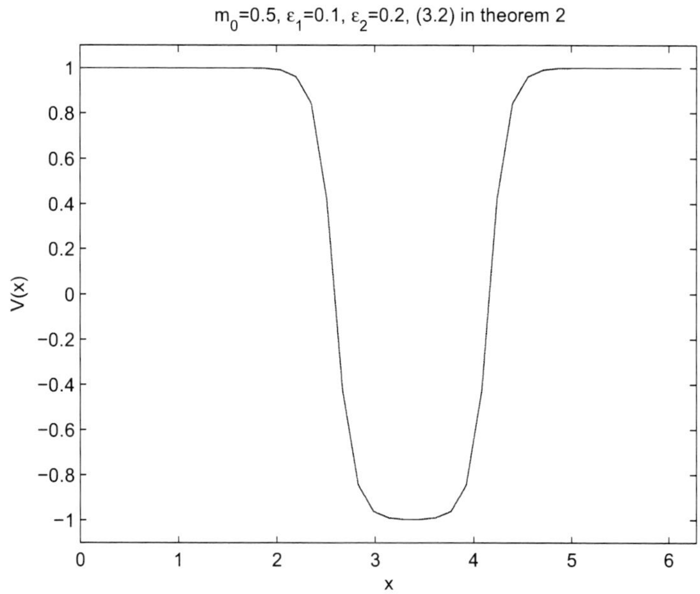

Figure 3.4. Case 4: $m_0 = 0.5$, $\varepsilon_1\varepsilon_2 = 0.02$, the second part of Item 3 in Theorem 1.2 with a unique non-trivial solutions.

References

[1] M. Affouf and R. E. Caflisch, A numerical study of Riemann problem solutions and stability for a system of viscous conservation laws of mixed type, *SIAM J. Appl. Math.* **51** (1991) 605-634.

[2] J. Carr, M. Gurtin and M. Slemrod, Structured phase transitions on a finite interval, *Arch. Rational Mech. Anal.*, **86** (1984) 317–351.

[3] J. Carr, M. Gurtin and M. Slemrod, One-dimensional structured phase transformations under prescribed loads, *J. Elasticity*, **15** (1985) 133–142.

[4] A. Eden, A. Milani and B. Nicolaenko, Local expential attractors for model of phase change for compressible gas dynamics, *Nonlinearity*, **6** (1993) 93–117.

[5] C. M. Elliott and S. Zheng, On the Cahn-Hilliard equation, *Arch. Rational Mech. Anal.*, **96** (1986) 339–357.

[6] H.-T. Fan, A limiting "viscosity" approach to the Riemann problem for materials exhibiting a change of phase. II, *Arch. Rational Mech. Anal.*, **116** (1992) 317–337.

[7] P. C. Fife and X.-P. Wang, Periodic structures in a van der Waals fluid, *Proc. Royal Soc. Edinburgh, Sect. A*, **128** (1998) 235–250.

[8] M. Grinfeld, J.E. Furter and J.C. Eilbeck, A monotonicity theorem and its application to stationary solutions of phase field model, *IMA J. Appl. Math.* **49** (1992) 61–72; **50** (1993) 203–204.

[9] M. Grinfeld and A. Novick-Cohen, Counting stationary solutions of the Cahn-Hilliard equation by transversality arguments, *Proc. Royal Soc. Edinburgh, Sect. A*, **125** (1995) 351–370.

[10] M. E. Gurtin and H. Matano, On the structure of equilibrium phase transitions within the gradient theory of fluids, *Quart. Appl. Math.*, **46** (1988) 301–317.

[11] M. E. Gurtin, On a theory of phase transitions with interfacial energy, *Arch. Rational Mech. Anal.*, **87** (1985) 187–212.

[12] R. Hagan and M. Slemrod, The viscous-capillarity criterion for shocks and phase transitions, *Arch. Rational Mech. Anal.* **83** (1983) 333–361.

[13] D.-Y. Hsieh and X.-P. Wang, Phase transition in van der Waals fluid, *SIAM J. Appl. Math.* **57** (1997) 871–892.

[14] R. D. James, The propagation of phase boundaries in elastic bars, *Arch. Rational Mech. Anal.* **73** (1980) 125–158.

[15] M. Mei, Y.-S. Wong and L. Liu, Phase transitions in a coupled viscoelastic system with periodic initial-boundary condition: (I) existence and uniform boundedness, *Discrete Continu. Dynam. Systems, Series B*, **7** (2007) 825–837.

[16] M. Mei, Y.-S. Wong and L. Liu, Phase transitions in a coupled viscoelastic system with periodic initial-boundary condition: (II) convergence, *Discrete Continu. Dynam. Systems, Series B*, **7** (2007) 839–857.

[17] B. Nicolaenko and W. Qian, Inertial manifolds for nonlinear viscoelasticity equations, *Nonlinearity*, **11** (1998) 1075–1093.

[18] A. Novick-Cohen, On Cahn-Hilliard type equations, *Nonlinear Analysis*, **15** (1990) 797–814.

[19] A. Novick-Cohen and L.A. Peletier, Steady states of the one-dimensional Cahn-Hilliard equation, *Proc. Royal Soc. Edinburgh, Sect. A*, **123** (1993) 1071–1098.

[20] A. Novick-Cohen and L.A. Segel, Nonlinear aspects of the Cahn-Hilliard equation, *Phsica D* **10** (1984) 277–298.

[21] M. Shearer, Dynamic phase transitions in a van der Waals gas, *Quart. Appl. Math.*, **46** (1988) 631–636.

[22] M. Slemrod, Admissibility criteria for propagating phase boundaries in a van der Waals fluid, *Arch. Rational Mech. Anal.*, **81** (1983) 301–315.

[23] S. Zheng, Asymptotic behavior of solution to the Cahn-Hillard equation, *Applicable Analysis*, **23** (1986) 164–184.

[24] S. Zheng, Nonlinear Parabolic Equations and Hyperbolic-Parabolic Coupled Systems, Pitman Monographs and Surverys in Pure and Applied Mathematics **76**, Longman Scientific & Technical, Harlow; copublished in the United States with John Wiley & Sons, Inc., New York, 1995.

[25] K. Zumbrun, Dynamical stability of phase transitions in the p-system with viscosity-capillarity, *SIAM J. Appl. Math.*, **60** (2000) 1913–1924.

In: Nonlinear Analysis Research Trends
Editor: Inès N. Roux, pp. 299-319
ISBN 978-1-60456-358-0

Chapter 10

IMMERSED BOUNDARY METHOD: THE EXISTENCE OF APPROXIMATE SOLUTION IN TWO-DIMENSIONAL CASE

Ling Rao[1,2] * ***and Hongquan Chen***[1]
[1]College of Aerospace Engineering
Nanjing University of Aeronautics and Astronautics
Nanjing 210016, China
[2]Department of Applied Mathematics
Nanjing University of Science and Technology
Nanjing 210094, China

Abstract

This paper deals with the two-dimensional Navier-Stokes equations in which the source term involves a Dirac delta function and describes the elastic reaction of the immersed boundary. We analyze the existence of the approximate solution with Dirac delta function approximated by differentiable function. We obtain the result via the Banach Fixed Point Theorem and the properties of the solutions to the Navier-Stokes equations of viscous incompressible fluids with periodic boundary conditions.

Key Words: Navier-Stokes Equations; Immersed boundary method; Nonlinear ordinary differential equations. AMS Subject Classification: 35K10, 46N20, 34A12.

1. Introduction

Problems involving the interaction between a fluid flow and elastic interfaces may appear in several branches of science such as engineering, physics, biology, and medicine. Regardless the field, as a rule, they share in common a high degree of complexity, often displaying intricate geometry or time-dependent elastic properties, turning the problem into a real challenge for applied scientists, from both the mathematical modeling and the numerical simulation points of view.

*E-mail address: lingrao@sina.com

In the early 70s, Peskin [2, 8] introduced a mathematical model and a computational method to study the flow patterns of the blood around the heart valves. Through years, Peskin's immersed boundary (IB) method was developed for the computer simulation of general problems [9–12] involving a transient incompressible viscous fluid containing an immersed elastic interface, which may have time-dependent geometry or elastic properties, or both.

The IB method is at the same time a mathematical formulation and a numerical scheme. The mathematical formulation is based on the use of Eulerian variables to describe the dynamic of fluid and of Lagrangean variables along the moving structure. The force exerted by the structure on the fluid is taken into account by means of a Dirac delta function constructed according to certain principles. The main idea is to use a regular Eulerian mesh for the fluid dynamics simulation, coupled with a Lagrangian representation of the immersed boundary. The advantage of this method is that the fluid domain can have a simple shape, so that structured grids can be used. The Lagrangian mesh is independent of the Eulerian mesh. The interaction between the fluid and the immersed boundary is modeled using a well-chosen discrete approximation to the Dirac delta function.

Although the immersed boundary method is a particularly effective approach in scientific computation, but very little theoretical analysis has been performed on either the underlying model equations or numerical methods. Daniele Boffi [3] gave a variational formulation of the problem and provided a suitable modification of the IB method which made use of finite elements method. Because the source term in the Navier-Stokes equations involves a Dirac delta function, the problem is highly nonlinear and presents several difficulties related with the lacking of regularity of the solution of the Navier-Stokes equations due to such source term. Daniele Boffi analyzed the existence of the solution in a very simple one-dimensional heat equation. In [13] we dealt with the two-dimensional heat equation. We analyzed the existence of the approximate solution with Dirac delta function approximated by differentiable function. In this paper will deal with two-dimensional Navier-Stokes equations. We analyze the existence of the approximate solution still with Dirac delta function approximated by differentiable function. We obtain the results via the Banach Fixed Point Theorem and a theorem in nonlinear ordinary differential equations in abstract space and the properties of the solutions to the Navier-Stokes equations of viscous incompressible fluids with periodic boundary conditions.

In section 2 we will present the mathematical model of the problem. In order to prove the existence of the approximate solution, some properties of the Navier-Stokes equations with the space-periodic boundary conditions are reviewed in section 3. In section 4 we will introduce the corresponding variational formulation of the problem. In section 5 we will prove the existence of the approximate solution of the problem via a fixed point argument.

2. Problem

The authors (see [3]) considered the model problem of a viscous incompressible fluid in a two-dimensional square domain Ω containing an immersed massless elastic boundary in the form of a curve. To be more precise, for all $t \in [0, T]$, let Γ_t be a simple curve, the configuration of which is given in a parametric form, $\mathbf{X}(s, t), 0 \le s \le L, \mathbf{X}(0, t) =$

$\mathbf{X}(L,t)$. The equations of motion of the system are

$$\frac{\partial \mathbf{u}}{\partial t} - \mu\Delta\mathbf{u} + \mathbf{u}\cdot\Delta\mathbf{u} + \Delta p = \mathbf{F} \qquad \text{in } \Omega\times(0,T), \tag{1}$$

$$\Delta\cdot\mathbf{u} = 0 \qquad \text{in } \Omega\times(0,T), \tag{2}$$

$$\mathbf{F}(\mathbf{x},t) = \int_0^L \mathbf{f}(s,t)\delta(x - \mathbf{X}(s,t))ds \qquad \forall\mathbf{x}\in\Omega,\ t\in(0,T), \tag{3}$$

$$\frac{\partial \mathbf{X}}{\partial t}(s,t) = \mathbf{u}(\mathbf{X}(s,t),t) \qquad \forall s\in[0,L],\ t\in(0,T). \tag{4}$$

Here $\mathbf{u}$ is the fluid velocity and p is the fluid pressure. The coefficient μ is the fluid viscosity constant. Eqs. (1) and (2) are the usual incompressible Navier-Stokes equations. $\mathbf{F}$ is the force density generated by the boundary on the fluid. The force exerted by the element of boundary on the fluid is $\mathbf{f}$. The function δ in the integrals is the two-dimensional Dirac delta function concentrate at the origin. Eq. (4) is equivalent to the no-slip condition that the fluid sticks to the boundary.

In [3] and [13] the problem was supplemented with the following boundary and initial conditions:

$$\mathbf{u}(\mathbf{x},0) = \mathbf{u}_0(\mathbf{x}) \qquad \forall\mathbf{x}\in\Omega, \tag{5}$$

$$\mathbf{X}(s,0) = \mathbf{X}_0(s) \qquad \forall s\in[0,L], \tag{6}$$

$$\mathbf{u}(\mathbf{x},t) = 0 \qquad \forall(\mathbf{x},t)\in\partial\Omega\times(0,T). \tag{7}$$

When we perform the numerical simulation with the IB method the fluid domain Ω often can be chosen to have a simple big enough shape such as square or rectangle, then we can make space-periodic extension of the problem and suppose that the fluid satisfies space-periodic boundary condition. Such assumption is also useful for idealizations and needed in the following proofs. In this paper, we use the space-periodic boundary condition instead of (7). We use Ω to denote the space-period:

$$\Omega = (-\frac{L_1}{2},\frac{L_1}{2})\times(-\frac{L_2}{2},\frac{L_2}{2}).$$

We assume that the fluid fills the entire space R^2 but with condition that

$\mathbf{u}, \mathbf{F}$ and p are periodic in each direction $ox_i,\ i=1,2,$
with corresponding periods $L_i > 0$.

It is sometimes useful and simpler to assume that average flow is zero, that is

$$\frac{1}{|\Omega|}\int_\Omega \mathbf{u}(\mathbf{x})d\mathbf{x} = 0.$$

Moreover, the general case-where the volume forces and the initial condition do not average to zero - can be reduced to this case. We may refer to [4, 6] for the method. We show it as follows.

First, averaging each term in (1). Using integration by parts and periodicity condition, several terms vanish, we get

$$\frac{d}{dt}\int_{\Omega}\mathbf{u}(\mathbf{x},t)d\mathbf{x} = \int_{\Omega}\mathbf{F}(\mathbf{x},t)d\mathbf{x}.$$

Therefore if we set

$$M_u(t) = \frac{1}{|\Omega|}\int_{\Omega}\mathbf{u}(\mathbf{x},t)d\mathbf{x} \tag{8}$$

and

$$M_F(t) = \frac{1}{|\Omega|}\int_{\Omega}\mathbf{F}(\mathbf{x},t)d\mathbf{x},$$

we obtain

$$\frac{dM_u(t)}{dt} = M_F(t),$$

thus

$$M_u(t) = \frac{1}{|\Omega|}\int_{\Omega}\mathbf{u}_0(\mathbf{x})d\mathbf{x} + \int_0^t M_F(t)dt.$$

Let

$$\begin{array}{l} \hat{\mathbf{u}} = \mathbf{u} - M_u, \\ \hat{\mathbf{F}} = \mathbf{F} - M_F, \\ I_u(t) = \int_0^t M_u(s)ds, \\ \hat{\hat{\mathbf{u}}}(\mathbf{x},t) = \hat{\mathbf{u}}(\mathbf{x} + I_u(t), t), \\ \hat{p}(\mathbf{x},t) = p(\mathbf{x} + I_u(t), t), \\ \hat{\hat{\mathbf{F}}}(\mathbf{x},t) = \hat{\mathbf{F}}(\mathbf{x} + I_u(t), t). \end{array} \tag{9}$$

Notice that $\hat{\hat{\mathbf{u}}}$, $\hat{\mathbf{p}}$, and $\hat{\hat{\mathbf{F}}}$ are also periodic with period Ω and $\hat{\hat{\mathbf{u}}}$, $\hat{\hat{\mathbf{F}}}$ have zero space average. With $\mathbf{u}$, p and $\mathbf{F}$ in (1)-(6) replaced by $\hat{\hat{\mathbf{u}}}$, $\hat{\mathbf{p}}$ and $\hat{\hat{\mathbf{F}}}$, we need to solve the following equations instead of (1)-(6) for $\hat{\hat{\mathbf{u}}}$, $\hat{\mathbf{p}}$ and $\mathbf{X}$:

$$\frac{\partial \mathbf{u}}{\partial t} - \mu\Delta\mathbf{u} + \mathbf{u}\cdot\Delta\mathbf{u} + \Delta p = \hat{\hat{\mathbf{F}}} \qquad \text{in } \Omega \times (0,T), \tag{10}$$

$$\Delta\cdot\mathbf{u} = 0 \qquad \text{in } \Omega \times (0,T), \tag{11}$$

$$\frac{\partial \mathbf{X}}{\partial t}(s,t) = \mathbf{u}(\mathbf{X}(s,t) - I_u(t), t) + M_u(t) \qquad \forall s \in [0,L], \quad t \in (0,T), \tag{12}$$

$$\mathbf{u}(\mathbf{x},0) = \hat{\mathbf{u}}_0(x), \qquad \forall x \in \Omega, \tag{13}$$

$$\mathbf{X}(s,0) = \mathbf{X}_0(s) \qquad \forall s \in [0,L], \tag{14}$$

$$\mathbf{X}(0,t) = \mathbf{X}(L,t) \qquad \forall t \in (0,T), \tag{15}$$

where for given f and $\mathbf{u}_0$, $\hat{\hat{\mathbf{u}}}_0$ and $\hat{\hat{\mathbf{F}}}$ may be calculated by (3) and (9),

$$\begin{aligned}\hat{\hat{\mathbf{u}}}_0(x) &= \mathbf{u}_0(x) - M_u(0) = \mathbf{u}_0(x) - \frac{1}{|\Omega|}\int_\Omega \mathbf{u}_0(x)d\mathbf{x},\\ \hat{\hat{\mathbf{F}}}(\mathbf{x},t) &= \mathbf{F}(\mathbf{x}+I_u(t),t) - M_F(t)\\ &= \int_0^L \mathbf{f}(s,t)\delta(\mathbf{x}+I_u(t)-\mathbf{X}(s,t))ds\\ &\quad -\frac{1}{|\Omega|}\int_0^L \mathbf{f}(s,t)(\int_\Omega \delta(x-\mathbf{X}(s,t))dx)ds.\end{aligned} \tag{16}$$

And with the solutions $\mathbf{u}(=\hat{\mathbf{u}})$, $p(=\hat{\mathbf{p}})$ of (10)-(16) we can recover the solutions $\mathbf{u}$, p of (1)-(6) by (9).

In order to obtain the existence of the approximate solution to equations (10)-(16), we review some properties of the Navier-Stokes equations with the space-periodic boundary conditions as follows.

3. Review of the Mathematical Theory of the Navier-Stokes Equations with the Space-Periodic Boundary Conditions

We introduce some basic mathematical properties about Navier-Stoks equations with the space-periodic boundary conditions. For more details of them, the readers may refer to [4, 6]. We shall be concerned with the spaces of two-dimensional vector functions. We use the notations

$$\mathbf{H}^m(\Omega) = \{H^m(\Omega)\}^2, \quad \mathbf{L}^2(\Omega) = \{L^2(\Omega)\}^2,$$

and we suppose that these product spaces are equipped with the usual product norm. The norm on $\mathbf{L}^2(\Omega)$ is denoted by $|\cdot|$ (also denoted by $\|\cdot\|_0$). The norm on $\mathbf{H}^m(\Omega)$ is denoted by $\|\cdot\|_m$. $(\cdot,\cdot)$ stands for the scalar product on $\mathbf{L}^2(\Omega)$. We denote by $\mathbf{H}^m_{per}(\Omega)$, $m \in \mathbb{N}$, the space of functions which are in $\mathbf{H}^m_{Loc}(R^n)$ (i.e. $\mathbf{u}|_{\mathscr{O}} \in \mathbf{H}^m(\mathscr{O})$ for every open bound set $\mathscr{O}$) and which are periodic with period Ω. For $m=0$, $\mathbf{H}^0_{per}(\Omega)$ is also denoted by $\mathbf{L}^2_{per}(\Omega)$ and coincides simply with $\mathbf{L}^2(\Omega)$ (the restrictions of the functions in $\mathbf{H}^0_{per}(\Omega)$ to Ω are the whole space $\mathbf{L}^2(\Omega)$). For an arbitrary $m \in \mathbb{N}$, $\mathbf{H}^m_{per}(\Omega)$ is a Hilbert space for the scalar product and the norm

$$(\mathbf{u},\mathbf{v})_m = \sum_{[\alpha]\le m}\int_\Omega D^\alpha\mathbf{u}(\mathbf{x})D^\alpha\mathbf{v}(\mathbf{x})d\mathbf{x}, \qquad \|\mathbf{u}\|_m = \{(\mathbf{u},\mathbf{u})_m\}^{\frac{1}{2}},$$

where $\alpha = (\alpha_1,\ldots,\alpha_n)$, $\alpha_i \in \mathbb{N}$, $[\alpha] = \alpha_1 + \ldots + \alpha_n$ and $D^\alpha = \dfrac{\partial^{[\alpha]}}{\partial x^{\alpha_1}\ldots\partial^{\alpha_n}}$.

We work with complex representation, for which we take $i = \sqrt{-1}$. Then a square integrable vector field $\mathbf{u} = \mathbf{u}(x)$ on Ω can be represented by the Fourier series expansion

$$\mathbf{u}(\mathbf{x}) = \sum_{\mathbf{k}\in\mathbb{Z}^n} c_{\mathbf{k}} e^{2\pi i\frac{\mathbf{k}}{\mathbf{L}}\cdot\mathbf{x}},$$

where

$$\frac{\mathbf{k}}{\mathbf{L}} = (\frac{k_1}{L_1},\frac{k_2}{L_2}).$$

The functions in $\mathbf{H}^m_{per}(\Omega)$ are easily characterized by their Fourier series expansion

$$\mathbf{H}^m_{per}(\Omega) = \{\mathbf{u},\ \mathbf{u} = \sum_{\mathbf{k}\in\mathbb{Z}^n} c_{\mathbf{k}} e^{2\pi i \frac{\mathbf{k}}{\mathbf{L}}\cdot\mathbf{x}}\quad \bar{c}_{\mathbf{k}} = c_{-\mathbf{k}},\ \sum_{\mathbf{k}\in\mathbb{Z}^n} |\mathbf{k}|^{2m}|c_{\mathbf{k}}|^2 < \infty\}, \tag{17}$$

and the norm $\|\mathbf{u}\|_m$ is equivalent to the norm $\{\sum_{\mathbf{k}\in\mathbb{Z}^n}(1+|\mathbf{k}|^{2m})|c_{\mathbf{k}}|^2\}^{\frac{1}{2}}$.

We denote by $\mathscr{V}$ the space of smooth($\mathcal{C}^\infty$) divergence-free vector fields on R^2 that are periodic with period Ω. Let H be the closure of $\mathscr{V}$ in $\mathbf{L}^2(\Omega)$ and let V be the closure of $\mathscr{V}$ in $\mathbf{H}^1(\Omega)$. The space H is equipped with the scalar product $(\cdot,\cdot)$ induced by $\mathbf{L}^2(\Omega)$; the space V with the scalar product

$$((\mathbf{u},\mathbf{v})) = \int_\Omega \sum_{i=1}^{2} \frac{\partial \mathbf{u}}{\partial x_i}\cdot\frac{\partial \mathbf{v}}{\partial x_i} d\mathbf{x}$$

and the associated norms, denoted by

$$|\mathbf{u}| = (\mathbf{u},\mathbf{u})^{\frac{1}{2}} \text{ for } \mathbf{u}\in H, \qquad \|\mathbf{u}\| = ((\mathbf{v},\mathbf{v}))^{\frac{1}{2}} \text{ for } \mathbf{v}\in V.$$

For the sake of simplicity, we restrict ourselves to the space-periodic case with vanishing space average. For the vanishing space average case, we have the additional condition

$$c_0 = \int_\Omega \mathbf{u}(\mathbf{x})d\mathbf{x} = 0.$$

Then

$$H = \{\mathbf{u}\in\mathbf{L}^2_{per}(\Omega);\ \nabla\cdot\mathbf{u} = 0,\ \int_\Omega \mathbf{u}(\mathbf{x}) = 0\}(\triangleq \dot{H});$$

$$V = \{\mathbf{u}\in\mathbf{H}^1_{per}(\Omega);\ \nabla\cdot\mathbf{u} = 0,\ \int_\Omega \mathbf{u}(\mathbf{x}) = 0\}(\triangleq \dot{V});$$

and

$$\mathbf{H}^m_{per}(\Omega) = \{\mathbf{u}\in\mathbf{H}^m_{per}(\Omega) \text{ of type } (17),\ c_0 = 0\}(\triangleq \dot{\mathbf{H}}^m_{per}(\Omega)).$$

The functions in $\dot{H}$ and $\dot{V}$ also can be characterized by there Fourier series expansion

$$\dot{H} = \{\mathbf{u},\ \mathbf{u} = \sum_{\mathbf{k}\in\mathbb{Z}^2\setminus\{0\}} c_{\mathbf{k}} e^{2\pi i\frac{\mathbf{k}}{\mathbf{L}}\cdot\mathbf{x}},\quad \bar{c}_{\mathbf{k}} = c_{-\mathbf{k}},\ \frac{\mathbf{k}}{\mathbf{L}}\cdot c_{\mathbf{k}} = 0,\ \sum_{\mathbf{k}\in\mathbb{Z}^2\setminus\{0\}} |c_{\mathbf{k}}|^2 < \infty\};$$

$$\dot{V} = \{\mathbf{u},\ \mathbf{u} = \sum_{\mathbf{k}\in\mathbb{Z}^2\setminus\{0\}} c_{\mathbf{k}} e^{2\pi i\frac{\mathbf{k}}{\mathbf{L}}\cdot\mathbf{x}},\quad \bar{c}_{\mathbf{k}} = c_{-\mathbf{k}},\ \frac{\mathbf{k}}{\mathbf{L}}\cdot c_{\mathbf{k}} = 0,\ \sum_{\mathbf{k}\in\mathbb{Z}^2\setminus\{0\}} |\frac{\mathbf{k}}{\mathbf{L}}|^2|c_{\mathbf{k}}|^2 < \infty\}.$$

For all $s\in R$ we may consider the space

$$V_s = \{\mathbf{u}\in\dot{\mathbf{H}}^s_{per}(\Omega),\ div\mathbf{u} = 0\} =$$
$$\{\mathbf{u},\ \mathbf{u} = \sum_{\mathbf{k}\in\mathbb{Z}^2\setminus\{0\}} c_{\mathbf{k}} e^{2\pi i\frac{\mathbf{k}}{\mathbf{L}}\cdot\mathbf{x}},\quad \bar{c}_{\mathbf{k}} = c_{-\mathbf{k}},\ \frac{\mathbf{k}}{\mathbf{L}}\cdot c_{\mathbf{k}} = 0,\ \sum_{\mathbf{k}\in\mathbb{Z}^2\setminus\{0\}} |\frac{\mathbf{k}}{\mathbf{L}}|^{2s}|c_{\mathbf{k}}|^2 < \infty\}.$$

Note that, V_s is a closed subspace of $\dot{\mathbf{H}}^s_{per}(\Omega)$. $V_{s_1} \subset V_{s_2}$ for $s_1 \geq s_2, V_1 = \dot{V}$ and $V_0 = \dot{H}$. Hence, $\dot{V} \subset V_s \subset \dot{H}$ for $0 \leq s \leq 1, V_s \subset \dot{V}$ for $s \geq 1$, and $V_s \supset \dot{H}$ for $s < 0$. It can be shown that V_s is a Hilbert space for the norm

$$\|\mathbf{u}\|_{V_s} = (\sum_{\mathbf{k}\in\mathbb{Z}^2\backslash\{0\}} |\frac{\mathbf{k}}{\mathbf{L}}|^{2s}|c_{\mathbf{k}}|^2)^{\frac{1}{2}}.$$

Of particular interest are the spaces V_2 and V_{-1}. The space V_{-1} is the dual space of $\dot{V}$, usually denoted V'; this is the space of linear continuous forms on $\dot{V}$. More generally, for all $s \geq 0, V_{-s}$ is the dual of V_s.

We have

$$H^{\perp} = \{\mathbf{u} \in \mathbf{L}^2(\Omega); \mathbf{u} = \nabla p,\ p \in \mathbf{H}^1_{per}(\Omega)\}.$$

We definite the so-called Leray projector by

$$P_L : \mathbf{L}^2(\Omega) \to H,$$

which is the orthogonal projector onto H in $\mathbf{L}^2(\Omega)$.

We definite the Stokes operator A by

$$A\mathbf{u} = -P_L\triangle\mathbf{u} \quad \forall\mathbf{u} \in D(A) = \dot{V} \cap \dot{\mathbf{H}}^2_{per}(\Omega),$$

where $\triangle$ is the Laplacian. And we can see that

$$-P_L\triangle\mathbf{u} = -\triangle\mathbf{u} \quad \forall\mathbf{u} \in D(A) = \dot{V} \cap \dot{\mathbf{H}}^2_{per}(\Omega).$$

The Stokes operator is a positive self-adjoint operator, so we can work with fractional powers of A. A is just the mapping

$$\mathbf{u} = \sum_{\mathbf{k}\in\mathbb{Z}^2\backslash\{0\}} c_{\mathbf{k}} e^{2\pi i\frac{\mathbf{k}}{\mathbf{L}}\cdot\mathbf{x}} \to A\mathbf{u} = \sum_{\mathbf{k}\in\mathbb{Z}^2\backslash\{0\}} |\frac{\mathbf{k}}{\mathbf{L}}|^2 c_{\mathbf{k}} e^{2\pi i\frac{\mathbf{k}}{\mathbf{L}}\cdot\mathbf{x}}.$$

$A^r\mathbf{u}$ is the operator

$$\mathbf{u} = \sum_{\mathbf{k}\in\mathbb{Z}^2\backslash\{0\}} c_{\mathbf{k}} e^{2\pi i\frac{\mathbf{k}}{\mathbf{L}}\cdot\mathbf{x}} \to A^r\mathbf{u} = \sum_{\mathbf{k}\in\mathbb{Z}^2\backslash\{0\}} |\frac{\mathbf{k}}{\mathbf{L}}|^{2r} c_{\mathbf{k}} e^{2\pi i\frac{\mathbf{k}}{\mathbf{L}}\cdot\mathbf{x}}.$$

It is straightforward to see that A^r maps V_{2s} continuously onto V_{2s-2r} $(s, r \in \mathbb{R})$. In particular, for $s \geq 0$, we have $A^s V_{2s} = \dot{H}$ and so $V_{2s} = D(A^s)$ is the domain of the (unbounded) operator A^s in $\dot{H}$. The norm $|A^s\mathbf{u}|(= \|\mathbf{u}\|_{V_{2s}})$ is equivalent to the norm induced by $\dot{\mathbf{H}}^{2s}_{per}(\Omega)$,

$$c\|\mathbf{u}\|_{2s} \leq |A^s\mathbf{u}| \leq c'\|\mathbf{u}\|_{2s} \quad \forall\mathbf{u} \in D(A^s), \tag{18}$$

with positive constants c, c' depending on L_1, L_2 and s.

We use some notations as follows. Let a, b be two extended real numbers, $-\infty \leq a < b \leq +\infty$, and let X be a Banach space. For given $\alpha,\ 1 \leq \alpha < +\infty,\ L^\alpha(a, b; X)$ denotes the space of L^α-integrable functions from [a, b] into X, which is equipped with the Banach norm

$$| \int_a^b \|f(t)\|_X^\alpha dt \,|^{\frac{1}{\alpha}} .$$

The space $\mathscr{C}([a,b];X)$ is the space of continuous functions from $[a,b]$ $(-\infty < a < b < \infty)$ into X and is equipped with the Banach norm

$$\sup_{t\in[a,b]} \|f(t)\|_X.$$

For $\mathbf{u},\ \mathbf{v},\ \mathbf{w} \in \mathbf{L}^1(\Omega)$, we set

$$b(\mathbf{u},\mathbf{v},\mathbf{w}) = \sum_{i,j=1}^{2} \int_\Omega \mathbf{u}_i(D_i\mathbf{v}_j)\mathbf{w}_j dx,$$

whenever the integrals make sense.

The trilinear operator $b = b(\mathbf{u},\mathbf{v},\mathbf{w})$ can be extended to a continuous trilinear operator defined on V. Moreover,

$$\begin{array}{ll} b(\mathbf{u},\mathbf{v},\mathbf{v}) = 0 & \mathbf{u},\mathbf{v} \in V. \\ b(\mathbf{u},\mathbf{v},\mathbf{w}) = -b(\mathbf{u},\mathbf{w},\mathbf{v}) & \mathbf{u},\mathbf{v},\mathbf{w} \in V. \end{array} \tag{19}$$

In the periodic case,

$$b(\mathbf{v},\mathbf{v},\mathbf{A}v) = 0 \quad \mathbf{v} \in D(A) = \dot{\mathbf{H}}^2_{per}(\Omega) \cap \dot{V}. \tag{20}$$

Now we consider the Navier-Stokes equations

$$\begin{cases} \dfrac{\partial \mathbf{u}}{\partial t} - \mu\Delta\mathbf{u} + \mathbf{u}\cdot\Delta\mathbf{u} + \Delta p = \mathbf{f}, \\ \Delta\cdot\mathbf{u} = 0, \end{cases} \tag{21}$$

in

$$\Omega = (-\frac{L_1}{2},\frac{L_1}{2}) \times (-\frac{L_2}{2},\frac{L_2}{2}), \quad L_1, L_2 > 0,$$

with $\mu > 0$ and require that

$$\mathbf{u} = \mathbf{u}(\mathbf{x},t),\ p = p(\mathbf{x},t),\ \text{and}\ \mathbf{f} = \mathbf{f}(\mathbf{x},t)\ \text{with}\ \mathbf{x} = (x_1,x_2), \\ \text{are } L_i\text{-periodic in each variable } x_i$$

and

$$\int_\Omega \mathbf{f}(\mathbf{x},t)d\mathbf{x} = \int_\Omega \mathbf{u}(\mathbf{x},t)d\mathbf{x} = 0.$$

Let $T > 0$ be given. Assume that $\mathbf{u} \in \mathscr{C}^2(R^2 \times [0,T]),\ p \in \mathscr{C}^1(R^2 \times [0,T])$ are the classical solutions of (21). For each $v \in \mathscr{V}$, by multiplying momentum equation in (21) by v, integrating over Ω and using the Stokes formula we find that (cf. [5] Chap.III for the details)

$$\frac{d}{dt}(\mathbf{u},\mathbf{v}) + \mu((\mathbf{u},\mathbf{v})) + b(\mathbf{u},\mathbf{u},\mathbf{v}) = \langle\mathbf{f},\mathbf{v}\rangle \qquad \forall\, \mathbf{v} \in \mathscr{V}. \tag{22}$$

By continuity, (22) holds also for each $v \in V$.

This suggests the following variational formulation of (21) (strong solutions).

For f and $\mathbf{u}_0$ given,

$$\mathbf{f} \in L^2(0,T;\dot{H}),$$

$$\mathbf{u}_0 \in \dot{V},$$

find $\mathbf{u}$ satisfying

$$\mathbf{u} \in L^2(0,T;D(A) \cap L^\infty(0,T;\dot{V})$$

such that

$$\begin{cases} \dfrac{d}{dt}(\mathbf{u},\mathbf{v}) + \mu((\mathbf{u},\mathbf{v})) + b(\mathbf{u},\mathbf{u},\mathbf{v}) = \langle \mathbf{f},\mathbf{v}\rangle & \forall\, \mathbf{v} \in V,\ t \in (0,T), \\ \mathbf{u}(0) = \mathbf{u}_0. \end{cases} \tag{23}$$

If f is square integrable but not in H then we can replace it by its Leray projection on H, so that f is always assumed to be in H. We may refer to [5], p.307 for the relation of equations (21) and its variational formulation (23). One can show that if $\mathbf{u}$ is a solution of (23) then there exists p such that (21) is satisfied in a weak sense.

Lemma 1. *(Existence and Uniqueness of Strong Solutions in Two Dimensions) [4, 6]*
(i) *Assume that* $\mathbf{u}_0, \mathbf{f}, T > 0$ *are given and satisfy*

$$\mathbf{u}_0 \in \dot{V}, \quad \mathbf{f} \in L^2(0,T;\dot{H}).$$

Then there exists a unique solution $\mathbf{u}$ *of (23), satisfying*

$$u_i, \frac{\partial u_i}{\partial t}, \frac{\partial u_i}{\partial x_j}, \frac{\partial^2 u_i}{\partial x_j \partial x_k} \in L^2(\Omega \times (0,T)), \quad i,j,k = 1,2,$$

and $\mathbf{u}$ *is a continuously function from [0,T] into* $\dot{V}$. *We also can see that*

$$\mathbf{u} \in L^2(0,T;D(A)) \cap \mathscr{C}([0,T],\dot{V}).$$

And $\mathbf{u}$ *satisfies the following priori estimates*

$$|\mathbf{u}(s)|^2 \le |\mathbf{u}(0)|^2 + \frac{1}{\mu}\int_0^s \|\mathbf{f}(t)\|_{V'}^2 dt \qquad \forall s \in [0,T],$$

$$\sup_{t\in[0,T]} \|\mathbf{u}(t)\|^2 \le K_1, \qquad \int_0^T |A\mathbf{u}(t)|^2 dt \le K_2,$$

where constants $K_1,\ K_2$ *are dependent on* $\|\mathbf{u}_0\|, \int_0^T |\mathbf{f}(t)|^2 dt,\ \mu, L_1, L_2$.
(ii) *For* $r \ge 1$, *if the initial data* $\mathbf{u}_0 \in V_r$ *and* $\mathbf{f} \in L^\infty(0,T;V_{r-1})$, *then the solution* $\mathbf{u} = \mathbf{u}(t)$ *belongs to* $\mathscr{C}([0,T];V_r)$. *And there exists constant* K *dependent on* $\mu,\ \|\mathbf{u}_0\|_r,\ |\mathbf{f}|_{L^\infty(0,T;V_{r-1})}$, *such that*

$$\sup_{t\in[0,T]} \|\mathbf{u}(t)\|_r \le K. \tag{24}$$

Gronwall's Lemma:

(A) [7] *Let $a \in L_1(\tau, T)$, b be absolutely continuous on $[\tau, T]$. If $x \in L^\infty(\tau, T)$ satisfies*

$$x(t) \le b(t) + \int_\tau^t a(s)x(s)ds,$$

then for $t \in (0, T)$

$$x(t) \le b(\tau)exp\int_\tau^t a(s)ds + \int_\tau^t b'(s)exp\left(\int_s^t a(\rho)d\rho\right)ds.$$

(B) [4] *Let $a, \theta \in L_1(0, T)$. If y satisfies*

$$y' \le a + \theta y,$$

then for $t \in (0, T)$

$$y(t) \le y(0)exp(\int_0^t \theta(\tau)d\tau) + \int_0^t a(s)exp(\int_s^t \theta(\tau)d\tau)ds. \quad \square$$

Lemma 2. *Assume that $0 < T \le \widetilde{T}$, c is a positive constant and $\mathbf{u}_0, \mathbf{f}_1,\ \mathbf{f}_2$ satisfy*

$$\mathbf{u}_0 \in \dot{V}, \quad \mathbf{f}_i, \in L^2(0, T; \dot{H}), \quad \int_0^T |\mathbf{f_i}(t)|^2 dt \le c \quad for\ i = 1, 2.$$

Suppose that $\mathbf{u}_1,\ \mathbf{u}_2$ are the two solutions to (23) corresponding forcing terms $\mathbf{f}_1,\ \mathbf{f}_2$ respectively. Then we have the following estimate

$$\int_0^T \|\mathbf{u}_1(t) - \mathbf{u}_2(t)\|_2^2 dt \le c_1 \int_0^T |\mathbf{f}_1(t) - \mathbf{f}_2(t)|^2 dt$$

where constant c_1 is dependent on $\mu,\ \|\mathbf{u}_0\|,\ \widetilde{T},\ c$.

Proof According to the assumptions and **Lemma 1**, we have

$$\mathbf{u}_1, \mathbf{u}_2 \in L^2(0, T; D(A))$$

and

$$\sup_{t\in[0,T]} \|\mathbf{u}_1(t)\|^2 \le K_1, \qquad \int_0^T |A\mathbf{u}_1(t)|^2 dt \le K_2, \tag{25}$$

where constants $K_1,\ K_2$ are dependent on $\mu,\ \|\mathbf{u}_0\|,\ c$.

By the assumption, we have

$$\begin{cases} \dfrac{d}{dt}((\mathbf{u}_1 - \mathbf{u}_2)(t), \mathbf{v}) + \mu((\mathbf{u}_1 - \mathbf{u}_2, \mathbf{v})) + b(\mathbf{u}_1, \mathbf{u}_1, \mathbf{v}) \\ \qquad -b(\mathbf{u}_2, \mathbf{u}_2, \mathbf{v}) = \langle \mathbf{f}_1 - \mathbf{f}_2, \mathbf{v}\rangle \qquad \forall\, \mathbf{v} \in V,\ t \in (0, T), \\ (\mathbf{u}_1 - \mathbf{u}_2)(0) = 0. \end{cases} \tag{26}$$

Let $\mathbf{u} = \mathbf{u}_1 - \mathbf{u}_2,\ \mathbf{f} = \mathbf{f}_1 - \mathbf{f}_2$. Using (19), we have

$$b(\mathbf{u}_1, \mathbf{u}_1, \mathbf{u}_1 - \mathbf{u}_2) - b(\mathbf{u}_2, \mathbf{u}_2, \mathbf{u}_1 - \mathbf{u}_2) = -b(\mathbf{u}_1, \mathbf{u}_1, \mathbf{u}_2) + b(\mathbf{u}_2, \mathbf{u}_1, \mathbf{u}_2)$$
$$= -b(\mathbf{u}, \mathbf{u}_1, \mathbf{u}_2) = b(\mathbf{u}, \mathbf{u}_1, \mathbf{u}_1) - b(\mathbf{u}, \mathbf{u}_1, \mathbf{u}_2) = -b(\mathbf{u}, \mathbf{u}, \mathbf{u}_1)$$

Replacing $\mathbf{v}$ by $\mathbf{u}$ in (26), we obtain

$$\frac{1}{2}\frac{d}{dt}|\mathbf{u}(t)|^2 + \mu\|\mathbf{u}\|^2 - b(\mathbf{u}, \mathbf{u}, \mathbf{u}_1) = \langle \mathbf{f}, \mathbf{u}\rangle \quad \forall t \in (0, T). \tag{27}$$

Using Poicare inequality (see [6]), we obtain

$$\langle \mathbf{f}, \mathbf{u}\rangle \leq |\mathbf{f}||\mathbf{u}| \leq \frac{1}{\lambda_1^{\frac{1}{2}}}|\mathbf{f}|\|\mathbf{u}\| \leq \frac{1}{\mu\lambda_1}|\mathbf{f}|^2 + \frac{\mu}{4}\|\mathbf{u}\|^2. \tag{28}$$

And by the following inequality (see [4], page 13)

$$|b(\mathbf{u}, \mathbf{u}, \mathbf{u}_1)| \leq k|\mathbf{u}|\|\mathbf{u}\|\|\mathbf{u}_1\|^{\frac{1}{2}}|A\mathbf{u}_1|^{\frac{1}{2}},$$

we obtain

$$|b(\mathbf{u}, \mathbf{u}, \mathbf{u}_1)| \leq \frac{\mu}{4}\|\mathbf{u}\|^2 + \frac{k^2}{\mu}\|\mathbf{u}_1\||A\mathbf{u}_1||\mathbf{u}|^2. \tag{29}$$

By (27),(28),(29) we obtain, for $t \in [0, T]$,

$$\frac{d}{dt}|\mathbf{u}(t)|^2 + \mu\|\mathbf{u}\|^2 \leq \frac{2}{\mu\lambda_1}|\mathbf{f}|^2 + \frac{2k^2}{\mu}\|\mathbf{u}_1\||A\mathbf{u}_1||\mathbf{u}|^2. \tag{30}$$

By **Gronwall's Lemma (B)** and (25), we deduce that, for $t \in [0, T]$,

$$|\mathbf{u}(t)|^2 \leq \int_0^t (\frac{2}{\mu\lambda_1}|\mathbf{f}(\mathbf{s})|^2)(exp\int_s^t \frac{2k^2}{\mu}\|\mathbf{u}_1(\tau)\||A\mathbf{u}_1(\tau)|d\tau)ds,$$

then

$$\sup_{0\in[0,T]} |\mathbf{u}(t)|^2 \leq K_3 \int_0^T |\mathbf{f}(\mathbf{s})|^2 ds, \tag{31}$$

where constant K_3 is dependent on $\mu,\ \|\mathbf{u}_0\|,\ c$.

Replacing $\mathbf{v}$ by $A\mathbf{u}$ in (26), we have, for $t \in (0, T)$,

$$\frac{1}{2}\frac{d}{dt}\|\mathbf{u}(t)\|^2 + \mu|A\mathbf{u}|^2 + b(\mathbf{u}_1, \mathbf{u}_1, A\mathbf{u}) - b(\mathbf{u}_2, \mathbf{u}_2, A\mathbf{u}) = \langle \mathbf{f}(t), A\mathbf{u}\rangle.$$

Since

$$b(\mathbf{u}_1, \mathbf{u}_1, A\mathbf{u}) - b(\mathbf{u}_2, \mathbf{u}_2, A\mathbf{u})$$
$$= b(\mathbf{u}_1, \mathbf{u}_1, A\mathbf{u}) - b(\mathbf{u}_2, \mathbf{u}_1, A\mathbf{u}) + b(\mathbf{u}_2, \mathbf{u}_1, A\mathbf{u}) - b(\mathbf{u}_2, \mathbf{u}_2, A\mathbf{u})$$
$$= b(\mathbf{u}, \mathbf{u}_1, A\mathbf{u}) + b(\mathbf{u}_2, \mathbf{u}, A\mathbf{u}),$$

then

$$\frac{1}{2}\frac{d}{dt}\|\mathbf{u}(t)\|^2 + \mu|A\mathbf{u}|^2 + b(\mathbf{u}, \mathbf{u}_1, A\mathbf{u}) + b(\mathbf{u}_2, \mathbf{u}, A\mathbf{u}) = \langle \mathbf{f}(t), A\mathbf{u}\rangle. \tag{32}$$

And using (20), we obtain the following equality (see [6], page 135)

$$\frac{1}{2}\frac{d}{dt}\|\mathbf{u}(t)\|^2 + \mu|A\mathbf{u}|^2 = b(\mathbf{u}, \mathbf{u}, A\mathbf{u}_1) + \langle \mathbf{f}(t), A\mathbf{u}\rangle. \tag{33}$$

Notice that

$$\langle \mathbf{f}, A\mathbf{u}\rangle \le |\mathbf{f}||A\mathbf{u}| \le \frac{1}{\mu}|\mathbf{f}|^2 + \frac{\mu}{4}|A\mathbf{u}|^2. \tag{34}$$

And by the following inequality (see [4], page 13)

$$|b(\mathbf{u}, \mathbf{u}, A\mathbf{u}_1)| \le k|\mathbf{u}|^{\frac{1}{2}}|A\mathbf{u}|^{\frac{1}{2}}\|\mathbf{u}\||A\mathbf{u}_1|,$$

we obtain

$$|b(\mathbf{u}, \mathbf{u}, A\mathbf{u}_1)| \le \frac{\mu}{4}|A\mathbf{u}|^2 + \frac{1}{2}|A\mathbf{u}_1|^2\|\mathbf{u}\|^2 + \frac{k^4}{4\mu}|\mathbf{u}|^2. \tag{35}$$

By (33),(34),(35), we deduce that

$$\frac{d}{dt}\|\mathbf{u}(t)\|^2 + \mu|A\mathbf{u}|^2 \le |A\mathbf{u}_1|^2\|\mathbf{u}\|^2 + \frac{k^4}{2\mu}|\mathbf{u}|^2 + \frac{2}{\mu}|\mathbf{f}|^2. \tag{36}$$

By **Gronwall's Lemma (B)**, we obtain, for $t \in [0, T]$,

$$\|\mathbf{u}(t)\|^2 \le (\int_0^t (\frac{k^4}{2\mu}|\mathbf{u}(s)| + \frac{2}{\mu}|\mathbf{f}(s)|^2)exp(\int_s^t |A\mathbf{u}_1(\tau)|^2 d\tau)ds.$$

Then

$$\sup_{t\in[0,T]} \|\mathbf{u}(t)\|^2 \le exp(\int_0^T |A\mathbf{u}_1(\tau)|^2 d\tau)(\int_0^T (\frac{k^4}{2\mu}|\mathbf{u}(s)| + \frac{2}{\mu}|\mathbf{f}(s)|^2)ds).$$

By (25), (31), there exists constant K_4 dependent on $\mu,\ \|\mathbf{u}_0\|,\ c,\ \widetilde{T}$, such that

$$\sup_{t\in[0,T]} \|\mathbf{u}(t)\|^2 \le K_4 \int_0^T |\mathbf{f}(\mathbf{s})|^2 ds. \tag{37}$$

Then by integration in t of (36) from 0 to T, after dropping unnecessary terms we obtain

$$\int_0^T |A\mathbf{u}(t)|^2 dt \le K_5 \int_0^T |\mathbf{f}(\mathbf{t})|^2 dt,$$

that is

$$\int_0^T |A\mathbf{u}_1(t) - A\mathbf{u}_2(t)|^2 dt \le K_5 \int_0^T |\mathbf{f}_1(t) - \mathbf{f}_2(t)|^2 dt, \tag{38}$$

where constant K_5 is dependent on $\mu,\ \|\mathbf{u}_0\|,\ c,\ \widetilde{T}$. Then using (18) with s=1, we obtain

$$\int_0^T \|\mathbf{u}_1(t) - \mathbf{u}_2(t)\|_2^2 dt \le c_1 \int_0^T |\mathbf{f}_1(t) - \mathbf{f}_2(t)|^2 dt$$

where constant c_1 is dependent on $\mu,\ \|\mathbf{u}_0\|,\ c,\ \widetilde{T}$.□

4. Variational Formulation of the Problem

$\mathbf{C}[0, L]$ denotes the space of continuous functions from $[0, L]$ into R^2 and is equipped with the norm

$$\|\mathbf{x}\|_c = \max_{0\le s\le L} |\mathbf{x}(s)|.$$

In what follows, we always suppose that $\widetilde{T} > 0$ is a given constant, and constant $T \in (0, \widetilde{T}]$. Let

$$G = \mathscr{L}([0,T]; \mathbf{C}[0,L]); \quad E = \{\mathbf{x} \in \mathbf{C}[0,L] : \mathbf{x}(0) = \mathbf{x}(L)\};$$
$$\mathscr{F} = \{\mathbf{x} \in E | \mathbf{x}(s) \in \Omega,\ 0 \le s \le L\},$$

where

$$\Omega = (-\frac{L_1}{2}, \frac{L_1}{2}) \times (-\frac{L_2}{2}, \frac{L_2}{2}).$$

Notice that

$$\overline{\mathscr{F}} = \{\mathbf{x} \in E | \mathbf{x}(s) \in \overline{\Omega},\ 0 \le s \le L\}.$$

For $\mathbf{x} \in G$, $\mathbf{x}(\cdot, t) \in \mathbf{C}[0, L]$, $\forall\, t \in [0, T]$, $\mathbf{x}(\cdot, t)$ is simply denoted by $\mathbf{x}(t)$.

Given $\mathbf{X}_0 \in \mathscr{F}$, let

$$\mathbb{X} = \{\mathbf{X} \in \mathscr{L}([0,T]; E) : \mathbf{X}(0) = \mathbf{X}_0\}.$$

$\mathbb{X}$ is a closed subset of Banach space G.

Given $\delta_h : R^2 \to R^2$, $\mathbf{f} : [0, L] \times [0, \widetilde{T}] \to R^2$, we assume that they satisfy *Condition **A***:

$$\delta_h \in \mathbf{H}_0^2(R^2), \quad \mathbf{f} \in \mathbf{C}([0,L] \times [0, \widetilde{T}]).$$

In this paper, we let Dirac delta function δ be approximated by differentiable function δ_h. In fact, some authors do so when performing the numerical simulation in order to get the approximate solution of the problem (1)-(7). For example, δ_h in [9] is chosen as follows

$$\delta_h(\mathbf{x}) = d_h(x) d_h(y), \qquad \forall \mathbf{x} \in \mathbf{R}^2$$

where

$$d_h(z) = \begin{cases} \dfrac{0.25}{h}\left[1 + \cos(\dfrac{\pi z}{2h})\right] & |z| \le 2h, \\ 0, & |z| > 2h. \end{cases}$$

It is clear that $\delta_h \in \mathbf{H}_0^2(R^2)$.

According to the theory and notations in **Section 2, Section3**, we introduce the corresponding variational formulation of the equations (10)-(16):

Problem 1. Given $\mathbf{u}_0 \in V$, $\mathbf{X}_0 \in \mathscr{F}$ and $\delta_h : R^2 \to R^2$, $\mathbf{f} : [0, L] \times [0, \widetilde{T}] \to R^2$ satisfying *Condition **A***, find $\mathbf{u} \in \mathbf{L}^2(0, T; D(A)) \cap \mathbf{L}^\infty(0, T; \dot{V})$, and $\mathbf{X} : [0, L] \times [0, T] \to \overline{\Omega}$, such that

$$\begin{aligned} &\frac{d}{dt}(\mathbf{u}(t), \mathbf{v}) + \mu((\mathbf{u}, \mathbf{v})) + b(\mathbf{u}, \mathbf{u}, \mathbf{v}) = \langle \hat{\hat{\mathbf{F}}}(t), \mathbf{v} \rangle && \forall\, \mathbf{v} \in V,\ t \in (0, T), \\ &\frac{\partial \mathbf{X}}{\partial t}(s, t) = \mathbf{u}(\mathbf{X}(s,t) - I_u(t), t) + M_u(t) && \forall\, s \in [0, L], t \in (0, T), \\ &\mathbf{u}(\mathbf{x}, 0) = \hat{\mathbf{u}}_0(\mathbf{x}) && \forall\, \mathbf{x} \in \Omega, \\ &\mathbf{X}(s, 0) = \mathbf{X}_0(s) && \forall s \in [0, L], \\ &\mathbf{X}(0, t) = \mathbf{X}(L, t) && \forall t \in (0, T), \end{aligned}$$

where

$$\hat{\mathbf{u}}_0(\mathbf{x}) = \mathbf{u}_0(\mathbf{x}) - \frac{1}{|\Omega|}\int_\Omega \mathbf{u}_0(\mathbf{x})d\mathbf{x},$$
$$\hat{\hat{\mathbf{F}}}(\mathbf{x},t)$$
$$= \int_0^L \mathbf{f}(s,t)\delta_h(\mathbf{x}+I_u(t)-\mathbf{X}(s,t))ds - \frac{1}{|\Omega|}\int_0^L \mathbf{f}(s,t)(\int_\Omega \delta_h(\mathbf{x}-\mathbf{X}(s,t))dx)ds.$$

5. Conclusion

In this section we shall prove the existence of the solution of **Problem 1** via a fixed point argument. We define an operator $\mathbf{T}$ on $\mathbb{X}$ as follows.

Given $\mathbf{u}_0 \in V$, $\overline{\mathbf{X}} \in \mathbb{X}$. $\forall x \in \Omega, t \in (0,T)$, let

$$\mathbf{F}_{\overline{X}}(\mathbf{x},t) = \int_0^L \mathbf{f}(s,t)\delta_h(\mathbf{x}-\overline{\mathbf{X}}(s,t))ds.$$

Hence

$$\hat{\hat{\mathbf{F}}}_{\overline{X}}(\mathbf{x},t) = \int_0^L \mathbf{f}(s,t)\delta_h(\mathbf{x}+I_{\overline{X}}(t)-\overline{\mathbf{X}}(s,t))ds - \frac{1}{|\Omega|}\int_0^L \mathbf{f}(s,t)(\int_\Omega \delta_h(x-\overline{\mathbf{X}}(s,t))dx)ds$$

where, according to (9),

$$I_{\overline{X}}(t) = \frac{t}{|\Omega|}\int_\Omega \mathbf{u}_0(\mathbf{x})d\mathbf{x} + \int_0^t\int_0^s M_{\mathbf{F}_{\overline{X}}}(\tau)d\tau ds, \tag{39}$$

$$M_{\mathbf{F}_{\overline{X}}}(t) = \frac{1}{|\Omega|}\int_\Omega \mathbf{F}_{\overline{X}}(\mathbf{x},t)d\mathbf{x} = \frac{1}{|\Omega|}\int_0^L \mathbf{f}(s,t)(\int_\Omega \delta_h(x-\overline{\mathbf{X}}(s,t))dx)ds. \tag{40}$$

If *Condition* **A** is satisfied, we can verify that $\hat{\hat{\mathbf{F}}}_{\overline{X}} \in \mathbf{L}^\infty(0,T;\dot{\mathbf{H}}^2_{per}(\Omega))$.

Then let $\overline{\mathbf{u}}$ (also denoted by $\mathbf{u}_{\overline{\mathbf{X}}}$) be the solution $\mathbf{u} \in \mathbf{L}^\infty(0,T;\dot{V})$ of

$$\begin{cases} \dfrac{d}{dt}(\mathbf{u}(t),\mathbf{v}) + \mu(\nabla\mathbf{u},\nabla\mathbf{v}) = \langle \hat{\hat{\mathbf{F}}}_{\overline{X}}(t),\mathbf{v}\rangle \quad \forall\mathbf{v}\in V,\ t\in(0,T), \\ \mathbf{u}(0) = \hat{\mathbf{u}}_0. \end{cases} \tag{41}$$

Finally, let $\overline{\overline{\mathbf{X}}}$ be the solution $\mathbf{X} \in \mathbb{X}$ of

$$\begin{cases} \mathbf{X}'(t) = \overline{\mathbf{u}}(\mathbf{X}(s,t) - I_{\overline{X}}(t),t) + M_{u\overline{X}}(t)) \quad \forall t \in (0,T), \\ \mathbf{X}(0) = \mathbf{X}_0, \end{cases} \tag{42}$$

where

$$M_{u\overline{X}}(t)) = \frac{1}{|\Omega|}\int_\Omega \mathbf{u}_0(\mathbf{x})d\mathbf{x} + \int_0^t M_{\mathbf{F}_{\overline{X}}}(\tau)d\tau, \tag{43}$$

and let $\mathbf{T}(\overline{\mathbf{X}}) = \overline{\overline{\mathbf{X}}}$.

We recall that the definition of $\mathbb{X}$ is

$$\mathbb{X} = \{\mathbf{X} \in \mathscr{L}([0,T];E) : \mathbf{X}(0) = \mathbf{X}_0\}.$$

For $t \in [0,T]$, $\mathbf{X}(t) = \mathbf{X}(\cdot,t) \in E$. (42) is an ordinary equation in the Banach space E and equivalent to

$$\begin{cases} \dfrac{\partial \mathbf{X}}{\partial t}(s,t) = \overline{\mathbf{u}}(\mathbf{X}(s,t) - I_{\overline{X}}(t),t) + M_{u\overline{X}}(t)) & \forall s \in [0,L], \forall t \in (0,T), \\ \mathbf{X}(s,0) = \mathbf{X}_0(s) & \forall s \in [0,L], \\ \mathbf{X}(0,t) = \mathbf{X}(L,t) & \forall t \in (0,T). \end{cases}$$

We observe that, if $\mathbf{T}$ has a fixed point $\widetilde{\mathbf{X}}$ in $\mathbb{X}$, then this fixed point and the solution $\widetilde{\mathbf{u}}$ to (41) corresponding to $\overline{\mathbf{X}} = \widetilde{\mathbf{X}}$, give the solutions to **Problem 1.**

The following lemma guarantees the existence of the solution to (42):

Lemma 3. *(see [1]) Let $x : [t_0, t_0 + a] \to Y$ be a mapping into the Banach-space Y. Consider the initial value problem*

$$\frac{dx}{dt} = f(x,t), \quad x(t_0) = x_0. \tag{44}$$

Let $Q = [t_0, t_0 + a] \times Y$. Suppose $f : Q \to Y$ is continuous and satisfies

$$\|f(t,x) - f(t,y)\| \le L\|x - y\|, \qquad for\ all \quad (t,x),(t,y) \in Q,\ and\ fixed\ L \ge 0.$$

Then initial problem (44) has exactly one continuously differential solution on $[t_0, t_0 + a]$ for each initial value $x_0 \in Y$. □

Theorem 1. *Assume that $T \in (0, \widetilde{T}]$, $\mathbf{u}_0 \in \mathbf{H}^3_{per}(\Omega) \cap H$, $\mathbf{X}_0 \in \mathscr{F}$ and Condition **A** holds.*

(a) *For given $\overline{\mathbf{X}} \in \mathbb{X}$, (41) has a unique solution $\overline{\mathbf{u}} \in \mathscr{L}([0,T], \mathbf{C}^1(\overline{\Omega}))$. Furthermore there exist constants c_1, c_2, c_3 dependent on μ, $\widetilde{T}$, δ_h, $|\mathbf{f}\|_{\mathbf{C}([0,L]\times[0,\widetilde{T}])}$, $\|\mathbf{u}_0\|_{V_3}$, such that*

$$\sup_{\substack{z\in\overline{\Omega} \\ 0\le t\le T}} |\overline{\mathbf{u}}(z,t)| \le c_1 < \infty, \tag{45}$$

$$\sup_{\substack{z\in\overline{\Omega} \\ 0\le t\le T}} |D\overline{\mathbf{u}}(z,t)| \le c_2 < \infty, \tag{46}$$

$$\sup_{\substack{z\in\overline{\Omega} \\ 0\le t\le T}} |\overline{\mathbf{u}}(z,t) + M_{u\overline{X}}(t)| \le c_3 < \infty.$$

(b) *Suppose that $\overline{\mathbf{u}}$ is the solution of (41). Then (42) has a unique solution $\overline{\overline{\mathbf{X}}} \in \mathbb{X}$. Furthermore, if $0 < T < \min\{\frac{L_m - \|\mathbf{X}_0\|_c}{c_3}, \widetilde{T}\}$ holds, where $L_m = min\{L_1, L_2\}$, then $\overline{\overline{\mathbf{X}}}(s,t) \in \Omega$, for all $(s,t) \in [0,L] \times [0,T]$.*

Proof:

(a) Since $\mathbf{u}_0 \in \mathbf{H}^3_{per}(\Omega) \cap H$, then $\hat{\mathbf{u}}_0 \in V_3$. Since *Condition **A*** is satisfied, then $\hat{\hat{\mathbf{F}}}_{\overline{X}} \in \mathbf{L}^\infty(0,T;\dot{\mathbf{H}}^2_{per}(\Omega))$ and $P_L\hat{\hat{\mathbf{F}}}_{\overline{X}} \in \mathbf{L}^\infty(0,T;V_2)$. We can verify that there exists constant C dependent on δ_h, such that

$$\|P_L\hat{\hat{\mathbf{F}}}_{\overline{X}}\|_{\mathbf{L}^\infty(0,T;V_2)} \le \|\hat{\hat{\mathbf{F}}}_{\overline{X}}\|_{\mathbf{L}^\infty(0,T;\dot{\mathbf{H}}^2_{per}(\Omega))} \le C\|\mathbf{f}\|_{\mathbf{C}([0,L]\times[0,\widetilde{T}])}. \tag{47}$$

By **Lemma 1 (ii)** with $r = 3$, there exists $\overline{\mathbf{u}}$ belonging to $\mathscr{C}([0,T];V_3)$ satisfying

$$\begin{cases} \frac{d}{dt}(\mathbf{u}(t),\mathbf{v}) + \mu(\nabla\mathbf{u},\nabla\mathbf{v}) = \langle P_L\hat{\hat{\mathbf{F}}}_{\overline{X}}(t),\mathbf{v}\rangle \quad \forall\mathbf{v}\in V,\ t\in(0,T), \\ \mathbf{u}(0) = \hat{\hat{\mathbf{u}}}_0. \end{cases}$$

Since P_L is the orthogonal projector onto H in $\mathbf{L}^2(\Omega)$, we have

$$\langle P_L\hat{\hat{\mathbf{F}}}_{\overline{X}}(t),\mathbf{v}\rangle = \langle\hat{\hat{\mathbf{F}}}_{\overline{X}}(t),\mathbf{v}\rangle \quad \forall\mathbf{v}\in V.$$

Then $\overline{\mathbf{u}}$ is the solution of (41). Since $\overline{\mathbf{u}}$ belonging to $\mathscr{C}([0,T];V_3)$ and V_3 is a closed subspace of $\dot{\mathbf{H}}^3_{per}(\Omega)$, we obtain that $\overline{\mathbf{u}} \in \mathscr{L}([0,T],\mathbf{C}^1(\overline{\Omega}))$ due to (18) and Sobolev embedding theorem (see [7]). Furthermore by (24) and (47) there exist constants c_1, c_2, c_3 dependent on $\mu,\ \widetilde{T},\ \delta_h,\ |\mathbf{f}\|_{\mathbf{C}([0,L]\times[0,\widetilde{T}])},\ \|\mathbf{u}_0\|_3$, such that

$$\sup_{\substack{z\in\overline{\Omega}\\0\le t\le T}} |\overline{\mathbf{u}}(z,t)| \le c_1 < \infty,$$

$$\sup_{\substack{z\in\overline{\Omega}\\0\le t\le T}} |D\overline{\mathbf{u}}(z,t)| \le c_2 < \infty,$$

$$\sup_{\substack{z\in\overline{\Omega}\\0\le t\le T}} |\overline{\mathbf{u}}(z,t) + M_{u\overline{X}}(t)| \le c_3 < \infty.$$

(b) We shall prove the conclusion by **Lemma 3** with $Y = E, f = \overline{\mathbf{u}}, x_0 = \mathbf{X}_0, t_0 = 0$. Below we verify that the conditions in **Lemma 3** hold.

First we prove that $\overline{\mathbf{u}} : E\times[0,T]\to E$ is continuous. Given $(x_0,t_0)\in E\times[0,T]$, $(x,t)\in E\times[0,T]$. It is easy to see $\overline{\mathbf{u}}(x,t)\in E,\ \overline{\mathbf{u}}(x_0,t_0)\in E$. We have

$$\begin{aligned} \|\overline{\mathbf{u}}(x,t)-\overline{\mathbf{u}}(x_0,t_0)\|_c &= \sup_{0\le s\le L} |\overline{\mathbf{u}}(x(s),t)-\overline{\mathbf{u}}(x_0(s),t_0)| \\ \le \sup_{z\in\overline{\Omega}} |\overline{\mathbf{u}}(z,t)-\overline{\mathbf{u}}(z,t_0)| &+ \sup_{z\in\overline{\Omega}} |\overline{\mathbf{u}}(z,t_0)-\overline{\mathbf{u}}(z_0,t_0)|. \end{aligned}$$

Then $\overline{\mathbf{u}}(x,t)$ is continuous at (x_0,t_0) due to $\overline{\mathbf{u}}\in\mathscr{L}([0,T],\mathbf{C}^1(\overline{\Omega}))$. For $t\in[0,T]$, $x,y\in E$, by (46) we have

$$\begin{aligned} \|\overline{\mathbf{u}}(x,t)-\overline{\mathbf{u}}(y,t)\|_c &= \sup_{0\le s\le L} |\overline{\mathbf{u}}(x(s),t)-\overline{\mathbf{u}}(y(s),t)| \\ \le \sup_{\substack{z\in\overline{\Omega}\\0\le t\le T}} |D\overline{\mathbf{u}}(z,t)|\|x-y\|_c &\le c_2\|x-y\|_c. \end{aligned}$$

Hence the conditions in **Lemma 3** hold. (42) has a unique solution $\overline{\overline{\mathbf{X}}}\in\mathbb{X}$.

By (42), we have

$$\overline{\overline{\mathbf{X}}}(t) = \mathbf{X}_0 + \int_0^t (\overline{\mathbf{u}}(\overline{\overline{\mathbf{X}}}(s,\tau) - I_{\overline{X}}(\tau),\tau) + M_{u\overline{X}}(\tau))d\tau.$$

Notice that $\mathbf{X}_0 \in \mathscr{F}$, then $L_m - \|\mathbf{X}_0\|_c > 0$. If $0 < T < \frac{L_m - \|\mathbf{X}_0\|_c}{c_3}$ holds, for $t \in [0, T]$, we have

$$\|\overline{\overline{\mathbf{X}}}(t)\|_c \le \|\mathbf{X}_0\|_c + t \sup_{\substack{z \in \overline{\Omega} \\ 0 \le \tau \le T}} |\overline{\mathbf{u}}(z, \tau) + M_{u\overline{X}}(\tau)|$$
$$< \|\mathbf{X}_0\|_c + c_3 \frac{L_m - \|\mathbf{X}_0\|_c}{c_3} = L_m.$$

Then $\overline{\overline{\mathbf{X}}}(t) \in \mathscr{F}$. That is, if $0 < T < \frac{L_m - \|\mathbf{X}_0\|_c}{c_3}$ holds, for all $(s, t) \in [0, L] \times [0, T]$, $\overline{\overline{\mathbf{X}}}(s, t) \in \Omega$. $\square$

Follow the above argument, we can prove the main conclusion:

Theorem 2. *Assume that* $\mathbf{u}_0 \in \mathbf{H}^3_{per}(\Omega) \cap H$, $\mathbf{X}_0 \in \mathscr{F}$ *and Condition A holds. Then there exists* $T'' \in (0, \widetilde{T}]$ *dependent on* μ, $\widetilde{T}, \delta_h, \mathbf{f}, \mathbf{u}_0, L_m, \|\mathbf{X}_0\|_c$, *such that if* $T \in (0, T'')$, ***Problem 1*** *has a unique pair of solutions* $\widetilde{\mathbf{u}}$ *and* $\widetilde{\mathbf{X}}$. *And* $\widetilde{\mathbf{X}}$ *may be solved by successive approximations*

$$\mathbf{Y}_{n+1} = \mathbf{T}\mathbf{Y}_n \quad \forall \mathbf{Y}_0 \in B \quad n = 0, 1, 2, \dots ,$$

where

$$B = \{\mathbf{X} \in \mathbb{X} \mid \mathbf{X}(t) \in \overline{\mathscr{F}} \quad \forall t \in [0, T]\}.$$

Proof: Below we take T satisfying $0 < T < \min\{\frac{L_m - \|\mathbf{X}_0\|_c}{c_3}, \widetilde{T}\}$. By **Theorem 1 (b)**, we can observe that $\mathbf{T}$ maps B into itself. The set B is a bounded closed convex subset of $\mathbb{X}$. We will prove that $\mathbf{T} : B \to B$ is a contractive map, and then by the Banach Fixed-Point Theorem (see [1]), $\mathbf{T}$ has a unique fixed point.

Arbitrarily given two elements $\mathbf{X}, \mathbf{Y} \in \mathbf{B}$, we denote the corresponding solutions of (41) by $\mathbf{u_X}$ and $\mathbf{u_Y}$ respectively. Then $\mathbf{u}_X - \mathbf{u}_Y$ solves the following equation

$$\begin{cases} \frac{d}{dt}((\mathbf{u_X} - \mathbf{u_Y})(t), \mathbf{v}) + \mu(\nabla(\mathbf{u_X} - \mathbf{u_Y})(t), \nabla \mathbf{v}) = \langle \hat{\hat{\mathbf{F}}}_{\mathbf{X}}(t) - \hat{\hat{\mathbf{F}}}_{\mathbf{Y}}(t), \mathbf{v} \rangle, \\ \qquad\qquad\qquad\qquad\qquad\qquad\qquad\qquad \forall v \in V,\ t \in (0, T), \\ (\mathbf{u_X} - \mathbf{u_Y})(0) = 0. \end{cases}$$

By the definitions of $\hat{\hat{\mathbf{F}}}_{\mathbf{X}}$, $\hat{\hat{\mathbf{F}}}_{\mathbf{Y}}$ and using *Condition A*, we can deduce that there exists a constant k dependent on $\mathbf{f}$, δ_h , such that

$$\int_0^T \|\hat{\hat{\mathbf{F}}}_X(t)\|_0^2 dt \le k, \quad \int_0^T \|\hat{\hat{\mathbf{F}}}_Y(t)\|_0^2 dt \le k.$$

Then according to **Lemma 2**, we have

$$\int_0^T \|\mathbf{u}_X(t) - \mathbf{u}_Y(t)\|_2^2 dt \le c_0 \int_0^T \|\hat{\hat{\mathbf{F}}}_X(t) - \hat{\hat{\mathbf{F}}}_Y(t)\|_0^2 dt \tag{48}$$

where constant c_0 is dependent on μ, $\|\mathbf{u}_0\|$, $\widetilde{T}$, $\mathbf{f}$, δ_h.

By the definitions of $\hat{\hat{\mathbf{F}}}_{\mathbf{X}}$, $\hat{\hat{\mathbf{F}}}_{\mathbf{Y}}$ and using *Condition A*, we can deduce that there exists a constant l_1 dependent on $\mathbf{f}$, δ_h, $\widetilde{T}$, such that, for each $t \in (0, T)$,

$$\|\hat{\hat{\mathbf{F}}}_{\mathbf{X}}(t) - \hat{\hat{\mathbf{F}}}_{\mathbf{Y}}(t)\|_0^2 \le l_1 \sup_{0 \le s \le L} |\mathbf{Y}(s, t) - \mathbf{X}(s, t)|^2 = l_1 \|\mathbf{Y}(t) - \mathbf{X}(t)\|_c^2. \tag{49}$$

By (48) and (49),

$$\int_0^T \|\mathbf{u}_X(t) - \mathbf{u}_Y(t)\|_2^2 dt \leq c_0 l_1 \int_0^T \|\mathbf{X}(t) - \mathbf{Y}(t)\|_c^2 dt. \tag{50}$$

Let $\widetilde{X} = \mathbf{TX}$ and $\widetilde{Y} = \mathbf{TY}$. According to (42), we have

$$\begin{cases} \widetilde{\mathbf{X}}'(t) = \mathbf{u}_X(\widetilde{\mathbf{X}}(t) - I_X(t), t) + M_{uX}(t) \quad \forall t \in (0, T), \\ \widetilde{\mathbf{X}}(0) = \mathbf{X}_0, \end{cases} \tag{51}$$

$$\begin{cases} \widetilde{\mathbf{Y}}'(t) = \mathbf{u}_Y(\widetilde{\mathbf{Y}}(t) - I_Y(t), t) + M_{uY}(t) \quad \forall t \in (0, T), \\ \widetilde{\mathbf{Y}}(0) = \mathbf{X}_0, \end{cases} \tag{52}$$

Subtracting (51) from (52) we obtain

$$\begin{aligned} &\widetilde{\mathbf{X}}'(t) - \widetilde{\mathbf{Y}}'(t) \\ &= \mathbf{u}_{\mathbf{X}}(\widetilde{\mathbf{X}}(t) - I_X(t), t) - \mathbf{u}_{\mathbf{Y}}(\widetilde{\mathbf{Y}}(t) - I_Y(t), t) + M_{uY}(t) - M_{uX}(t) \\ &= \mathbf{u}_{\mathbf{X}}(\widetilde{\mathbf{X}}(t) - I_X(t), t) - \mathbf{u}_{\mathbf{Y}}(\widetilde{\mathbf{X}}(t) - I_X(t), t) \\ &\quad + \mathbf{u}_{\mathbf{Y}}(\widetilde{\mathbf{X}}(t) - I_X(t), t) - \mathbf{u}_{\mathbf{Y}}(\widetilde{\mathbf{Y}}(t) - I_Y(t), t) \\ &\quad + M_{uY}(t) - M_{uX}(t) \end{aligned}$$

Integrating from 0 to t in the two sides of above equation, we have

$$\begin{aligned} \widetilde{\mathbf{X}}(t) - \widetilde{\mathbf{Y}}(t) &= \int_0^t [\mathbf{u}_{\mathbf{X}}(\widetilde{\mathbf{X}}(\tau) - I_X(\tau), \tau) - \mathbf{u}_{\mathbf{Y}}(\widetilde{\mathbf{X}}(\tau) - I_X(\tau), \tau)] d\tau \\ + \int_0^t [\mathbf{u}_{\mathbf{Y}}(\widetilde{\mathbf{X}}(\tau) &- I_X(\tau), \tau) - \mathbf{u}_{\mathbf{Y}}(\widetilde{\mathbf{Y}}(\tau) - I_Y(\tau), \tau)] d\tau + \int_0^t [M_{uY}(\tau) - M_{uX}(\tau)] d\tau. \end{aligned}$$

Hence

$$\begin{aligned} \|\widetilde{\mathbf{X}}(t) - \widetilde{\mathbf{Y}}(t)\|_c &\leq \int_0^t \|\mathbf{u}_{\mathbf{X}}(\widetilde{\mathbf{X}}(\tau) - I_X(\tau), \tau) - \mathbf{u}_{\mathbf{Y}}(\widetilde{\mathbf{X}}(\tau) - I_X(\tau), \tau)\|_c d\tau \\ &+ \int_0^t \|\mathbf{u}_{\mathbf{Y}}(\widetilde{\mathbf{X}}(\tau) - I_X(\tau), \tau) - \mathbf{u}_{\mathbf{Y}}(\widetilde{\mathbf{Y}}(\tau) - I_Y(\tau), \tau)\|_c d\tau \\ &+ \int_0^t |M_{uY}(\tau) - M_{uX}(\tau)| d\tau. \end{aligned} \tag{53}$$

By Sobolev embedding theorem, there exists a constant l_2 such that

$$\begin{aligned} &\|\mathbf{u}_{\mathbf{X}}(\widetilde{\mathbf{X}}(\tau) - I_X(\tau), \tau) - \mathbf{u}_{\mathbf{Y}}(\widetilde{\mathbf{X}}(\tau) - I_X(\tau), \tau)\|_c \\ &= \sup_{0 \leq s \leq L} |\mathbf{u}_{\mathbf{X}}(\widetilde{\mathbf{X}}(s, \tau) - I_X(\tau), \tau) - \mathbf{u}_{\mathbf{Y}}(\widetilde{\mathbf{X}}(s, \tau) - I_X(\tau), \tau)| \\ &\leq \sup_{z \in \overline{\Omega}} |\mathbf{u}_{\mathbf{X}}(z, \tau) - \mathbf{u}_{\mathbf{Y}}(z, \tau)| \\ &\leq l_2 \|\mathbf{u}_{\mathbf{X}}(\tau) - \mathbf{u}_{\mathbf{Y}}(\tau)\|_2. \end{aligned} \tag{54}$$

By (50) and (54), there exists a constant l_3, dependent on $\mu,\ \mathbf{f}, \delta_h, \widetilde{T},\ \|\mathbf{u}_0\|$, such that for each $t \in [0, T]$,

$$\begin{aligned}
&\textstyle\int_0^t \|\mathbf{u}_{\mathbf{X}}(\widetilde{\mathbf{X}}(\tau) - I_X(\tau), \tau) - \mathbf{u}_{\mathbf{Y}}(\widetilde{\mathbf{X}}(\tau) - I_X(\tau), \tau)\|_c d\tau \\
&\le l_2 \textstyle\int_0^t \|\mathbf{u}_{\mathbf{X}}(\tau) - \mathbf{u}_{\mathbf{Y}}(\tau)\|_2 d\tau \\
&\le l_2 t^{1/2} \left(\textstyle\int_0^T \|\mathbf{u}_{\mathbf{X}}(\tau) - \mathbf{u}_{\mathbf{Y}}(\tau)\|_2^2 d\tau\right)^{1/2} \\
&\le l_3 t^{1/2} \left(\textstyle\int_0^T \|\mathbf{X}(t) - \mathbf{Y}(t)\|_c^2 dt\right)^{\frac{1}{2}} \\
&\le l_3 T \sup_{0 \le t \le T} \|\mathbf{X}(t) - \mathbf{Y}(t)\|_c.
\end{aligned} \tag{55}$$

By (46), there exists c_2 dependent on $\mu,\ \widetilde{T}, \delta_h, \mathbf{f}, \mathbf{u}_0$, such that

$$\sup_{\substack{z \in \overline{\Omega} \\ 0 \le t \le T}} |D\mathbf{u}_{\mathbf{Y}}(z, t)| \le c_2.$$

Hence for $\tau \in [0, T]$,

$$\begin{aligned}
&\|\mathbf{u}_{\mathbf{Y}}(\widetilde{\mathbf{X}}(\tau) - I_X(\tau), \tau) - \mathbf{u}_{\mathbf{Y}}(\widetilde{\mathbf{Y}}(\tau) - I_Y(\tau), \tau)\|_c \\
&= \sup_{0 \le s \le L} |\mathbf{u}_{\mathbf{Y}}(\widetilde{\mathbf{X}}(s, \tau) - I_X(\tau), \tau) - \mathbf{u}_{\mathbf{Y}}(\widetilde{\mathbf{Y}}(s, \tau) - I_Y(\tau), \tau)| \\
&\le \sup_{\substack{z \in \overline{\Omega} \\ 0 \le \tau \le T}} |D\mathbf{u}_{\mathbf{Y}}((z, \tau)|(\|\widetilde{\mathbf{X}}(\tau) - \widetilde{\mathbf{Y}}(\tau)\|_c + |I_Y(\tau) - I_X(\tau)|) \\
&\le c_2 \|\widetilde{\mathbf{X}}(\tau) - \widetilde{\mathbf{Y}}(\tau)\|_c + c_2 |I_Y(\tau) - I_X(\tau)|.
\end{aligned} \tag{56}$$

By (53), (55) and (56), for $t \in [0, T]$, we have

$$\begin{aligned}
\|\widetilde{\mathbf{X}}(t) - \widetilde{\mathbf{Y}}(t)\| \le{}& l_3 T \sup_{0 \le t \le T} \|\mathbf{X}(t) - \mathbf{Y}(t)\|_c + c_2 \int_0^t \|\widetilde{\mathbf{X}}(\tau) - \widetilde{\mathbf{Y}}(\tau)\| d\tau \\
&+ c_2 \int_0^t |I_Y(\tau) - I_X(\tau)| d\tau + \int_0^t |M_{uY}(\tau) - M_{uX}(\tau)| d\tau.
\end{aligned}$$

By (39, (40),(43) and using *Condition A*, we can deduce that there exists constant l_4 dependent on $\mathbf{f},\ \delta_h, \widetilde{T}$, such that, for $\tau \in [0, T]$,

$$c_2 |I_Y(\tau) - I_X(\tau)| + |M_{uY}(\tau) - M_{uX}(\tau)| \le l_4 \sup_{0 \le t \le T} \|\mathbf{X}(t) - \mathbf{Y}(t)\|_c.$$

Then for $t \in [0, T]$,

$$c_2 \int_0^t |I_Y(\tau) - I_X(\tau)| d\tau + \int_0^t |M_{uY}(\tau) - M_{uX}(\tau)| d\tau \le l_4 T \|\mathbf{X}(t) - \mathbf{Y}(t)\|_c.$$

Hence there exists constant l_5 dependent on $\mu,\ \mathbf{f},\ \delta_h, \widetilde{T},\ \mathbf{u}_0$, such that

$$\|\widetilde{\mathbf{X}}(t) - \widetilde{\mathbf{Y}}(t)\| \le l_5 T \sup_{0 \le t \le T} \|\mathbf{X}(t) - \mathbf{Y}(t)\|_c + c_2 \int_0^t \|\widetilde{\mathbf{X}}(\tau) - \widetilde{\mathbf{Y}}(\tau)\| d\tau.$$

By **Gronwall's Lemma (A)**,

$$\|\widetilde{\mathbf{X}}(t) - \widetilde{\mathbf{Y}}(t)\| \leq l_5 T e^{c_2 T} \sup_{0 \leq t \leq T} \|\mathbf{X}(t) - \mathbf{Y}(t)\|_c. \tag{57}$$

The inequality (57) implies for each constant $l_6 \in (0,1)$, there exists T' dependent on $\mu,\ \widetilde{T}, \delta_h, \mathbf{f}, \mathbf{u}_0$, such that when $0 < T \leq T'$,

$$l_5 T e^{c_2 T} \leq l_5 T e^{c_2 \widetilde{T}} \leq l_6,$$

$$\|\widetilde{\mathbf{X}}(t) - \widetilde{\mathbf{Y}}(t)\| \leq l_6 \sup_{0 \leq t \leq T} \|\mathbf{X}(t) - \mathbf{Y}(t)\|_c \qquad 0 \leq t \leq T.$$

Let $\|\mathbf{x}\|_{\mathscr{C}}$ denote the norm in $\mathbb{X}$,

$$\|\mathbf{x}\|_{\mathscr{C}} = \sup_{0 \leq t \leq T} \max_{0 \leq s \leq L} |\mathbf{x}(s,t)|.$$

Then

$$\sup_{0 \leq t \leq T} \|\widetilde{\mathbf{X}}(t) - \widetilde{\mathbf{Y}}(t)\| \leq l_6 \|\mathbf{X} - \mathbf{Y}\|_{\mathscr{C}},$$

that is

$$\|\mathbf{TX} - \mathbf{TY}\|_{\mathscr{C}} \leq l_6 \|\mathbf{X} - \mathbf{Y}\|_{\mathscr{C}}. \tag{58}$$

Let $T'' = min\{T', \frac{L_m - \|\mathbf{X}_0\|_c}{c_3}, \widetilde{T}\}$. Notice that T'' is dependent on $\mu,\ \widetilde{T}, \delta_h, \mathbf{f}, \mathbf{u}_0, L_m, \|\mathbf{X}_0\|_c$. The inequality (58) implies that if $T \in (0, T'')$, $\mathbf{T} : B \to B$ is a contractive map. By the Banach Fixed-Point Theorem (see [1]), $\mathbf{T}$ has a unique fixed point $\widetilde{\mathbf{X}} \in B$, and $\widetilde{\mathbf{X}}$ may be solved by successive approximations

$$\mathbf{Y}_{n+1} = \mathbf{TY}_n, \quad \forall \mathbf{Y_0} \in B, \quad n = 0, 1, 2, \ldots .$$

Let $\widetilde{\mathbf{u}}$ be the unique solution to (41) corresponding $\widetilde{\mathbf{X}}$. Then $\widetilde{\mathbf{u}}$ and $\widetilde{\mathbf{X}}$ is the unique pair of solutions we need. □

Remark: In our conclusion we only obtain the existence of the approximate solution with Dirac delta function approximated by differentiable function. How the solution of the original problem is approximated by this approximate solution is still an open problem.

Acknowledgement

This work is partially supported by National Natural Science Foundation of China(No: 10671092).

References

[1] Ebhard Zeidler. *Nonlinear functional analysis and its applications, I: Fixed-point theorems*. Berlin: Springer-Verlag, 1986.

[2] Peskin C.S. *The immersed boundary method, Acta Numerica*. Cambridge University Press, 2002.

[3] Daniele Boffi, Lucia Gastaldi. A finite element approach of the immersed boundary method. *Computers and Structures*, 2003, Vol.81, 491-501.

[4] Roger Temam. *Navier-Stokes equations and nonlinear functional analysis* (CBM-SNSF Regional Conf. Ser. in Aplied Mathematics 66), SIAM, Philadelphia, 1983.

[5] Roger Temam. *Navier-Stokes equations, theory and numerical analysis, Studies in Mathematics and its applications*. volume 2, North-Holland Publishing Company 1977.

[6] C.Foias, O.Manley, R.Rosa, Roger Temam. *Navier-Stokes equations and turbulence*. Cambridge University Press 2001.

[7] Ruth F.Curtain, A.J.Pritchard. *Functional Analysis in Modern Applied Mathematics*. Academic Press INC. (London) LTD. 1977.

[8] C.S. Peskin. Numerical analysis of blood flow in the heart, *J. Comp. Phys.* 1977, Vol.25, 220-252.

[9] Santos Alberto Enriquez-Remigio and Alexandre Megiorin Roma. Incompressible flows in elastic domains: an immersed boundary method approach. *Applied Mathematical Modeling*, January 2005, Volume 29, Issue 1, 35-54.

[10] Lima, A.L.F., Silva, E., Silveira-Neto A.,& Damasceno J.J.R. Numerical simulation of two-dimensional flows over a circular cylinder using the immersed boundary method. *Journal of Computational Physics*, 2003, Vol.189, 351-370.

[11] Arthurs, K.M., Moore, L.C., Peskin, C.S., Pitman, E.B., &Layton, H.E. Modeling arteriolar flow and mass transport using the immersed boundary method. *J. Comp. Physiol.* 1998, Vol.147, 402-440.

[12] Roma, A.M., Peskin, C.S., & Berger, M.J. An adaptive version of the immersed boundary method. *J. Comp. Physiol.*, 1999, Vol.153, 509-534

[13] Ling Rao and Hongquan Chen. Immersed boundary method: The existence of approximate solution of the two-dimensional heat equation. *Nonlinear Analysis: Real World Applications*, April 2008, Volume 9, Issue 2, 384-393.

Reviewed by Professor Yuanguo Zhu, Department of Applied Mathematics, Nanjing University of Science and Technology.

INDEX

D

E

F

G

Q

R

S

W

Y

Z